ENERGY SCIENCE, ENGINEERING AND TECHNOLOGY

GASIFICATION

CHEMISTRY, PROCESSES AND APPLICATIONS

Energy Science, Engineering and Technology

Additional books in this series can be found on Nova's website under the Series tab.

Additional E-books in this series can be found on Nova's website under the E-books tab.

ENERGY SCIENCE, ENGINEERING AND TECHNOLOGY

GASIFICATION

CHEMISTRY, PROCESSES AND APPLICATIONS

MICHAEL D. BAKER
EDITOR

Nova Science Publishers, Inc.
New York

For permission to use material from this book please contact us:
Telephone 631-231-7269; Fax 631-231-8175
Web Site: http://www.novapublishers.com

Library of Congress Cataloging-in-Publication Data

Gasification : chemistry, processes, and applications / editor, Michael D. Baker.
p. cm.
Includes bibliographical references and index.
ISBN 978-1-61209-681-0 (hardcover : alk. paper)
1. Coal gasification. I. Baker, Michael D.
TP759.G327 2011
665.7'72--dc22

2011001552

Published by Nova Science Publishers, Inc. †New York

CONTENTS

PREFACE

This new book presents current research in the study of the chemistry, processes and application of gasification processes. Topics discussed include experimental activities in a flexible gasification pilot plant; water gas shift and membrane reactor studies for the production of a hydrogen-rich stream in steam-oxygen gasification processes; the development of a dynamic model for a stratified downdraft wood gasifier; gasification of biomass for the production of syngas with reduced tar formation; the System Biomass Gasifier and its operation and the study of the catalytic effect of magnetite in the gasification of chars from lignite/rubber mixtures

Chapter 1 - World's energy demand is expected to continue to grow at large scale in emerging economies, due to the increase of world population and also due to increasing claims for high life quality. Though, in developed countries the rise of energy demands are predicted to grow at much lower levels, world's carbon dioxide (CO_2) emissions are expected to grow at an extent that may put in danger 'life as we know it.' Therefore, it is urgent to develop, demonstrate and implement alternative fuels sources and cleaner energy production technologies. It is predictable that developing countries, like China and India will increase the utilization of coal for energy production, thus the implementation of clean coal technologies, especially carbon capture and storage (CCS) are most needed.

Gasification is considered as a pre-combustion capture process, as by this process fuel is converted into syngas, which is afterwards converted into CO_2 and H_2. Thus, a H_2 rich fuel and free of carbon is produced while simultaneously CO_2 is captured producing a stream ready for storage. Gasification operating conditions may be optimized with the aim of reducing CO_2 emissions, while maintaining the heating value and the quality of syngas produced. Several materials with CO_2 capture ability may be added to the gasification bed to study their action. Another option is the use of oxygen instead of air producing a syngas without the dilution effect of nitrogen, which also simplifies CO_2 separation process.

The main challenges of current research work are to achieve high conversion of low grade coals and different types of wastes into syngas, with low contents of undesirable components, to simplify and decrease the cost of syngas cleaning and upgrading processes. Another important issue is syngas conversion into H_2 and the development of effective H_2 separation processes, which are both technically and economically viable. However, before these processes are ready to be implemented at industrial scale, further research and development activities are still needed and the demonstration at large scale of the known processes is fundamental. The integration of technologies under development in IGCC

(Integrated Gasification Combined Cycle) will be the main focus of the near future R, D&D (Research, Development & Deployment), as IGCC has high energy conversion efficiency.

This Chapter reviews the latest achievements in gasification processes and syngas cleaning and upgrading processes to produce a syngas with a wide range of applications, including effective H_2 separation processes to produce H_2 rich fuel and CO_2 capture. The stage of development of these subjects and the R, D&D needs before the full integration of the most promising ones are also analyzed.

Chapter 2 - In the field of hydrogen production through coal gasification for distributed power generation, Sotacarbo, an Italian limited company which operates in the field of clean coal technologies, is developing different research and development projects for the tuning of a coal-to-hydrogen process configuration. Toward this goal, a flexible pilot platform was built in 2007-2008, and it is currently in operation. In particular, the platform includes demonstrative and pilot air-blown fixed-bed gasifiers, the latter equipped with a flexible syngas treatment line for a combined power generation and CO_2-free hydrogen production.

This chapter presents a detailed description of all the experimental equipment and the main results obtained during the first 700 hours of tests in the pilot unit. In particular, the optimum operating conditions with a low sulphur South African coal have been defined, with a syngas production of about 43 Nm^3/h (from the gasification of 8 kg/h of coal), characterized by a lower heating value of 4.5 MJ/kg, and a high cold gas efficiency, of about 97%.

A series of experimental tests has also been carried out in order to assess the plant performance under different operating conditions and the effects of the main parameters, such as air/fuel and steam/fuel mass ratios and oxidant composition (air, eventually enriched in oxygen, or mixtures of oxygen and carbon dioxide). Moreover, a series of preliminary tests has been carried out in order to evaluate the gasification performance with a portion of high sulphur Sulcis coal or biomass (wood pellets) in the fuel blend.

The experimental results obtained in syngas treatment processes show a very efficient removal of pollutants (in particular, a final concentration of sulphur compounds lower than 10 ppm, in volume, can be obtained, even with high sulphur coal in the fuel blend) and the possibility to reduce carbon dioxide emissions of about 90% by using a single-stage amine-based carbon capture system, which assures a CO_2 absorption efficiency of about 99%.

Finally, the guidelines of the future research activities are described in this chapter, with particular reference to the further optimization of both gasification and syngas treatment processes in different operating conditions

Chapter 3 - Gasification of different types of fuels with steam and oxygen produces a synthesis gas (syngas) which consists mainly of CO, H_2, CO_2 and H_2O. Depending on the fuel and the operating conditions, the contents of CO and H_2 in the syngas usually range from 35 % to 55 % and from 20 % to 45 % respectively. If higher hydrogen contents are desired, the Water Gas Shift (WGS) reaction allows the conversion of CO to H_2 and CO_2 in presence of excess steam and a specific catalyst. This reaction is industrially carried out in two consecutive stages, first a high-temperature step for rapid CO conversion and then a low-temperature step for achievement of lower equilibrium CO contents.

A novel approach which is being investigated to increase the production of hydrogen in gasification processes is the use of a Water Gas Shift (WGS) membrane reactor in which a single high-temperature WGS catalyst is combined with a hydrogen selective membrane. In this system, the removal of hydrogen from the reaction chamber through the membrane drives the WGS reaction to the products side and thermodynamic equilibrium can be exceeded.

Catalyst and membrane must be carefully chosen and special attention must be paid to operating conditions in order to guarantee the correct performance of the system. In this work the performance of semi-commercial palladium membranes and two different high temperature WGS catalysts are described and the advantages of using a WGS membrane reactor for the production of H_2 in gasification processes versus a conventional WGS reactor are presented.

Chapter 4 - Gasification of different types of fuels with steam and oxygen produces a synthesis gas (syngas) which consists mainly of CO, H_2, CO_2 and H_2O. Depending on the fuel and the operating conditions, the contents of CO and H_2 in the syngas usually range from 35 % to 55 % and from 20 % to 45 % respectively. If higher hydrogen contents are desired, the Water Gas Shift (WGS) reaction allows the conversion of CO to H_2 and CO_2 in presence of excess steam and a specific catalyst. This reaction is industrially carried out in two consecutive stages, first a high-temperature step for rapid CO conversion and then a low-temperature step for achievement of lower equilibrium CO contents.

A novel approach which is being investigated to increase the production of hydrogen in gasification processes is the use of a Water Gas Shift (WGS) membrane reactor in which a single high-temperature WGS catalyst is combined with a hydrogen selective membrane. In this system, the removal of hydrogen from the reaction chamber through the membrane drives the WGS reaction to the products side and thermodynamic equilibrium can be exceeded. Catalyst and membrane must be carefully chosen and special attention must be paid to operating conditions in order to guarantee the correct performance of the system. In this work the performance of semi-commercial palladium membranes and two different high temperature WGS catalysts are described and the advantages of using a WGS membrane reactor for the production of H_2 in gasification processes versus a conventional WGS reactor are presented.

Chapter 5 - This paper describes the development and the predictions of a dynamic model for a stratified downdraft wood gasifier. The numerical model, which incorporates a two-step pyrolysis mechanism, uses mass and energy balances, together with information about rates of chemical reactions and physical transport processes, to describe the complex physical and chemical processes taking place in the gasifier. The model is capable of predicting the gas composition, primary and secondary tar contents, and gas and solid surface temperature profiles along the axis of the gasifier. The model predicts an outlet gas tar content of between 20 and 200 mg/Nm^3 for the range of input parameters used in the model simulation, assuming that prior to ignition the section of the gasifier below the ignition ports contains charcoal rather than wood. The stability of the reaction zones was also investigated and for stable operation it was found that the mass flux of supplied air needed to be in the range 0.158-0.167 kg/m^2s.

Chapter 6 - An analysis was carried out to determine the sensitivity of a dynamic model developed for a stratified downdraft wood gasifier to changes in input parameter values. The sensitivity of the model to small increments and decrements in three operating parameters and eight model parameters was investigated. The results show that a higher air supply rate increases the combustion zone temperature and reduces outlet tar levels. Changes of ±20% in air supply rate, initial particle diameter and solid-to-gas heat transfer coefficient led to changes in predicted reaction zone velocity of more than 50%. The final mass of condensable tars in the producer gas was predicted to change by more than 300% as a result of 20%

changes in the values of the following model parameters: solid-to-gas heat transfer coefficient, mass transfer coefficient and tar cracking reaction rates.

Chapter 7 - A pilot-scale downdraft gasifier was designed, fabricated and tested. The experimental data obtained from it were compared with the predictions of a dynamic model the development and analysis of which are described in previous papers. River red gum (*Eucalyptus camaldulensis*) wood blocks of two different sizes were used as gasifier feedstock. The gasifier was operated at inlet airflow rates of 7.5 to 27.6 kg/hr, with corresponding energy outputs ranging from 54 to 168 MJ/hr. An average cold gas efficiency of 70% was obtained. Two modes of operation were used. In the first mode, the section of the gasifier below the ignition port was filled with charcoal prior to start up and the remainder of the gasifier filled with wood; in the second mode, no initial charge of charcoal was used. Operating the reactor with wood fuel alone produced an outlet gas with a final tar content that was around 20 times higher than the average tar content values in gas produced when using an initial charge of charcoal. Transient temperature profiles, outlet gas compositions, and outlet tar contents predicted by the model were in generally good agreement with experimental values. However, the model predicted a more rapid increase in concentrations of H_2 and CO along the gasifier axis than was observed experimentally. There was good agreement between the predicted and experimentally observed rates and directions of reaction zone movement within the gasifier over the range of inlet airflow rates studied.

Chapter 8 - Among all the biomass conversion technologies, gasification has attracted increasing research interest in recent decades, as biomass gasification combined with gas turbine, internal combustion engines and gas burners has great potential for large-scale green power generation (through IGCC, integrated gasification combine cycle) at higher efficiencies than the conventional pulverized coal-fired power generation. Gasification of biomass can be achieved by partial oxidation of the feedstock using air, O_2 and steam or CO_2, producing a low to medium-BTU syngas gas (4-15 MJ/Nm^3), consisting primarily of CO, H_2, CH_4, CO_2 and trace higher hydrocarbons (C_2+C_3). However, a major challenge for biomass gasification is the high yields of tars that are composed mainly of aromatics and complex poly-aromatic hydrocarbons (PAHs). In an air-blown fluidized bed gasifier for instance, typical tar contents in the producer gas are between 0.5 to 100 g/m^3. For many applications, with the exception of direct and immediate syngas combustion for heat or electricity production, these tar levels must be reduced, often to below 50 mg/m^3. Tars can have significant negative effects on gasification with respect to efficiency and operation. Specifically, the production of tars instead of combustible gases reduces the gasification efficiency, and condensation and deposition of tars at temperatures below 350°C can lead to fouling and potential blockage of downstream equipment and piping. Although the biomass-derived syngas with tar can be effectively cleaned up with the wet scrubbing technology, hot gas clean-up technology, preferably employing using catalysts, to remove these contaminants at the "hot" state is advantageous with respect to energy efficiencies as it eliminates the need to cool the product gas and re-heat again for the syngas applications (internal combustion, gas engines, fuel cells and methanol synthesis, etc.). The hydrocarbons and tars formed during the gasification process could be removed/suppressed at a hot state through steam/dry reforming or cracking/hydro-cracking reactions using catalysts. This chapter overviews various biomass gasification technologies and tar formation during gasification, tar reduction technologies, and catalysts for biomass gasification and tar reduction.

Chapter 9 - Among alternative energy resources, biomass, such as wood wastes and sewage sludge, and waste plastics are very attractive due to its abundance, accessibility, and ease of use. The conversion of lignocellulosic biomass or waste plastics to a fuel or chemicals usually can take place with thermo-chemical conversion technologies. Among the thermo-chemical conversion processes, gasification is a complex combination of pyrolysis and oxidation reactions in the condensed and vapor phases. Under typical gasification conditions, oxygen levels are restricted to less than 30 % of that required for complete combustion. Gasification converts biomass or waste plastics to a uniform gaseous mixture that mainly comprises H_2, CO, and CH_4. The gaseous mixture is called producer gas. Producer gas can be used to run internal combustion engines to generate electricity, can be used as substitute for fuel oils in direct heat applications and can be used to produce chemicals which are useful as fuel, as well as chemical feedstock for industries.

Gasification itself is very old technology, which flourished during the 2nd world war. Today, there is renewed interest in the gasification, due to the increased petroleum prices. The main advantage of gasification lies in that it is decentralized conversion system which can be operated even in a small scale.

In Korea, an average of 5 million tons/yr of wood waste is being generated of late, much of which has been treated simply through incineration and recycling, such as into wood pellets for boiler fuel. Therefore, the need for more efficient treatment of wood wastes has recently gained strong national recognition.

Meanwhile, plastics are widely used in everyday life, as well as in high-tech industries, as an indispensable ingredient. Consequently, the amount of waste plastics has also increased. Therefore, there is a growing need for waste plastics management. The traditional ways for the disposal of waste plastics have been either to bury it in landfills, or to burn it in incinerators. Landfill and incineration, however, are associated with serious environmental problems. Therefore, there is an increasing need to utilize waste plastics environmental friendly and economically.

In this regard, the present work gives detailed experimental results of gasification of forest wood waste (FWW), construction wood waste (CWW), dried sewage sludge (DSS), and post-consumed mixed waste plastics (MPW) in a newly developed two-stage gasifier. The aim of the research is to directly reduce the tar yield in the gasifier and to produce a high caloric producer gas. In accordance with the aim, this article reports the development of a producer gas composition according to experimental conditions, such as the reactor temperatures and the equivalence ratio, the role of activated carbon in tar removal, and the chemical and thermal characteristics of the tar

Chapter 10 - The chapter presents a detailed description of the System Johansson Biomass Gasifier and its operation; it also deals with the factors influencing the efficiency of fixed bed biomass gasifier systems since the latter gasifier is a fixed bed type. Basically, the fuel is fed into the reactor where the gasification process takes place resulting in the formation of syngas. Syngas is channeled through a system of pipes to the downstream processes that consists mainly of the purification unit where it is cleaned of impurities such as fine carbon particles. The gas is then used to drive the engine that drives the generator and generates electricity. The gasifier is mainly used for electricity generation. A 150kVA pilot project aimed at improving the socio-economic status of a rural community has been implemented. The project aims to provide low-cost electricity to stimulate the establishment of small businesses in the area.

Chapter 11 - The utilization of waste rubber is still a topical issue. It is proven by a number of studies devoted to its treatment, especially thermal, to the process conditions, the products as well as a description of the behavior of this material in pyrolysis and gasification, and kinetics of reaction of the rubber char with CO_2. A realistic solution could be the co-gasification of waste rubber with lignite in the Integrated Gasification Combined Cycle. Recent experiments on the laboratory but predominantly commercial scale have demonstrated that waste rubber can without any technological difficulties thus be gasified at a content of 10% in a gasified mixture, in the short term at 20%. Higher content however causes an increase in the unburned carbon, whose content in ash grows significantly with the rising share of rubber in the mixture, as demonstrated in Fig. 1. The reason is apparently the significant difference between the reactivities of the lignite and rubber chars in the gasification zone of the generator, where the reactivity of the rubber char is low in comparison with the reactivity of the lignite char. The removal of the negative phenomenon of the unburned carbon would contribute to the more effective treatment of waste rubber, avert the losses of energy inserted into the process, improve the quality of the ash for its further use and facilitate the maintenance of the surface of the disposal site of the generator ash. From the environmental perspective, the possibility to treat greater volumes of waste rubber (e.g. waste-tire) than before is significant, whereas from the economic perspective it is important to avert energy losses. The author's operational experiments on the Lurgi gasifiers (Sokolov Coal A.S., Czech Rep.) have shown that if the content of the unburned carbon in the generator ash is for example 10.5%, then the loss of energy is 1.26% of the higher heating value introduced in the system at the input of the generator and that at a content of unburned carbon of 28.4% is already 4.39%, which are high to extreme losses. The question thus is how to lower the unburned carbon at higher shares of rubber in the gasified mixtures to an acceptable level or to the level reached at the gasification of the lignite alone. A possible solution is to lower the size of the particles of the treated rubber, e.g. from 40 mm to 15-20 mm (in the moving-bed gasifiers), and thus to increase the surface of the gasified particles, but this solution is not essential.

Another approach is to find the effective catalyst of the gasification reaction. Alkali metal salts, especially potassium salts, including those presented in a high content potassium biomass, are considered as effective catalysts for gasification by steam and CO_2. The most recently tested catalyst requires coal modified by a solvent extraction (HyperCoal), because this catalyst reacts with the mineral components of the coal. It is, however, an unfeasible variant for the given case. According to the authors experience with the Lurgi gasification process, it is possible to consider iron-based catalysts, which are effective, available and industrially used. For the authors experiments as model Fe catalyst a finely ground magnetite was selected for the laboratory tests. The experiments were conceived so that the chars from lignite and rubber without magnetite and with magnetite were prepared first in a rotary kiln. The comparative chars from rubber alone, lignite/polymers, lignite/celluloses and lignite/wood mixtures, and lignite alone were prepared in a similar way. They served for the comparison of their reactivity with the reactivity of the lignite/rubber chars and further for the more general determination of the impact of magnetite on the rate of the gasification reactions. The prepared materials were then gasified by steam and CO_2 and the rates of gasification measured at isothermal conditions. With lignite/rubber chars, the effective activation energy and pre-exponential factor of the gasifying reaction with CO_2 were further evaluated using non-isothermal isoconversional kinetics, and these parameters were compared for the samples with and without magnetite.

The objective of the treatise is to describe the effect of magnetite on the rate of the gasification of the chars from the mixture of lignite and rubber, compare it with the effect determined with comparative chars and explain the impact of magnetite on the basis of non-isothermal isoconversional kinetics.

Chapter 12 - Carbon, slag chemistry, temperature, and partial pressure of oxygen affect material reactions in a gasification chamber. Corresponding thermodynamic parameters are critical as they determine the types of crystals that form in slags, weight fractions of the phases, and corrosive interactions between slags and refractory materials. The crystal formation then effectively changes physical characteristics of slags, impacting slag viscosity. In order to study interactions between slags and refractory materials, it is important to assess thermodynamic data for the slags and the timing of crystal formation as opposed to slag residence time in the gasification chamber. The current study focuses on the thermodynamic analysis of impurities contained in carbon feedstock used in gasification, its kinetic behavior, and the effects of carbon feedstock chemistry on the degradation of gasifier refractory materials. Utilizing qualitative and quantitative techniques (XRD, SEM-EDX/WDX, TEM-EDX and ICP-OES); a thermodynamic phase diagram encompassing industrial slags based upon coal/petcoke ash mixture compositions was constructed using equilibration conditions obtained in samples after 72 hours of exposure at 1500 °C in an atmosphere of 64%CO/36%CO_2. This CO/CO_2 ratio corresponded to a simulated gasification environment of $Po_2 = 10^{-8}$ atm. Mullite crystal was found in Al_2O_3 rich slags, while V_2O_3 crystal was present in slags with high V_2O_3 (content corresponding to slags with high petcoke content). No crystalline phases were observed in SiO_2 rich slags at low V_2O_3 content (slag with high content of coal ash).

The formation kinetics of crystals in coal/petcoke slags was also studied from 1200 – 1700 °C in an atmosphere of 64%CO/36%CO_2, and a Time-Temperature-Transformation diagram constructed. This diagram exhibited two distinct crystallization events; with higher than 10 wt.% additions of petcoke slag to coal slag promoting V_2O_3 crystallization in the slag above 1350 °C, and with additional crystalline phases such as V-rich spinel forming at temperatures lower than 1350 °C. Experimental results indicated that the V_2O_3 crystal would form in the slag as fast as 35 seconds.

Finally, an evaluation of the interfacial reactions between synthetic slags (average coal and petcoke slag chemistry) and refractory materials (90 wt.% Cr_2O_3 – 10 wt.% Al_2O_3 and 100 wt.% Al_2O_3) was conducted utilizing a Confocal Scanning Laser Microscope. Ground slag samples at specific locations on refractory coupons were heated at 1500 °C in a 64%CO/36%CO_2 gas mixture using a gold-image heating chamber. Characterization of the slag/refractory interface by the SEM-EDX indicated that slag penetration occurred into porous areas of the refractory microstructure, and that refractory dissolution into slag occurred through the refractory surface and in the porous areas penetrated by slag, which promoted de-bonding of refractory material. It was noted that VO_x-containing crystalline phases precipitated at the slag/refractory interface in coal/petcoke slag mixtures. Chemical spalling of the Cr-containing refractory grains along with chemical dissolution of the grain in the slag facilitated surface degradation of the refractory.

In: Gasification: Chemistry, Processes and Applications
Editor: Michael D. Baker
ISBN: 978-1-61209-681-0

Chapter 1

ANALYSIS OF GASIFICATION VIABILITY AS A PRE-COMBUSTION CCS OPTION, VIA THE PRODUCTION OF A H_2 RICH FUEL WITH CO_2 CAPTURE

Filomena Pinto*

Rui Neto André, Gulyurtlu INETI, Estrada do Paço do Lumiar, Lisboa, Portugal

ABSTRACT

World's energy demand is expected to continue to grow at large scale in emerging economies, due to the increase of world population and also due to increasing claims for high life quality. Though, in developed countries the rise of energy demands are predicted to grow at much lower levels, world's carbon dioxide (CO_2) emissions are expected to grow at an extent that may put in danger life as we know it. Therefore, it is urgent to develop, demonstrate and implement alternative fuels sources and cleaner energy production technologies. It is predictable that developing countries, like China and India will increase the utilization of coal for energy production, thus the implementation of clean coal technologies, especially carbon capture and storage (CCS) are most needed.

Gasification is considered as a pre-combustion capture process, as by this process fuel is converted into syngas, which is afterwards converted into CO_2 and H_2. Thus, a H_2 rich fuel and free of carbon is produced while simultaneously CO_2 is captured producing a stream ready for storage. Gasification operating conditions may be optimized with the aim of reducing CO_2 emissions, while maintaining the heating value and the quality of syngas produced. Several materials with CO_2 capture ability may be added to the gasification bed to study their action. Another option is the use of oxygen instead of air producing a syngas without the dilution effect of nitrogen, which also simplifies CO_2 separation process.

* E-mail: filomena.pinto@lneg.pt; Tel: 351 21 092 4789

The main challenges of current research work are to achieve high conversion of low grade coals and different types of wastes into syngas, with low contents of undesirable components, to simplify and decrease the cost of syngas cleaning and upgrading processes. Another important issue is syngas conversion into H_2 and the development of effective H_2 separation processes, which are both technically and economically viable. However, before these processes are ready to be implemented at industrial scale, further research and development activities are still needed and the demonstration at large scale of the known processes is fundamental. The integration of technologies under development in IGCC (Integrated Gasification Combined Cycle) will be the main focus of the near future R, D&D (Research, Development & Deployment), as IGCC has high energy conversion efficiency.

This Chapter reviews the latest achievements in gasification processes and syngas cleaning and upgrading processes to produce a syngas with a wide range of applications, including effective H_2 separation processes to produce H_2 rich fuel and CO_2 capture. The stage of development of these subjects and the R, D&D needs before the full integration of the most promising ones are also analyzed.

Keywords: pre-combustion CCS, gasification, hot syngas cleaning, steam reforming, water gas shift (WGS) reaction, membrane H_2 separation, chemical looping (CL).

1. Introduction - The Role of Syngas Towards ZET

International Energy Agency (IEA) predicts that the world's energy demand will continue to grow, between 2005 and 2030 by about 55% (IEA, 2009), due to the growth of world population and also because of increasing demand for greater life quality, especially in developing countries, like China and India. Consequently, carbon dioxide (CO_2) emissions will continue to grow, especially in emerging economies, whose contributions to CO_2 release is predicted to increase from 39%, in 2005, to over 50%, in 2030 (IEA, 2009). If the right measures are not taken, CO_2 emissions from energy sector will be, in 2050, around 130% higher than those verified in 2005 (IEA, 2008). Due to the negative consequences in climate change and in worldwide economics, CO_2 emissions from energy production have to be reduced. Hence, clean coal technologies, particularly carbon capture storage (CCS) are most needed, especially in countries where coal use is growing faster.

Pre-combustion capture or gasification is a possible option to capture carbon from the fuel before combustion and to produce a hydrogen rich fuel and a CO_2 stream for storage. Moreover oxy-gasification uses oxygen instead of air to produce a syngas without the dilution effect of nitrogen, which also simplifies CO_2 separation process. Some research gasification studies have been concentrated in developing gasification processes with lower CO_2 emissions, either by optimizing gasification conditions, such as gasification temperature, gasifying agent type and composition, especially the ratios between gasifying agents and solid fuel. Another option consists of adding to the gasification bed CO_2 capture materials.

The production of syngas without or with minor contents of undesirable compounds, like tar, NH_3, H_2S and HCl, is one of the great challenges of nowadays gasification processes, especially when low grade coals or wastes are co-gasified. The need of using alternative fuels to decrease the dependency on imported oil has led several countries to study the possibility of using poor quality fuels and wastes, which has imposed new challenges to co-gasification and syngas cleaning processes. The most recent issue is the development of effective syngas cleaning processes to remove undesirable compounds and to produce a syngas with a wide range of end-uses.

Syngas produced from coal and natural gas may be converted into transportation fuels, through chemical synthesis by Sasol and Mobil processes. Another possibility is the production of syngas from biomass wastes, but the use of wastes that may contain undesirable elements will require an improvement of clean-up processes. The European Union has encouraged the production and use of biofuels in transportation sector and has been financing research project in these areas, which has motivated R&D (Research and Development) activities.

After syngas cleaning and improvement processes, its main components are CO, H_2 and CO_2. CO can be converted into H_2 and CO_2, by water gas shift (WGS) reaction. Although WGS reaction is a well known process, nowadays researchers have the challenge of applying known processes to syngas with different origins that consequently, may have adverse components. Another area of interest is the development of new catalysts that may increase WGS reaction conversion at milder conditions of temperature and pressure and with lower cost catalysts. Due to the potentialities of WGS reaction to achieve CO_2 capture, a huge amount of publications have appeared lately, reporting the performance of new catalysts.

Another challenge of current gasification processes studies is to achieve high conversion of syngas into CO_2 and H_2 and to develop an effective H_2 separation processes, both technically and economically viable. Hence, a H_2 rich stream is produced, while a CO_2 flow may go to storage. Consequently, coal and different types of wastes may be converted into H_2, through gasification and syngas, thus these technologies may be considered as ZET (zero emissions technologies).

H_2 demand has increased, due to its industrial utilization and also due to its potentialities for energy production, either through combustion or as raw material in fuel cells. Therefore many researchers have concentrated their efforts in developing and testing new membranes for pure hydrogen production with high efficiencies. In this chapter the main developments in this area are reviewed. The integration of these technologies under development in IGCC (Integrated Gasification Combined Cycle) will be the main focus of the near future R, D&D (Research, Development & Deployment), as IGCC has high energy conversion efficiency.

The cost of power generation with CCS is higher than those of existing processes, when CO_2 is not captured. Hence, today´s challenge is the development of processes that are less energy demanding and with lower costs. Thus, a great effort has been done in improving chemical looping processes and many researchers have focused their activities in applying the chemical looping process to gasification and CO_2 capture. This issue is also analyzed in this chapter.

One easier way of achieving CO_2 emission reduction is by increasing biomass use as fuels to enlarge biomass contribution to world's energy production. Biomass co-gasification technologies may play an important role in the future, due to their capacity to convert biomass waste into bioenergy and biofuels. Many researchers have stated the potentialities of gasification process and some developed technologies have been tested proved in operating industrial installations. However, more research is still needed in this area and mainly the demonstration of some of the developed processes at large scale is much required.

2. THE EFFECT OF GASIFICATION CONDITIONS ON CO_2 CAPTURE

Standard gasification technologies where developed and optimized to increase the energy output of the system while maintaining the best possible gas composition in terms of the desired heating value, H_2 and CO fractions and with minimal amounts of tars and gaseous contaminants.

Whichever is the chosen path for the conversion of a solid fuel, e.g. direct combustion or a two step approach including gasification and combustion of the obtained gas, the final environmental impact should be equivalent in terms of CO_2 emissions. It is possible to optimize CO_2 reduction and capture in combustion facilities but with some important effects on the combustion process. It is easier to act on the gasification process to reduce CO_2 emissions, without an adverse impact on the obtained gas, or even with very important benefits on its composition, properties and possible uses.

Most part of the experimental work to decrease CO_2 emissions has been done on the optimization of the gasification conditions, regarding namely the process temperature, the gasifying agent composition and the gasifying agent/solid fuel ratios. Some of the most promising results include the use as gasifying agent of oxygen enriched air or even pure oxygen mixed with steam. Other possibilities include the use of CO_2 capture agents with the side-benefit of some catalytic effect on the gasification process itself or the recycling of a CO_2 stream as one of the components of the gasifying agent mixture.

2.1 Effect of gasification agent

Early gasification work dealt mainly on the optimization of the characteristics of the produced syngas namely its energetic content and presence of contaminants. It was clear that, besides temperature, one of the major factor on syngas characteristics was the composition of the used gasification agent. Gil et al., 1997 studied the effect of different steam-oxygen mixtures on biomass gasification. The use of higher (H_2O+O_2)/biomass ratios led to a higher fuel conversion and a higher gas yield. This higher ratio also led to changes on composition, like a decrease on H_2, CO and hydrocarbons and an increase of CO_2 concentrations corresponding to a syngas with a decreasing lower heating value (LHV) originating a combined decrease on the apparent thermal efficiency of the gasifier. The optimal conditions found by the author, for a gasification temperature between 800 and 860°C included a (H_2O+O_2)/biomass of 0.8–1.2 kg/kg daf and a H_2O/O_2 molar ratio of 3.0. The obtained syngas would present a LHV between 13.5–15.0 MJ/Nm^3 of dry gas. The resulting CO_2 concentration on the syngas would lie between the 15–25% range.

Kim et al., 2000 studied coal gasification using air and steam, finding that increasing O_2/coal ratio between 0.3–0.53 kg/kg daf increased CO_2 concentration between 15–23%. Studying the effect of steam/coal ratios led to CO_2 concentrations in the same range and an increase of gasification temperature caused reductions on CO_2 concentrations. This higher CO_2 concentration would be favorable for a further capture phase.

Studies on the potential CO_2 capture and reduction resulting from the use of IGCC technology to obtain both H_2 and electricity were done by Shim et al., 2010. Simulations were based on the flowchart of an IGCC power station and low cost fuels like coal, heavy residue oil or bitumen using a gasifier, water-gas shift reactor. It was used monoethanolamine to capture both CO_2 and H_2S. A pressure swing adsorption membrane was used to obtain 99% pure H_2. The results show that bitumen has the highest H_2 production while pulverized dry coal has the best CO_2 avoided emissions with the capture of more than 93% of the carbon from the fuel.

Valero and Usón, 2006 studied the oxy-co-gasification of biomass, coal and coke mixtures in an IGCC power plant. Data obtained from an existing plant and several simulations have shown that by optimization of the steam/fuel and O_2/ fuel mass ratios it is possible to minimize CO_2 production in the gasifier. However the optimization shows that the gasification of mixtures is a non-linear process, as different fuel mixtures will present different LHV values and even maintaining the same H/C and O/C ratios on the gasifier will lead to different operating temperatures and cold gas efficiencies (CGE). The best result obtained allowed an increase of CGE, to more than 77%, while decreasing CO_2 concentrations from 5 to 2%.

Campoy et al., 2009 studied the effect of using mixtures of steam and O_2–enriched air to gasify biomass in a fluidized bed. The use of enriched air, while reducing the nitrogen dilution effect, increased the gasification temperature, allowing the use of lower quality steam while maintaining the thermal level input to the gasifier. The appropriate combination of temperature and steam/fuel ratio leads to higher CO and H_2 yields, heating value and carbon conversion. It was verified that by increasing the O_2 concentrations up to 40% it was possible to improve the total carbon conversion to more than 97% and increase the cold gas efficiency from 54–62% to 63–69%. Depending on the experimental conditions CO_2 concentrations showed some reduction, but it was clear that the CO_2 yield, decreased by more than 50%, from values with more than 700 g/kg daf biomass to less than 350 g/kg daf biomass.

Wahlund et al., 2004 evaluated the potential to achieve the largest CO_2 reduction by switching from fossil fuels to biomass. The results show that the best and sustainable CO_2 reduction would be achieved by using biomass to obtain fuel pellets for coal substitution in IGCC installations. These pellets can either be stored or transported over long distance with a substantial increase on CO_2 emissions. Depending on the energy output considered e.g. with only electricity production or with an additional use of all available heat the use of biomass pellets will lead to a reduction of about 76 to 100 Mg CO_2 for each TJ of used biomass. Indirect use of biomass to generate liquid fuels, like ethanol, methanol or dimethylether (DME) generates only half of CO_2 emission reductions, but with a higher specific cost.

One of the best examples of the effect of the gasifying agent on coal gasification was presented by Cormos et al., 2006 who studied entrained-coal gasification using either N_2, CO_2 or water as transporter gases required for fuel feeding. The base scenario, Figure 1, is the behavior of conventional water-slurry coal feed gasifiers where the entrained flow oxygen promoted gasification reaches temperatures around 1370°C, at a pressure of 70 bar. The supplied steam is enough for the operation of a secondary water-gas shift reactor, being obtained a raw syngas with around 42.2% CO, 25.3% H_2, 10.5% CO_2, 20.8% H_2O and 0.6% N_2.

The use of N_2 as coal transport gas, Figure 2, will require two additional supply lines of steam for both the gasifier and the water-gas shift reactor. In this configuration the gasifier operates at 40 bar reaching a temperature of 1560 °C. The raw syngas composition is 61.5% CO, 26.5% H_2, 2.6% CO_2, 4.6% H_2O and 4.3% N_2.

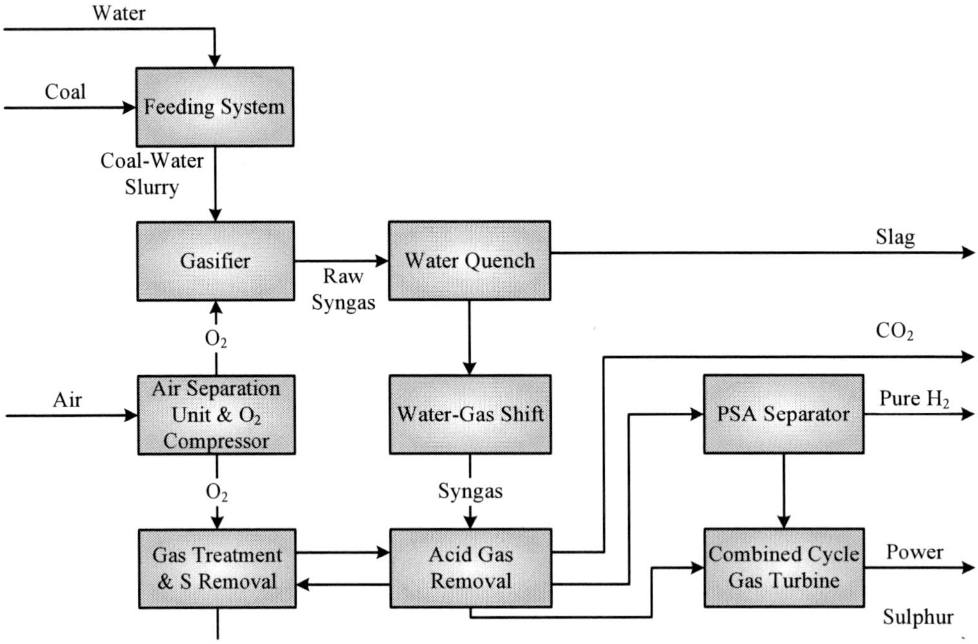

Figure 1. Hydrogen production by slurry feed gasification, with water-gas shift conversion and sulfur and CO_2 removal. Adapted from Cormos et al., 2006.

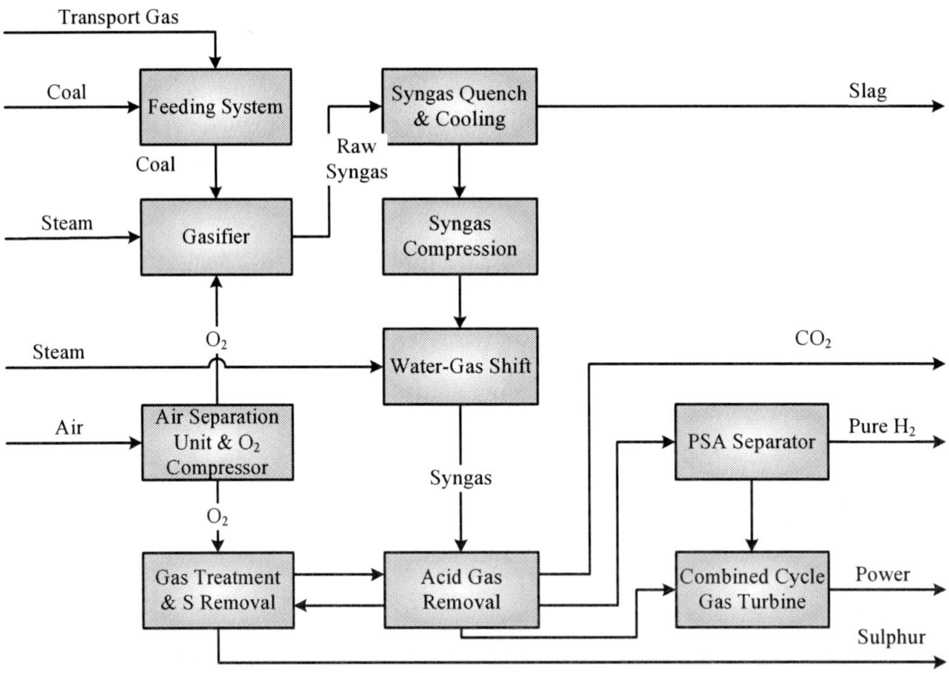

Figure 2. Hydrogen production by dry feed gasification, with syngas compression, water-gas shift conversion and Sulfur and CO_2 removal. Adapted from Cormos et al., 2006.

Using CO_2 as coal transport gas, Figure 2, enables the operation of the gasifier at 40 bar reaching a temperature of 1550 °C. In this case the raw syngas composition is 63.6% CO, 25.6% H_2, 3.6% CO_2, 5.9% H_2O and 0.8% N_2. Either of these gases will pass through a water-gas shift reactor, where the CO is almost totally converted to CO_2. The gas is then cooled and sent to an acid gas removal system (Selexol®) for the removal and capture of H_2S and CO_2. The obtained gas will present H_2 concentrations between 91 to 96%, which can be improved to 99.99% using a Pressure Swing Adsorption (PSA) stage. In terms of plant efficiency parameters best results were obtained using CO_2 as coal transport agent, while comparing with the water-slurry coal feeding. Using CO_2 generates 7% more electricity, 9% more pure H_2, increasing the plant efficiency from 53.5% to 58.3%, while maintaining equivalent CO_2 capture rates of over 96%.

2.2 Effect of using limestone and other low cost CO_2 absorbents

While it is possible to partially decrease CO_2 contents on syngas through some adjustments on the gasifying agent composition, an easy method for early CO_2 removal is by the use of limestone or other inexpensive Ca containing minerals. This method will product solid residue that could be used or stored elsewhere, without major negative drawbacks. In the case of the gasification of sulfur containing fuels, calcium may act as an agent to effectively remove sulfur compounds from the syngas. An additional catalytic activity results from the metallic elements existing in the used minerals, acting on the promotion of the gasification reactions and on the destruction and the removal of some syngas contaminants like tars or other undesirable compounds.

Mahishi and Goswami, 2007 tested the gasification of biomass in the presence of calcium oxide, acting as a CO_2 sorbent. The experimental set-up included a dual system reactor where the gasification gas passed through a sorbent bed. Steam gasification of pine bark was carried out at atmospheric pressure in the temperature range 500–700 °C. The use of the calcium oxide increased the hydrogen yield, total gas yield and carbon conversion efficiency. The results show that calcium oxide played the dual role of sorbent and catalyst, promoting the reforming of tars and hydrocarbons in the raw gasification gas. Before saturation of the sorbent, CO_2 concentrations on the exit gas decreased from 27.7% to 9.2%, with appreciable reductions in both CO and CH_4 concentrations. Final H_2 concentration in the product gas increased from 60 to 83%. The authors suggest that the use of a sorbent will allow operating the gasifier at lower temperatures while still maintaining the same H_2 output.

Felice et al., 2009 tested the use of dolomite and a commercial nickel catalyst to promote tar destruction and steam reforming of hydrocarbons in biomass gasification studies. Using model species like CH_4, toluene and 1-methyl naphthalene it was found that the use of dolomite clearly helps on their complete conversion. As dolomite also absorbs CO_2, the final gas composition is almost pure H_2, in a dry and N_2 free basis. After some reaction time, as dolomite CO_2 absorption capacity reaches saturation, the exit gas will contain significant amounts of CO and CO_2. The use of Ni does not affect the initial behavior of dolomite but in the latter phases there is a promotion of water-gas shift reaction increasing the CO_2/CO ratio.

Koppatz et al., 2009 and Pfeifer et al., 2009 tested an internally circulating fluidized bed with a steam promoted gasification zone and an air feeding combustion zone. The use of limestone as a reactive fluidized bed material promotes a continuous CO_2 removal by cyclic carbonation of CaO and calcination of $CaCO_3$. Gasification of biomass (wood chips) was possible at lower temperatures both in the gasification zone, decreasing by 200°C to 650–700°C while the temperature in the combustion zone was reduced from around 950°C to 850°C. The gasifier promoted a significant CO_2 removal from the producer gas resulting in an increased conversion of CO toward H_2. The product gas composition was upgraded with an increased H_2 content of about 50 vol.-%dry. The tar content was quantified to about 1 g/Nm^3 (d.b.). The system is also versatile enough to easily obtain the desired CO/H_2 ratio according to the desired syngas final use.

Pecho et al., 2008 studied the optimization of the biomass gasification in a internally circulating fluidized bed using different reactive bed materials. Besides natural minerals like calcite ($CaCO_3$) and olivine [$(Fe_xMg_{1-x})_2$ SiO_4] other synthesized perovskite-type oxides like $Ba_{0.3}Sr_{0.7}Fe_{0.9}Mn_{0.1}O_{3-\delta}$, $La_{0.65}Sr_{0.35}Cr$ $0.5Mn_{0.5}O_{3-\delta}$ and $Gd_{0.1}Ce_{0.9}O_2$ were used. It was evaluated each mineral effect on oxygen capacity and release, the catalytic activity towards toluene reforming as well as mechanical and chemical stability in studies with either sulfur-free and H_2S enriched conditions. Typical results included the formation of gas mixture with high CO_2/CO and H_2/CO_2 ratios that were explained by the oxygen transfer capacities of the used bed materials. It is reported that under sulfur–free conditions the synthesized materials are the best reactive bed materials, where $Gd_{0.1}Ce_{0.9}O_2$ presents better catalytic activity. In the presence of H_2S the best results were obtained with the natural minerals, where calcite and to a lower extent olivine, have clear advantages with respect to price, catalytic activity and oxygen transfer capacity.

2.3 Effect of syngas recycling on CO_2 enrichment

Ahmed and Gupta, 2009 compared CO_2 gasification of paper and cardboard finding that rises in temperature from 800°C to 1000°C increased CO concentration from less than 20% to more than 50%. This was explained by an increase on the extent of Boudouard reaction. Kinetical studies revealed that at short times final composition is limited by the extent of pyrolysis of the fuel, with high initial peaks of H_2 and hydrocarbons. Only at longer reaction times, more than 5 min, it was found an effective CO_2 consumption, indicating the suitability of this agent to promote biomass and wastes gasification.

Butterman and Castaldi, 2007 studied the effect of CO_2 feed on biomass steam gasification. It was found that the addition of up to 50% CO_2 into the feed increased the CO formation and decreased H_2 and CH_4 concentration, especially at higher temperatures. These results indicated a positive correlation on the effects of recirculation of CO_2 that could be used as an additional gasification agent. Butterman and Castaldi, 2009 also studied biomass gasification with steam and CO_2 mixtures, ranging from 0–100%. It was found that carbon conversion was complete with 25% CO_2/75% steam while pure steam only gave 90% conversion in the temperature range of 800–1000 °C. Changes on CO_2 amount in the steam mixture from 0% to 50% caused a decrease on H_2/CO ratio from 5.5 to 0.25 allowing a comprehensive adjustment of the desired final H_2/CO ratio. Biomass devolatilization using CO_2 was improved using low temperatures and low heating rates. It was found that this agent promotes the separation of lignine from cellulose of the original biomass originating a highly porous lignin based residue, facilitating the solid-gas reactions occurring during the gasification phase. In Table 1 is presented the comparative effect of using either steam or CO_2 in biomass gasification.

Table 1. Comparison of biomass gasification using CO_2 or steam as gasifying agents. Adapted from Butterman and Castaldi, 2009

Using CO_2 as gasifying agent	Using Steam as gasifying agent
Uses CO_2 (generally an unwanted end product of various industrial processes)	Uses H_2O
No heat required for phase change	Requires 2.26 kJ g^{-1} to vaporize the water
Produces higher CO levels	Produces higher H_2 levels
Produces a more reactive char, with higher porosity and easy diffusion to micropores and macropores	Produces a less reactive char, with high porosity and lower diffusion to micropores
Produces a more reactive char	Produces a less reactive char
Higher conversion of char to volatiles corresponding to a more efficient gasification	Lower conversion of char, that can be used as feedstock in other processes
Lowers the H_2/CO ratio to a range (1.5-2.5) more suitable for Fischer-Tropsch synthesis	Raises the H_2/CO ratio to a range (5-8) more suitable for us in solid oxide fuel cells

Cao et al., 2006 proposed an innovative method two-stage biomass gasifier, using a conventional fluidized bed but with fuel feeding at the top of the reactor, Figure 3. By these means volatile compounds will be released into an oxygen-lean or reducing environment to prevent their combustion. Some of the fuel gas is recirculated to the upper part of the reactor, but is totally combusted using the required amount of secondary air, providing an additional heat source for the gasification reactions and for tar destruction. This innovative concept design allows for a large residence time of the solid residues obtained from the devolatilization of biomass while the gaseous products leave the reactor without contact to the oxidizing zones of the reactor. The temperatures used on the top of the reactor, between 860–890°C allow for an extensive reduction on tar content, reaching about 10 mg/Nm^3.

Zhu et al., 2002 compared carbon gasification with O_2 and CO_2 finding that activation energies for CO_2 gasification would be about 30% higher than for O_2 gasification. It was also found that CO_2 has a much lower chemisorption than O_2.

Li et al., 2010 evaluated the techno-economical performances of various coal gasification-based hydrogen and electricity cogeneration processes. The results indicate that the use a membrane and syngas chemical looping strategies can reduce the energy and cost penalties associated to CO_2 capture in coal conversion processes. The membrane-based systems studied integrate the functions of water-gas shift and CO_2 separation in a single unit using selective membranes. Both scenarios consider either H_2 or CO_2 selective membranes in the water-gas shift reactor, using steam as the sweeping gas. The use of steam allows for an easy separation of water by cooling, originating almost pure H_2 or CO_2.The results indicate a carbon capture efficiency of 90%, with a higher H_2 production and plant efficiency in the case of using the H_2 selective membrane. Total plant cost will be lower when a CO_2 selective membrane is used.

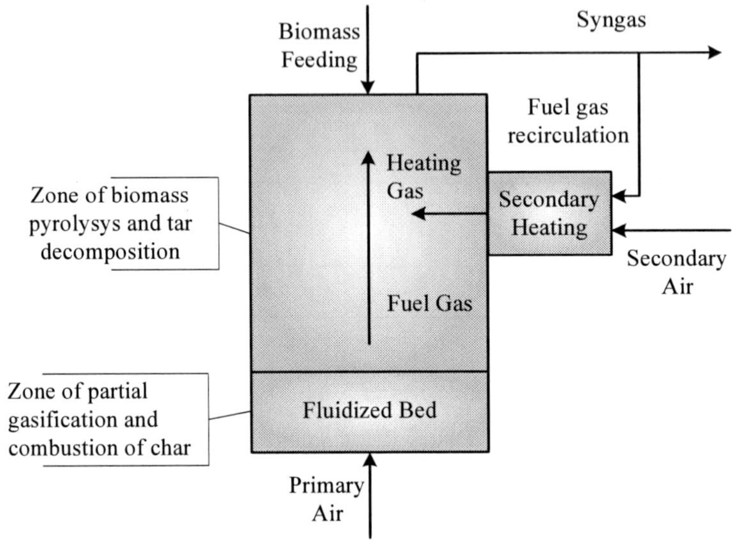

Figure 3. Model of a two-stage biomass gasifier with partial fuel gas recirculation. Adapted from Cao et al., 2006.

3. Hot Syngas Cleaning Processes

3.1. Tar removal

Gasification may play an important role in CCS, it is considered as a pre-combustion process, as it may produce a carbon free fuel, allowing CO_2 capture. However, before achieving these goals, it is necessary to up-grade syngas cleaning processes, as the syngas that comes out from the gasifier usually contains several undesirable components that difficult both the production of a carbon free fuel and CO_2 capture. One of the most problematic components is tar. Tar deposits are responsible for installation shutdowns for maintenance and repair, therefore to overtake this problem some installations may duplicate gas cleaning equipments, but this option increases greatly the investment costs. On the other hand, most syngas utilizations require very low contents of tar.

Tar definition is not consensual, but in general tar includes the condensable organic components present in syngas. Tar is responsible for serious operating problems, such as plugging, fouling and clogging. There are five tar classes as defined by Devi et al., 2005, and Pinto et al., 2009 a). Analytical methods for tar determination and quantification are not consensual, as several researchers use different analytical procedures. Nowadays the most used procedure is that defined by "Tars Protocol" guidelines, 2005.

The formation of tar can be controlled and reduced during gasification process, both by optimizing gasification conditions and by using low cost catalysts, usually minerals that promote tar destruction, as reported by Pinto et al., 2009 a) and 2009 b). Keeping a high temperature all along the gasification reactor promotes the destruction of tar initial formed.

The same objective can be also achieved by introducing secondary air or steam inside the reactor above bed height, but this option has the disadvantage of decreasing syngas heating value.

Whenever syngas contains tar molecules, they have to be removed before syngas end-use. During syngas cleaning and upgrading processes, tar can be physically removed or chemically converted into lighter gaseous compounds by thermal and/or catalytic processes. Tar physical removal is usually done by wet scrubbers and devices used for removal of particulates, including fly ash and dust particles. It is a better option to remove particulates at high temperatures to prevent tar condensation on particulate surfaces, which may difficult the operation of particulates removal devices. There are several mechanical devices for particulates removal, such as cyclones, bag filters, baffle filters, ceramic filters, fabric filters, rotating particle separators, wet electrostatic precipitators and water scrubbers. Pinto et al., 2009 a) describes the main features of these devices and enlightens the main disadvantages of each one.

Electrostatic filters are very expensive and packed bed filters have limited applications, due to waste filter disposal and operability problems. Other devices, like fabric filters, rotating particle separators and water scrubbers require quenching syngas to low temperatures, usually below 200 °C, which is a disadvantage when syngas needs to be heated again, either for further cleaning or upgrading. Cyclones are the most used equipments for particulates removal, because they are easy to build and operate, can operate over a wide range of temperatures and are commercially available at a lower cost than other devices. However, though they are suitable to remove particulates with larger particle sizes, they are ineffective for removing sub-micron particulates. For particulates with particle sizes below 5 μm ceramic candle or sintered metal filters could be a possible choice, as they can operate at temperatures close to those of syngas leaving the gasifier.

Wet scrubbers for tar removal are efficient and largely employed and besides tar and particulate some other undesirable syngas components may also be removed, like acids and alkali contaminants, but wet scrubbers are not environmental friendly, due to the need to treat large volumes of liquid effluents, which is very expensive. On the other hand, syngas has to be cooled down to temperature below 60°C, before it enters a wet scrubber, which decreases the overall energy efficiency of the process, whenever syngas needs to be heated afterwards.

Cold gas is a better option when syngas has to be cooled for further processing or to be transported. However, hot gas cleaning processes are more convenient than cold cleaning methods, as they improve energy efficiency and lower operational costs when syngas utilization requires its heating, as it happens for hydrogen production through steam reforming and water–gas and shift reactions or for heat and power generation in a high-temperature fuel cell. According to Wang et al., 2008, hot gas cleaning processes have a lower environmental impact and are normally safe.

Tar can be chemically removed by thermal and/or catalytic processes, which promote tar cracking reactions (1). Steam reforming reactions, (2) and (3), and CO_2 reforming reactions, (4) to (5), might also contribute to tar destruction, leading to the formation of H_2 and CO, whose contents are expected to increase.

$$p\ C_nH_m \rightarrow q\ C_xH_y + r\ H_2 \qquad (x<n \text{ and } y<m) \tag{1}$$

$$CH_4 + H_2O \leftrightarrows CO + 3\ H_2 \tag{2}$$

$$C_nH_m + nH_2O \leftrightarrows n\,CO + (n+m/2)\,H_2 \tag{3}$$

$$CH_4 + CO_2 \leftrightarrows 2\,CO + 2\,H_2 \tag{4}$$

$$C_nH_m + n\,CO_2 \leftrightarrows 2n\,CO + (m/2)\,H_2 \tag{5}$$

In thermal processes, syngas is heated till temperatures around 1000°C to crack tar molecules into lighter compounds, gaseous in nature. Morf et al., 2002 reported that when syngas was heated between 700° and 1000°C, both CO, H_2 and CH_4 concentrations rose. However, tar chemical conversion also increased the release of Polycyclic aromatic hydrocarbons (PAHs) and soot. On the other hand, the operation at such high temperatures implied an energy penalty and required that reactors had to be built with expensive materials to resist heat. The use of suitable catalysts for tar destruction could decrease operation temperature and pressure and increase tar conversion into gases and decrease the formation of soot.

Many researchers have concentrated their effort and activities in developing and testing different types of catalysts to achieve high tar reductions at the lowest possible cost and consequently a huge number of publications have appeared recently. Pinto et al., 2009 a) and 2009 b) reviewed the main types of catalysts that may be used for tar reduction either during gasification or for hot syngas cleaning. Catalysts suitable for tar abatement may be grouped as: natural minerals, like limestone, dolomite or olivine, metal alkalis such as: KOH, $KHCO_3$ and K_2CO_3, stable metal oxides such as nickel oxide and noble metals based compounds. Natural minerals have the advantages of low cost and availability in most countries, but show an intermediate performance. Catalytic action of natural minerals depends on their origin, porosity and composition, for instance dolomites iron content have proved to increase tar destruction. Dolomite is soft and easy to erode and break into small solid particles that may be entrained with the syngas and may cause plugging in cyclones, filters or pipes. Olivine has a higher attrition resistance than dolomite and thus may be a better option, as olivine and dolomite have similar tar reduction capacities. Some authors have reported that olivine led to higher tar reductions than dolomite, whilst other researchers have stated the opposite, depending on each mineral contents of elements with catalytic action.

Nickel based catalysts are more effective for tar reduction than natural minerals, but are also more expensive and they may deactivate easily by the presence of sulfur compounds in syngas. More research is still needed to develop nickel based catalysts more resistant to poisoning, with longer lifetimes, higher mechanical strength, lower cost and more effective for tar abatement. An alternative is the use of a guard bed made of dolomite, which will act mainly on sulfur removal, to protect the more exponsive nickel catalyst. Noble metals based catalysts present the highest effectiveness, but they are quite prohibitive, due to their high cost, Pinto et al., 2009 b).

Some authors have developed bi and tri-metallic catalysts to combine the catalytic action of different metals in the same catalyst with the aim of obtaining a better performance at a lower cost. Wang L. et al., 2008 prepared a bi-metallic catalyst with ruthenium and nickel supported in titania and zirconia to be used for steam reforming of tar at low temperature. Rapagnà et al., 2002 studied the performance of several tri-metallic and ternary oxide structures for syngas methane reforming with steam. The results obtained showed that $LaNi_{0.3}Fe_{0.7}O_3$ had the highest activity at 800°C, as around 90% of the tar present in the syngas was converted and also about 90% of CH_4 was converted, without leading to carbon formation.

3.2. Sulfur and halogens based compounds abatement

Due to gasification reduction conditions sulfur and halogens are usually released into syngas as hydrogen compounds like H_2S and HCl that are responsible for corrosion problems in pipes and in equipment where syngas is used and are precursors to acid rain and photochemical smog. Apart from this, whenever, syngas is cleaned or upgraded by thermal catalytic processes, sulfur and halogens compounds need to be eliminated, because catalysts may be deactivated or poisoned by the presence of such compounds in syngas.

The release of sulfur and halogens based compounds into syngas may be controlled during gasification by optimizing experimental conditions and by adding to gasification bed material sorbents, containing Fe, Ca and other elements that retain sulfur and chlorine inside the bed through the formation of sulfides and halogenides of these elements, reactions (6) to (8). The presence of H_2O and CO_2 in reaction medium may also affect H_2S reduction, by reactions (9) and (10).

$$CaO + H_2S \leftrightarrows CaS + H_2O \quad (6)$$
$$CaO + 2\,HCl \leftrightarrows CaCl_2 + H_2O \quad (7)$$
$$Ca(OH)_2 + 2\,HCl \leftrightarrows CaCl_2 + 2\,H_2O \quad (8)$$
$$CaO + H_2 + CO \leftrightarrows Ca + H_2O + CO_2 \quad (9)$$
$$Ca + H_2S \leftrightarrows CaS + H_2 \quad (10)$$

The rise of air or oxygen flow rate usually leads to a decrease of H_2S content in the syngas, probably due to the formation of SO_x. The rise of temperature usually leads to an increase in H_2S in the syngas. However, H_2S formation is affected by many experimental parameters: steam and oxygen flow rates, temperature, feedstock nitrogen content, fuel inorganic matter and even reactor material. As some of these parameters may act in opposite directions, it is not easy to predict the effect of all of them on H_2S formation and destruction.

Even with gasification operating at optimized conditions and especially when feedstocks with high contents of sulfur are gasified, syngas still contains higher sulfur contents than those required by some end-uses. Some syngas utilization require very low contents of H_2S and HCl or even the total absence of these compounds, like for instance fuel cells or chemical synthesis, Pinto et al., 2009 a). Thus, syngas needs to be treated to ensure the abatement of these compounds.

Sulfur and halogens based compounds reduction may be achieved by low temperature processes, like: physical absorption, chemical absorption, oxidation and dry bed processes. Physical absorption processes use sorbents like: methanol, dimethyl ether, polyethylene glycol, sulfolane or carbon dioxide. In chemical absorption processes are usually used potassium carbonate and different types of amines (MEA, DEA, MDEA, MGA or DIPA). Some of these processes are well know, but despite the disadvantage of requiring the moving and regeneration of large solvents volumes, they are not advisable when syngas is going to be used in any thermal process.

When syngas needs to be hot cleaned or upgraded and/or when it is going to be used in CO shift reaction or for chemical synthesis, it is a better option to keep syngas at high temperature and use one or more guard beds or fixed bed reactors with natural minerals to retain sulfur and halogens. Several natural minerals, like: limestone (CaO), dolomite (CaO•MgO) or olivine ((Mg, Fe)•2SiO_4) may be used. They have proven to be suitable to retain sulfur, chlorine and other halogenides and to promote some tar reduction. Natural minerals are also cheap and available in many regions.

Instead of natural minerals, ZnO has proven to be a good sorbent to retain sulfur compounds. However, ZnO may also react with CO and H_2 by reactions (11) and (12) and originate metallic zinc that may diffuse to the surface and vaporize during the sulfidation/regeneration process. To avoid zinc vaporization, ZnO may be stabilized by the addition of natural zeolites or by the presence of inorganic oxides, like TiO_2 to increase ZnO stability, mechanical strength and capacity of retention, Pinto et al., 2009 b). The metal-based sorbents, such as zinc titanate ($ZnO.xTiO_2$), zinc ferrites ($ZnO.xFe_2O_3$) and copper and manganese oxides are effective for H_2S reduction, but are expensive and may lose their activity at temperatures below 550°C.

$$ZnO + H_2 \rightarrow Zn + H_2O \quad (11)$$

$$ZnO + CO \rightarrow ZnC + O_2 \quad (12)$$

Sorbents are regenerated by reversing the retention reaction (6), therefore, sorbents have to remain stable under repeated sulfidation/regeneration cycles. Both zinc titanate and zinc ferrite have shown to be versatile and stable sorbents for desulphurization, Zevenhoven and Kilpinen, 2004. On the other hand, MnO supported in γ-Al_2O_3 was reported to have a high capacity, but poor stability, due to self-poisoning by soot formation, Atakul and Wakker, 1995.

3.3. Nitrogen-based compounds abatement

Nitrogen present in the materials to be gasified is generally released as NH_3, due to reduction conditions used during gasification. NH_3 is a precursor to acid rains, as it forms NO_x by combustion reactions. The release of NH_3, into syngas may be controlled by adjusting gasification experimental conditions, for instance the rise of air or oxygen flow rates usually decrease NH_3 formation, but increases the release of NO_x and the rise of temperature generally promotes NH_3 destruction. Nevertheless, NH_3 formation and destruction depends on several parameters, such as: temperature, steam and oxygen contents in the gasification medium, feedstock nitrogen content, fuel inorganic matter and reactor material, thus the effect of some of these parameters is not always easy to predict, as they may be affected by some others.

Depending on nitrogen content of feedstock to be gasified and on gasification experimental parameters syngas may present NH_3 up to 6000 ppmv. Some syngas utilizations require low contents of NH_3, for instance gas engines require concentrations below 50 mg/Nm^3, while fuel cells impose contents lower than 5000 ppm and Fischer-Tropsch and methanol synthesis only admits values around 1 ppmv, Pinto et al., 2009 b). Therefore, it is important to promote Nitrogen-based compounds abatement.

Molecular sieves made of highly porous aluminum silicates, zeolites, may be used for NH_3 removal through adsorption/desorption cycles. Another option is the catalytic decomposition of NH_3 into N_2 by reaction (13) during syngas hot cleaning and upgrading processes. Several authors have stated that nickel-based catalysts and other-metal based catalysts that are capable of tar destruction are also suitable for NH_3 abatement, Devi et al., 2003 and Pinto et al., 2009 b).

$$2\ NH_3 \leftrightarrows N_2 + 3\ H_2 \quad (13)$$

Wang W. et al., 1999 reported that the use of nickel based catalysts during the hot cleaning process of syngas could decrease the initial values of NH_3 to values up to 95%. Pfeifer et al., 2008 also stated that in presence of a monolith type catalyst made from nickel-based catalysts and at temperatures in the range from 850 to 900°C, tar conversion of around 98% and ammonia destruction of about 40% were obtained. These results agree with those obtained by Pinto et al., 2010 who reported NH_3 reductions higher than 97% after the syngas has been treated with a nickel based catalysts at 850°C. Lower reductions were reported by Leppälahti et al., 1997 who obtained NH_3 reductions from 50 to 60 % in syngas gas hot treated with aluminum oxides and aluminum silicates at the temperatures in the range of 600–700°C.

3.4. Carbon-based compounds abatement

During tar destruction by thermal and/or catalytic processes, some tar molecules may be converted into CH_4 and heavier hydrocarbons, through reaction (1). As some tar destruction catalysts are not powerful enough to destroy completely these hydrocarbons, especially CH_4, syngas may contain appreciable amounts of CH_4 after hot cleaning processes. Thus, syngas usually needs to be further treated by catalytic steam reforming reactions, (2), (3), (14) and (15) to convert CH_4, and other heavier gaseous hydrocarbons into CO, CO_2 and H_2.

$$CH_4 + 2\ H_2O \leftrightarrows CO_2 + 4\ H_2 \quad (14)$$

$$C_nH_m + n/2\ H_2O \leftrightarrows n/2\ CO + (m-n)\ H_2 + n/2\ CH_4 \quad (15)$$

CO_2 reforming reactions, (4), (5) and (16) may also occur, as syngas is usually rich in CO_2.

$$C_nH_m + n/4\ CO_2 \leftrightarrows n/2\ CO + (m-3n/2)\ H_2 + (3n/4)\ CH_4 \quad (16)$$

Several researchers have studied carbon-based compounds abatement through cracking and reforming reactions and consequently many catalysts have been developed and tested. CH_4 reforming reactions is a well known subject, however, research work is still going on with the aim of reducing operational costs, including decreasing experimental parameters severity and catalyst price. Some authors stated that the rise of temperature favoured CH_4 reforming reaction, Choudhary and Rajput 1996 and Galluci et al., 2006. Wu et al., 2009 reported that it was possible to attain CH_4 conversion around 95% at 850°C and in presence of metal monolith catalysts with Ni. Several authors observed that the rise of temperature increased CH_4 conversion, because steam reforming reactions are endothermic, thus they are favoured by high temperatures. Galluci et. al., 2006 reported that with H_2O/CH_4 feed ratios between 3 and 4, it was possible to attain methane conversion around 80% at 850°C. Some other authors also suggested the use H_2O/CH_4 ratios around 4 to achieve CH_4 conversion higher than 80%, Hou and Hughes, 2001, Maluf and Assaf, 2009 and Rakass et al., 2006.

Choudhary and Rajput, 1996 obtained good CH_4 conversions in presence of a NiO–CaO catalyst, with a Ni/Ca ratio of 3.0. The rise of temperature and of residence time decrease CH_4 content, however, at 850°C the effect of residence time was lower than that observed at lower temperatures. The increase of temperature and of residence time also improved CO selectivity and thus, H_2/CO mole ratio decreased on final gas composition.

Huang et al., 2006 also studied the catalytic effect of nickel based catalysts by preparing and testing catalysts with 2 % of Ni supported on samaria-doped ceria (SDC), on gadolinia-doped ceria and on α-Al_2O_3. SDC-supported Ni-Cu catalysts were also developed, which contained 0.5 wt % of Ni or Cu and addition of 0.01-0.1 wt % of Cu or Ni. It was observed that CH_4 steam reforming was favored at higher temperature, while WGS reaction decreased. These authors observed that the CH_4-H_2O reaction was much favored in presence of the Ni catalyst doped with Ce. This catalyst also led to a higher production of CO_2, because WGS reaction was favored. In general, the addition of Cu to Ni catalysts increased the WGS reaction activity, which enhanced with the amount of the bimetallic Cu-Ni species.

The effect of a NiO/MgO catalyst on CH_4 steam reforming was studied by Yang et al., 2010 who also reported that the catalyst was efficient and stable, though CH_4 conversions were much affected by reaction temperature and steam/carbon mole ratio. CH_4 conversion was around 97% at 875°C and when the steam/carbon mole ratio was around 1 and the hydrogen volume content was enhanced 1.5 times.

Wu et al., 2009 tested several Ni/SBA-15/Al_2O_3/FeCrAl metal monolith catalysts with different Ni contents (5, 10 and 15 wt%) and reported that the monolith catalysts with 10 wt% of Ni loading had good activity. Other metal monolith catalyst with 10% Ni/$Ce_xZr_{1-x}O_2$/SBA-15/Al_2O_3/FeCrAl ($x = 0$, 0.5 and 1) with $Ce_xZr_{1-x}O_2$ of 1.7 wt% were also prepared and tested. The results obtained by these authors showed that the use of CeO_2 and $Ce_{0.5}Zr_{0.5}O_2$ as promoters improved the activity and stability of metal monolith catalysts, whilst opposite effect were obtained by the addition of ZrO_2, as catalyst activity and stability decreased. The effect of Ni loadings (0.2-10 wt %) on the performance of α-alumina supported Ni catalysts was also studied by Li et al., 2006. CH_4 conversion increased with the rise of Ni loading amount and the catalyst bed temperature during the reaction was dependent on the Ni loading amount. However, the bed temperature profile was much dependent on the ratio of catalyst amount/gas flow rate. At low ratios, around 0.12 g h/mol, the highest bed temperature increased considerably with Ni loading, at medium ratios, about 0.40 g h/mol, the highest bed temperature decreased with Ni loading in the range of 0.2-0.9 wt %, while beyond this value, the highest bed temperature rose only moderately. According to the authors a catalyst with 0.9 wt % of Ni showed good performance for CH_4 steam reforming at a lower bed temperature.

The effect of using different supports like silica, γ-alumina, and zirconia for nickel based catalysts was studied by Matsumura and Nakamori, 2004. These catalysts were reduced with H_2 before use. Nickel supported on γ-alumina was inactive for reforming at 500°C, probably because it was not much reduced with hydrogen at 500°C, but the catalyst reduced at 700°C showed to be rather active with CH_4 conversion around 17.4% and with 91.8% of CO_2 selectivity, while nickel was partially oxidized during the reaction. The activity of nickel supported on silica and reduced with hydrogen at 500°C decreased with oxidation of nickel particles by steam during the reaction. The most active catalyst for steam reforming at 500 °C was nickel supported on zirconia. Surface hydroxyl groups on these catalysts reacted readily with CH_4 to produce H_2 and CO, probably because of the role of hydroxyl groups in the mechanism of the steam reforming. The accumulation of water on the surface of zirconia-supported nickel catalyst seemed to have favored the formation of H_2 and CO_2 in the subsequent reaction with CH_4.

Oliveira et al., 2010 studied catalyst extrudates for steam methane reforming and stated that in presence of these catalysts steam reforming reaction presented high mass and heat limitations: diffusion of reactants/products and heat transfer in the gas/solid interface. Thus, these authors prepared a Ni-Al_2O_3 catalyst extrudate promoted with potassium and obtained a catalyst with large pores that allowed reducing diffusional limitations. The nickel based catalyst needed to be reduced to become active to steam methane reforming. Experiments done with powder catalyst at a temperature between 701 and 800K and at a pressure of 2 bar showed that the this catalyst is very selective for CO_2, as the conversion was below 10%, while the CO_2 selectivity (CO_2/CO) was always above 10.

Hou and Hughes, 2001 studied methane steam reforming and the reverse of WGS reaction (17) using a commercial Ni/α-Al_2O_3 catalyst and observed that CO and CO_2 were formed as primary products and that CH_4 conversion was proportional to the partial pressure of methane at low product concentrations. Temperature and steam/CH_4 ratio had a great effect on product distribution, while no significant effect of pressure was observed. The rate controlling steps for steam reforming reaction were surface reactions between adsorbed species under conditions of no diffusion limitation. The experimental results showed that the formation of H_2 was favored by low temperature and by a high steam/CH_4 ratio and that the rate of CO_2 formation was much faster than that of CO at low temperature. These results might mean that CH_4 decomposition into H_2 and CO_2, through reaction (14) was dominant over methane steam reforming on the catalyst used at low temperature. However, the results obtained also showed that reaction (2) happened simultaneously, thus, the domination of the former reaction were expected to decrease with temperature

Murata et al., 2004 studied steam reforming of heavier hydrocarbons (isooctane and methylcyclohexane) in presence of Fe and Ni based catalysts, as FeMg/Al_2O_3 catalyst was effective for CH_4 decomposition in the presence of O_2/CO_2. Though, this catalyst was active for steam reforming of isooctane, cracking C-C bond, H_2 production was low, but it could be improved by Rh modification. This modification also increased the formation of aromatic byproducts. However, H_2 formation of Ni/ZrO_2 catalyst was around 7 times higher than that of FeMg/Al_2O_3.

3.5. Integration of syngas cleaning processes

Nowadays challenges is the development of catalysts that could resist to some of syngas undesirable components, like sulfur compounds and that could have several functions in syngas hot cleaning processes. Some authors have developed catalysts that could simultaneously reduce NH_3 and H_2S. Gangwal et al., 1993 reported that a number of catalysts containing Ni-based and MoS_2-based materials were able to remove simultaneous NH_3 and H_2S from syngas at temperatures around 800ºC or even higher. With the aim of reducing H_2S and NH_3 contents in syngas, Gangwal et al., 1998 developed and tested at 725ºC several sorbents/catalysts containing Ni, Co, Mo, and W with Al_2O_3, TiO_2, and other oxides as supports, combined with a zinc titanate sorbent. These materials were named as HART-xx, for example HART-49 reveled to be a promising material, due to attrition resistance in fluidized beds, HART-56 led to NH_3 reductions around 90%, while H_2S contents went down to 20 ppmv. However, some of these materials still need to improve their reactivity and attrition resistance and to lower their cost, before they may be used at larger scale.

CC&S (Catalysts & Chemical Specialities) commercializes catalysts that allow reducing simultaneously H_2S, NH_3 and tar. These catalysts contain different contents of nickel and CaO supported in Al_2O_3 and are available with different shapes and sizes. One of these catalysts, G-90 B5, is commercialized in cylindrical pieces with 16 x 8 mm and 5 holes with 11% of Ni, 6-9% of CaO and 76-82% of Al_2O_3 and showed a good performance for hot syngas leaning.

Pinto et al., 2009 c) and 2010 studied the activity of G-90 B5 catalyst during syngas hot treatment. Syngas produced by co-gasification of coal and different types of wastes went through a fixed bed reactor with dolomite and afterwards it passed through another fixed bed reactor with G-90 B5. After the dolomite reactor, syngas presented methane content reductions between 20 and 47%, while heavier gaseous hydrocarbons concentrations were around 80 to 90 % lower, depending on syngas initial composition. As a result, hydrogen contents in syngas increased to values around 40 to 45%. No tar and gaseous hydrocarbons with more than one carbon atom were ever detected in syngas sampled after the fixed bed reactor with G-90 B5. Consequently, H_2 contents in syngas increased to values between 50 and 55%, depending on the composition of the feedstock gasified, Pinto et al., 2009 c). In Figure 4 may be analysed the effect of syngas hot treatment with dolomite and G-90 B5 catalyst on syngas composition.

Syngas was also sampled to determine the evolution of H_2S and NH_3 contents along the syngas cleaning process tested. Syngas sampled after dolomite fixed bed reactor showed H_2S reductions between 68% and 74%, depending on H_2S concentration in syngas, while the reductions observed in NH_3 were between 30 and 50 %, also depending on feedstock nitrogen content. Further reduction in H_2S and NH_3 contents were observed after the syngas had gone through the G-90 B5 catalyst. The overall H_2S and NH_3 reductions were higher than 97%, thus, final H_2S and NH_3 concentration in syngas were usually below 20 ppmv and 30 ppmv, respectively, which allowed syngas utilization in boilers and gas engines, Pinto et al., 2010. In Figure 5 are shown the great reductions in H_2S concentration when syngas went through a fixed bed reactor with dolomite. H_2S contents were further reduced inside the fixed bed reactor with G-90 B5 catalyst. Some decreases in NH_3 concentration were also observed during syngas treatment with dolomite, however, the highest reductions were obtained inside the reactor with G-90 B5 catalyst.

In Figure 6 is shown a possible configuration for hot cleaning a syngas with initial appreciable contents of several undesirable components. Due to the high H_2S content in initial syngas composition it was decided to decrease the concentration of this component by passing syngas through a fixed bed reactor with dolomite. Besides achieving a H_2S reduction of about 68%, NH_3 decreased around 55%. Also hydrocarbons and tar contents showed important reductions. After tar and ammonia abatement with a nickel based catalyst (stream

4) the presence of tar and gaseous hydrocarbons with more than one carbon atom was not detected, CH_4 concentration decreased around 80% in relation to initial composition in syngas and both NH_3 and H_2S concentrations had reductions of around 98%. After final gaseous hydrocarbons abatement with a steam reforming catalyst (stream 5) gaseous main components were CO_2 and H_2, whose concentration presented an increase of around 65%.

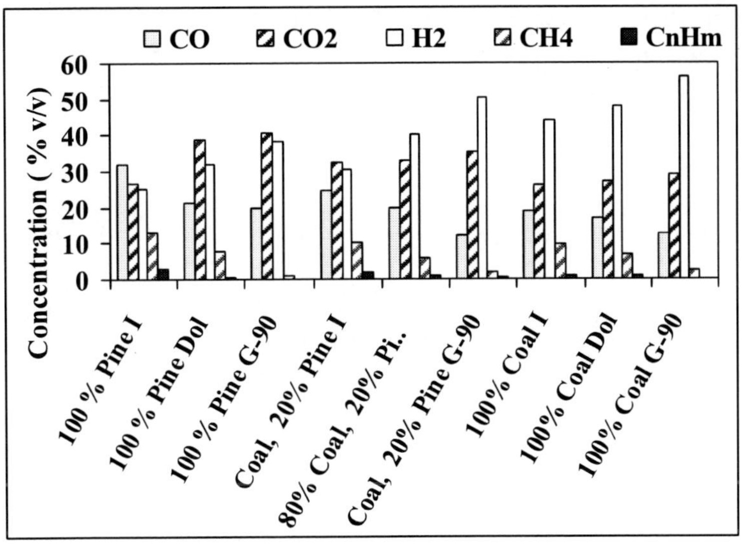

Figure 4. Effect of syngas hot treatment on syngas composition. I - initial composition, Dol - syngas composition after dolomite reactor, G-90 - syngas composition after G-90 B5 catalyst reactor.

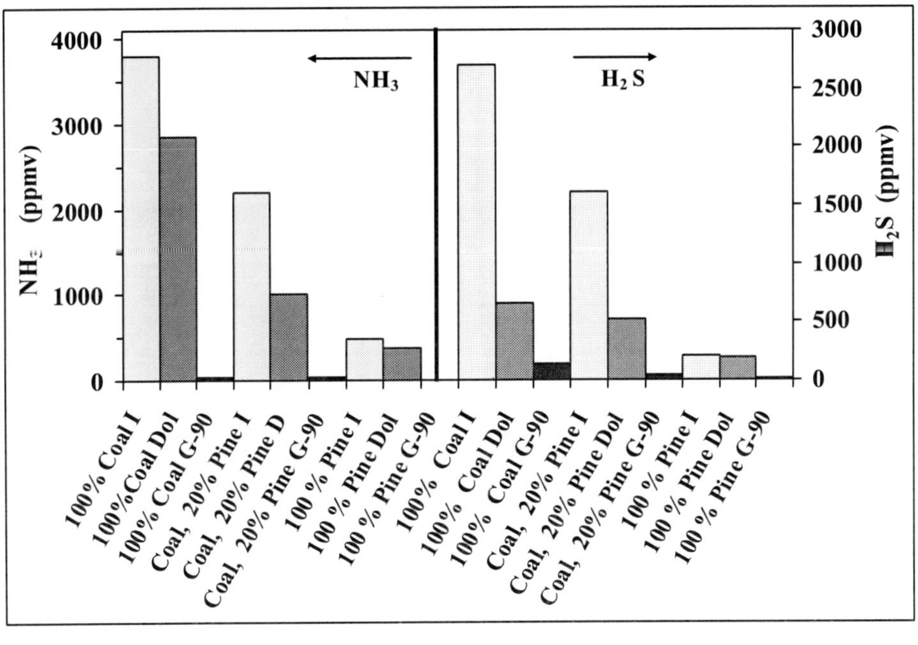

Figure 5. Effect of syngas hot treatment on H_2S and NH_3 contents. I - initial composition, Dol - syngas composition after dolomite reactor, G-90 - syngas composition after G-90 B5 catalyst reactor.

The configuration proposed in Figure 6 can deal with a wide range of syngas compositions with the most common undesirable components, however it could be simplified according to syngas initial composition. For instance the sulphur and halogens abatement reactor may be eliminated if the contents of these compounds are low enough to justify such change. On the other hand, the catalyst to be used inside the tar and ammonia abatement reactor should be chosen according to the concentration of these compounds and syngas final utilization. Also, the use of multi-function catalyst could also reduce the number of operations proposed in Figure 6 for syngas hot cleaning. It is of great importance to optimize gasifier operating conditions in order to decrease the contents of all undesirable components, as the syngas cleaning operation may be greatly simplified according to syngas initial composition and syngas end-use.

As most syngas hot cleaning processes are endothermic, it is fundamental have good equipment insulation and to achieve a good integration of downstream gas cleaning reactors to improve the overall process energy conversion efficiency and reduce operating costs, Brown et al., 2009. It is also advisable to have hot cleaning operations close to the gasifier to avoid the need for extensive tar control and to improve energy conversion efficiency.

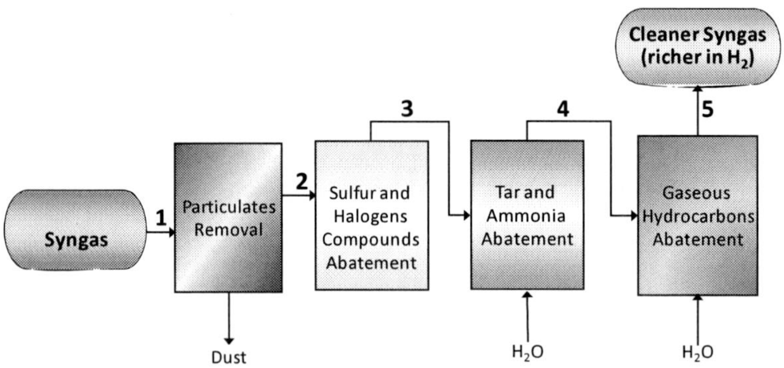

	2	3	4	5
H_2 (% v/v)	31	40	51	60
CO_2 (% v/v)	32	33	35	39
CH_4 (% v/v)	10	6	2	n.d.
C_nH_m (% v/v)	1.7	0.8	n.d.	n.d.
Tar (g/Nm^3)	11	5	n.d.	n.d.
H_2S (ppmv)	1610	517	35	n.d.
NH_3 (ppmv)	2206	1005	37	n.d.

Figure 6. Possible configuration for syngas hot cleaning (n.d. – not detected).

4. WATER GAS SHIFT REACTION

After syngas cleaning and upgrading, its H_2 content may be further increased by water gas shift reaction (WGS) that converts CO into H_2 and CO_2 in presence of steam and specific catalysts, reaction (17). WGS reaction is usually promoted after CH_4 steam reforming. WGS reaction also allows achieving the right H_2/CO ratio for chemical synthesis and the removal of CO to produce a gas suitable for fuel cells applications. On the other hand, it is possible to achieve to CO_2 capture to prevent global warming and climate change through WGS reaction, as CO is converted into CO_2, which after capture may go to storage, while a H_2 stream is obtained, to be used afterwards as fuel or in several applications.

$$CO + H_2O \leftrightarrows CO_2 + H_2 \quad (17)$$

$$CH_4 + 2\,CO_2 \leftrightarrows 2\,CO + 2\,H_2 \quad (18)$$

The WGS reaction is equilibrium limited and CO conversion depends on the shift reactor temperature, Carbo et al., 2009. Though CO conversion is thermodynamically favored at lower temperatures, catalyst activity is usually higher at higher temperatures. When syngas has high CO concentration, large steam amounts are needed.

WGS reaction is favored by the presence of a suitable catalyst, depending on catalyst type and performance different temperature and pressure conditions are needed. WGS reaction is well known for a long time and has been used to produce H_2 from natural gas. There are two main processes for WGS reaction with industrial application: a) at higher temperature (320–450°C) with High-Temperature Catalyst (HTC) like iron-chromium catalysts and b) at low temperature (200–250°C) in presence of Low-Temperature Catalyst (LTC) like copper-zinc based catalysts for further decreasing CO content. HTC requires sulphur content in syngas below 20 ppm, while LTC requires syngas to have sulphur content below 0.1 ppm, Carbo et al., 2009. There is also another type of catalysts for WGS reaction commercially available, sour shift catalysts. Active components of these catalysts are sulfided Co and Mo (CoMoS). They operate at temperatures in the range of 250 to 500°C and sulphur contents in syngas higher than 1000 ppm, the minimum sulfur concentration at 480°C to keep catalyst sulfide, lower temperatures correspond to lower minimum sulfur content, Frank, 2003.

Usually, syngas is first introduced into a reactor with HTC and afterwards it goes into a reactor with a LTC, as the HTC reactor converts the bulk of CO and the LTC reactor converts lower amounts of CO, but ensures that only small CO concentrations are present in outlet gas. When sour shift catalysts are used, reactors also operate in sequence for bulk CO conversion, since the CO final contents are higher than those obtained with LTC, Carbo et al., 2009.

A high-temperature catalyst and a low-temperature catalyst for WGS reaction were studied by Chen et al., 2007, who reported substantially different behaviors for HTC and LTC. The results obtained suggested that for HTC steam/CO ratio of 4 and temperature between 400 and 500°C should be used, while for LTC steam/CO ratio of 4 and temperature of 200°C were the optimal operational conditions.

Hla et al., 2009 a) studied two iron–chromium based commercial catalysts, High-Temperature Catalyst (HTC1 and HTC2) at 450°C. Both catalysts are copper promoted iron–chromium oxide based, they have high iron content and have only a slight difference in copper and chromium concentration in the oxides. No great differences between the performances of these catalysts were reported, which is not surprising as they both have similar compositions. Four different gas inlet compositions were tested and the highest WGS reaction rate was achieved for the syngas with the maximum CO concentration for both catalysts.

Guo et al., 2009 prepared $Cu/ZnO/Al_2O_3$ catalysts by the coprecipitation method and tested their performance for WGS reaction. After several shut-down and start-up operations, $Cu/ZnO/Al_2O_3$ catalysts presented continuous deactivation, probably due to blocking or deterioration of the active sites by $Zn_6Al_2(OH)_{16}CO_3.4H_2O$ resulting from the degeneration of the oxides under cyclic operations. When $Cu/ZnO/Al_2O_3$ catalysts were calcined in air at 450°C, followed by re-reduction hydroxycarbonate species were removed. Thus the catalyst achieved the best activity and stability, though continuous deactivation in shut-down/start-up operation was still observed.

Besides Fe–Cr-based catalysts and the Cu–Zn-based catalysts, gold-based catalysts and platinum group metals have been tested for WGS reaction, despite the high cost of these catalysts. Other materials such as zeolites, Souza et al., 1999 and metals like Sn, Ce, Ru and Rd, Basinska et al., 2004 or B, Cu, Ba, Pb, Hg and Ag Rhodes et al., 2002 impregnated in the catalyst have also been studied to increase the performance of WGS reaction.

Souza et al., 1999 prepared and tested Cu- and Cu,Zn-ZSM-5 zeolite catalysts for WGS reaction. As zinc changed copper distribution species in the zeolite, the addition of zinc ions enhanced the activity of the Cu-ZSM-5. Thus, these catalysts showed considerable activity and selectivity for WGS reaction at 300—400°C and at steam/gas ratio of 3. Rodríguez et al., 2008 tested the performance of Cu/MoO_2, as WGS catalyst, because the Cu/ZnO catalyst, commonly used in industrial WGS reaction requires long and complex activation steps before use. These authors also tested Au/MoO_2 performance in WGS reaction, MoO_2 was chosen as an oxide support, because Mo cations were not fully oxidized and thus could take part in different steps of WGS. The results reported by these authors refer that Au/MoO_2 and Cu/MoO_2 had a catalytic activity five to eight times higher than that of Cu, being the activity of Cu/MoO_2 similar to that of Cu/CeO_2 and higher than that of Cu/ZnO. However, the presence of CO_2 and H_2 affected the stability and activity of Au/CeO_2 catalyst. The rise of CO_2 content increased catalysts deactivation, though the effect on the initial reaction rate was small, while H_2 significantly reduced the initial reaction rate but had a small effect on deactivation, Denkwitz et al., 2007.

Many other catalysts have been developed and tested for WGS reaction with the aim of increasing CO conversion. Catalyst performance may be increased by the presence of oxygen vacancies, traces of other metals or formation of proper alloys. Still, the nature of the catalyst support and the catalyst preparation process may also affect catalyst activity. Basinska et al., 2004 studied catalysts with Sn-Ce-Rh-O and its oxide components SnO_2 and CeO_2 were also tested. These catalysts showed an activity similar to that of low activity iron oxides. On the other hand, Rhodes et al., 2002 used about 2 wt% of several elements Cu, Ba, Pb, Hg and Ag as promoters for Fe_3O_4/Cr_2O_3, using a co-precipitation procedure. The results reported by these authors indicate an activity increase by the order Hg > Ag ≈ Ba > Cu > Pb.

The effect of WGS reaction temperature on the stability and deactivation behavior of a 4.5 wt% Au/CeO_2 catalyst was studied by El-Moemen et al., 2009. Reaction temperature of around 300°C, instead of the commonly used 180°C, was investigated. These authors observed that the rise of temperature led to considerable increase of reaction rate. On the other hand, the deactivation of the Au/CeO2 catalyst did not decrease with temperature, but was more pronounced at 300°C than at 180°C. After oxidative pretreatments only 35% of the XPS accessible Au was present as Au^0, while the remaining fraction was present as Au^0 (45%) or Au^{3+} (20%), respectively. On the other hand, after reductive pretreatment, the amount of Au^0 nanoparticle was the highest, while the amounts of $Au^{0\prime}$ and Au^{3+} were the lowest. On the contrary Ce^{3+} content was only little affected by the different pretreatment conditions. Oxidative pretreatments lead to the highest activity and highest tendency for deactivation (lowest stability), while the opposite was observed for reductive treatment in H_2.

Several researchers have focused their work in improving WGS reaction with the aim of increasing CO conversion and of decreasing temperature and pressure conditions. Thus, several catalysts have been tested. Gunawardana et al., 2009 prepared and tested several copper–ceria catalysts with copper loading in the range of 20–90 at% Cu at temperatures in the range of 150 to 360°C. It was found that these catalysts activity increased with copper loading and the best catalyst performance was obtained with 80 at% copper–ceria with which proved the synergy between copper and ceria. The catalysts showed stable activities at 360°C, but at 400°C significant deactivation was observed.

The performance of industrial catalysts can also be improved by the addition of noble metal or its alloys (Pd, Au, Pt), though these catalysts may deactivate at WGS experimental conditions, which together with their high cost are important disadvantages. Platinum based catalysts (Pt/ZrO_2), Tibiletti et al., 2006, and gold based ones have also been tested, Cameron et al., 2003. The main disadvantage of these catalysts was the high price, as the results obtained by Cameron et al., 2003 showed that gold supported on iron oxide ($Au/\alpha\text{-}Fe_2O_3$) was more active at lower temperatures than either $\alpha\text{-}Fe_2O_3$ or the mixed CuO/ZnO, currently used at commercial scale.

Thus, Liu et al., 2009 prepared a different catalyst, $Ni_2P(001)$, and found that it presented a higher activity than Ni(100) or even Cu(100). They also observed that the addition of a small amount of cesium to the phosphide surface increased even better the catalytic activity of $Ni_2P(001)$.

Lian et al., 2009 prepared Co-Mo-based catalysts supported on mixed oxide supports $MgO\text{-}Al_2O_3$ with different Mg/Al atom ratios. It was found that the octahedral Mo species in oxidized $Co\text{-}Mo/MgO_{(x)}\text{-}Al_2O_3$ catalyst and the contents of Mo^{5+}, Mo^{4+}, S^{-2} and S_2^{-2} species increased with increasing the Mg/Al atom ratio of the support. The formation of such species favoured the formation of the active Co-Mo-S phase of the catalysts. According to these authors, the Mg/Al atom ratio in the range of 0.475−0.525 led to an optimal catalytic activity for WGS reaction. In a more recent publication Lian et al., 2010 reported the performance of a cobalt-molybdenum/alumina catalyst ($Co\text{-}Mo/MgO\text{-}Al_2O_3$) which was impregnated with ammonium sulfide solution followed by H_2 activation, thus a large amount of Co-Mo sulfides was formed after activation, containing a well dispersed MoS_2 and CoMoS phase, which was formed formed due to the redispersion of Co_9S_8 particles decorated on the MoS_2 edges. The presulfided $Co\text{-}Mo/MgOAl_2O_3$ catalyst presented a higher activity and stability for WGS reaction with CO conversions around 95% at 250°C.

Liang et al., 2009 prepared and tested a monolithic macroporous $Pt/CeO_2/Al_2O_3$ catalyst that showed to have a similar reducibility property to that of the particle PtO_x/CeO_2 reported in literature. In presence of monolithic macroporous $PtO_x/CeO_2/Al_2O_3$, CO conversion was was one order of magnitude higher than that observed for the micro $Pt/CeO_2/Al_2O_3$.

Kim et al., 2009 tested ceria-promoted Pt catalysts in various supports such as $g\text{-}Al_2O_3$, SiO_2, TiO_2 (P-25), CeO_2, $SiO_2\text{-}Al_2O_3$, yttria stabilized zirconia (YSZ) and ZrO_2 and found that $Pt\text{-}Ce/TiO_2$ has the highest WGS activity particularly at low temperatures showed the most stable catalytic activity, thus the performance of this catalyst was better than that of conventional $Pt\text{-}Ce/CeO_2$.

Most of published information about WGS catalyst refers to synthetic gas mixtures with only one or two compounds and sometimes with only CO in an inert gas, as the presence of gaseous components commonly present in real syngas mixture may change catalysts activity and CO conversion (Watanabe et al., 2009, Hla et al., 2009 a), Kalamaras et al., 2009 and Shangguan et al., 2007). CO may react with H_2 to form CH_4 by the reverse of reforming reactions, (2) and (4). These reactions may occur without any catalyst and in presence of some catalysts that may also catalyse CH_4 formation. They need to be avoided, due to the consumption of H_2.

Aznar et al., 2006 used two shift reactors in series with several ICI and BASF catalysts to study WGS reaction. CO reductions higher than 90%, hydrogen contents as high as 73 vol % and CO remaining contents of 2.6 vol % were reported. These authors also stated that in presence of metallic iron catalyst and at temperatures below 500 °C, occurred the reverse of reaction (4), thus CH_4 formation and H_2 consumption were observed. Aznar et al., 2006 reported that when there was not enough steam in the reaction medium, Fe_3O_4 might lead to the formation of metallic iron, due to reactions (19) or (20).

$$Fe_3O_4 + 3\ H_2 \leftrightarrows 3\ Fe + 4\ H_2O \qquad (19)$$
$$Fe_3O_4 + 4\ CO \leftrightarrows 3\ Fe + 4\ CO_2 \qquad (20)$$

Kalamaras et al., 2009 studied a catalyst with 0.5 wt% Pt in TiO_2 in the range of 210 to 270°C and found that the reaction rate of WGS reaction increased with the rise of CO or H_2O concentrations in the feed stream, while the addition of H_2 in the feed stream led to significant decreases. On the contrary, the kinetic reaction rate was nearly independent on the concentration of CO_2. Hla et al., 2009 a) reported that the rate of CO conversion decreased as reaction (17) proceeds to the right leading to the decrease of CO ant to the rise of CO_2 and H_2 concentrations, which led to the backend of a WGS reactor. The reverse WGS reaction depended on the catalyst used. According to these authors, the removal of H_2 from reaction medium through a specific membrane or other device, would prevent or at least reduce the reverse WGS reaction.

Watanabe et al., 2009 reported that when Ni–Al oxide composite catalyst was used with a syngas mixture containing 10% of CO, 20% of CO_2 and 60% of H_2, besides WGS reaction, undesired CO methanation also took place at a large scale, which consumed significant amounts of hydrogen. However, if appropriate amounts of Fe were added to the Ni–Al oxide composite catalyst during the catalyst preparation plasma process, methanation was suppressed remarkably. On the other hand, when the gas mixture only contained CO mixed with N_2, at 310°C and with a steam/CO ratio of 1.5, methanation reaction was not observed and CO conversion into CO_2 reached 98%, for higher temperatures CO conversion was near equilibrium value.

El-Moemen et al., 2010 reported that gas mixture composition considerably affected the activity and stability of the catalysts, which also depended on the pre-treatment of the catalysts. The presence of H_2 or CO_2 commonly present in real syngas mixture significantly lowered the activity or the stability of the Au/CeO_2 catalyst, because of changes in catalyst properties, due to the formation of thermally stable monodentate carbonates that worked as reaction–inhibiting species. These authors also observed that the presence of high H_2 partial pressures partially reduced the support, modifying the adsorption sites for surface species adsorbed on the support. Thus, the stability of the Au/CeO_2 catalyst was considerably higher in presence of H_2-rich atmospheres than in absence of H_2.

Shangguan et al., 2007 studied water-gas shift reaction catalyst at lower temperature (200—400°C) in presence of B303Q (sulfided cobalt-molybdenum/alumina catalyst). The effect of temperature, space velocity and gas components on the formation of carbonyl sulfide (COS) were studied in a tubular fixed-bed quartz-glass reactor. COS was formed due to the reaction of CO with H_2S. Thus, the yield of COS over B303Q catalyst increased with temperature, reaching a maximum at 220°C, while it decreased sharply with the rise of space velocity and of steam content. In relation to syngas composition, Shangguan et al., 2007 observed that COS increased with the rise of CO and CO_2 and the yield of COS attained a stable value with the increase of H_2 and H_2S contents. The formation of COS could be avoided by the use of high ratio of water vapor to shift gas, short residence time and suitable temperature.

A new concept for preparing WGS catalysts was reported by Chen et al., 2009 that used atomic layer epitaxy (ALE) technique to prepare uniform copper nanoparticles dispersed on a silica support (ALE-Cu/SiO_2 with 2.85±0.32 nm), which showed to be highly active for WGS reaction. These authors suggested that nano Cu particles could be involved in strong interactions with the SiO_2 support, leading to a partially electropositive state, due to interactions with oxygen atoms at the surface of the support, even if the copper was reduced. The high efficiency of this catalyst for WGS reaction was due to its strong activity in promoting H_2O dissociation.

Several researchers still continue to prepare and test new catalysts to achieve higher WGS reaction conversion. Nowadays main challenge is the development of catalysts that may be used with syngas produced by wastes gasification, as syngas may contain some undesirable compounds that may poison the catalysts used for WGS reaction, such as sulfur compounds. Catalysts should be resistant to the contaminants present in syngas after gas cleaning. Several authors have tested different catalysts to analyse their activity, selectivity and sulfur resistance. Hla et al., 2009 tested syngas with different H_2S concentrations on the performance of Fe/Cr based catalyst for WGS reaction at 450°C. For H_2S concentration below 50 ppm the effect on catalyst activity was slight, but the catalyst activity was reduced to half of its initial performance when H_2S concentrations were higher than 250 ppm. However, if the H_2S content increased till around 700 ppm no further significant alterations were observed in catalyst activity.

CO conversion and H_2 formation trough WGS reaction is much affected by catalyst composition, thus most researchers have focused their activities in developing new catalysts and less attention have been paid on the performance of WGS reaction. However, some authors have concentrated their efforts on developing new WGS concepts.

Galvita et al., 2008 produced H_2 by cyclic water gas shift (CWGS) reactor, using Fe_2O_3–CeO_2–ZrO_2 as catalyst, which was converted into a reduced form by syngas. Next, steam was fed into the reactor for re-oxidation of the reduced material and H_2 was also released. Some coke was formed on the solid surface, due to Bouduard reaction (21), but is was destroyed during the re-oxidation step with steam, leading to the formation of carbon monoxide. H_2 gas with low CO content was produced by this process.

$$C + CO_2 \leftrightarrows 2\,CO \qquad (21)$$

Carbo et al., 2009 developed an advanced process with four staged reactors with distributed feeds of synthesis gas and quench water in between the reactors. Conventional shift reactor configurations use two reactors one with a HTC, followed by another reactor with LTC, with intermediate cooling, where the entire synthesis gas stream passes both reactors. In Carbo et al., 2009 process synthesis gas is split and introduced into the four reactors, the first split stream is mixed with intermediate pressure steam. The hot outlet stream of the first reactor is quenched with water, next mixed with the second synthesis gas split stream and went into the second shift reactor. A similar procedure was followed for the other reactors. According to Carbo et al., 2009, this configuration allowed reducing the overall steam requirement and increasing CO conversion, due to staged addition of reactants together with the corresponding temperature quenches. The use of HTC or sour shift catalysts was advantageous, as unlike LTC further purity requirements for sulfur components were not needed. Besides, steam requirements reductions in about 70%, the four staged reactors process also led to CO_2 capture ratios of around 85%. Consequently, lower electric efficiency penalties and higher net electric outputs for IGCC power plants with CO2 capture were obtained, which could counterbalance the increase of capital costs due to the use of a higher number of reactors.

A new concept of WGS reaction was reported by Stevens et al., 2010. The WGS reaction took place in presence of CO_2 sorbents, either in presence or in absence of catalyst and at mild reaction conditions with temperatures between 300 and 600°C and pressure values between 1 and 11.2 atm. These authors treated CaO sorbents with sodium hydroxide (NaOH) and with steam, which led to the formation of $Ca(OH)_2$. The presence of $Ca(OH)_2$ in the sorbent improved its performance, by increasing CO_2 sorption, to values near 100%, and also by rising WGS reaction conversion, as CaO showed catalytic properties towards WGS reaction. Moreover, the promotion of CaO with NaOH also allowed reducing significantly the regeneration temperature of the former.

5. H_2 Separation Through Membranes

Hydrogen separation from gaseous streams may be performed by different methods. A very well known process is chemical or physical absorption using solvents. These processes have high heat losses, great energy consumption and high solvent degradation rates. They also have low efficiency and negative environmental impact, due to the huge amounts of solvents that need to be circulated and regenerated and to the large equipments needed. Cryogenics separation has the advantage of making easier the transport of liquid CO_2, but as this process has high energy needs and high costs, it is used only in large-scale facilities. Another option is adsorption processes by different methods. Pressure swing adsorption (PSA) is the most used process, as it works at room temperature and at high pressure and recovers less of the feed-stream hydrogen. Other adsorption techniques, temperature swing adsorption (TSA), and electrical swing adsorption (ESA), are not so used, McKee, 2002. According to some authors the use of membranes for hydrogen separation from gaseous streams may be a viable alternative to PSA and cryogenic separation, depending on scale and on the purity required for the H_2 stream. Bounaceur et al., 2006 stated that membranes can provide significant improvements in relation to the best available technology, amine absorption. The main advantages of membranes are: compactness and light weight, simplicity of operation and maintenance and small size. Membranes also have low energy requirements, are environmentally friendly and need low labor effort

First membranes were developed for liquid phase separation processes in the XVIII century. Only during the forty decade of XX century were developed membranes for gas separation processes, in which there are gaseous phases on both sides of the membrane. These are the membranes suitable for H_2 separation form syngas. Since 1979 selective polymeric membranes have been used for separation of H_2 from off-gases in the ammonia industry and in the last decade of XX century, selective hydrogen palladium membranes have been used to produce pure H_2 in the electronics industry. A wide range of research work has been done by several research teams to develop selective membranes suitable for hydrogen separation from other gases, mainly for application within IGCC-processes. A membrane to be selective has to permit the mass transport through it of some components more easily than others. However, some polymer membranes have limited temperature tolerance and are more vulnerable to chemical damage from aromatic compounds and H_2S that may be present in syngas, thus they have limited applications for coal-based hydrogen production, Basu et al., 2004.

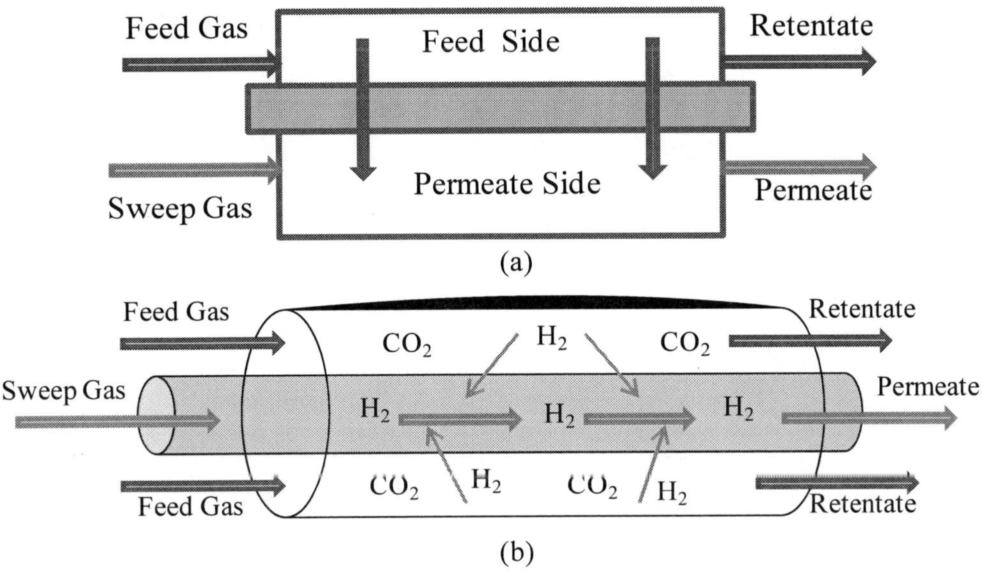

Figure 7. Co-current flows through a simple membrane. (a) The feed gas enters at one side of the membrane, while the sweep gas is on the other side of the membrane (b) The membrane is a cylindrical tube inside the outer tube where the feed gas enters.

A membrane can be settled in a device with only one flow input in the feed side of the membrane and one way out at the other side of the membrane (permeate side). Thus, the recovery is high at be beginning of operation, but is decreases with time, because the concentration of non-permeating species increase in the feed side, decreasing the permeating species, which reduces the driving force and the transport through the membrane. A better option is to remove the flow after permeation (called retentate). In the permeate side, the input stream is called sweep flow and the outlet stream is called permeate flow. In Figure 7 is shown a possible configuration for co-current flow, besides this one, there are other operation options: counter-current, cross-flow with perfect permeate mixing and perfect mixing, Kluiters, 2004. Cross-flow options are usually better, because as the flows run along the membrane, decline of membrane flux in time is reduced.

Membranes can be dense or porous materials. Porous membranes usually present low selectivity and high flux, while dense and small-pore membranes have low flux, but high selectivity. In porous membranes, molecules can move by different diffusion mechanisms: (a) Knudsen diffusion, (b) surface diffusion, (c) capillary condensation, (d) molecular sieving or by more than one of the previous mechanisms. These mechanisms depend on temperature and pressure conditions, membrane characteristics and gas composition. Knudsen diffusion usually happens with larger porous sizes, thus lower selectivity is obtained than that of capillary condensation and molecular sieving, Kluiters, 2004. In dense membranes the gaseous molecules are adsorbed on one side of the membrane, diffuse through the membrane and are desorbed on the other side of the membrane. Sometimes molecules decompose into ions, electrons or atoms, which diffuse through the membrane, thus these species have to recombine while they are desorbed.

Membranes can be either organic or inorganic. Organic membranes are made of polymers, they have many applications and a relative low cost, due to their advanced stage of development, but they are usually sensitivity to process conditions and trace contaminants. Operating temperatures are usually not higher than 100 °C and they have low chemical resistance to HCl, SO_x, and CO_2, and because of that they have limited applications for coal-based hydrogen production. Other drawbacks of polymeric membranes are limited mechanical strength and high sensitivity to swelling and compaction, Kluiters, 2004. Polymeric membranes are usually classified as glassy or rubbery, both are usually prepared by the phase inversion method. Rubbery membranes are prepared at temperatures above the glass transition temperature, while glassy ones produced at temperatures beneath the glass transition temperature. Rubbery membranes have higher flux and lower selectivity than glassy membranes.

Table 2. Main types of H_2 selective membranes

Main Membrane Types	Group	Sub-Group
Organic Membranes	Polymer	
Inorganic Membranes	Glass	Microporous Mesoporous
	Metal Membranes	
	Carbon Membranes	
	Ceramics	Porous Ceramic Dense Ceramic

Inorganic membranes appear to be more suitable for gasification gas applications, they can be made of glass, metal, carbon and ceramics (including zeolites). Inorganic membranes can operate at higher temperatures and have higher chemical stability than polymeric membranes. They can be either porous or dense, porous ones allow mass transport through their pores, while dense membranes enable mass transport through the bulk of the material, these can be metallic or solid electrolytes (ceramic). Membranes may also be classified according to their structure as symmetrical or composite membranes. In symmetrical membranes there is a homogeneous structure and in composite ones there are several layers with different characteristics, which usually change regularly from a dense to a porous material. In Table 2 is presented a possible classification for main types of H_2 selective membranes.

Glass membranes

Glass membranes are generally prepared by leaching manufacturing process, using silica. They can present several pore sizes: microporous, with pore sizes lower than 2 nm, and mesoporous, when pore sizes are in the range between 2-5 nm. Microporous membranes have higher selectivity, but lower fluxes than mesoporous membranes. Glass membranes can work at temperatures between 400-500 °C, but mainly due to their low selectivity, they are not very important for hydrogen separation.

Carbon membranes

Carbon membranes can be used at temperatures between 500 and 900°C and with non-oxidising environments, thus they are advisable for syngas. However, the cost of these membranes is high, its manufacturing still needs to be improved and they show brittleness, which difficult the package of larger membrane surfaces. These membranes may have different transport mechanisms: molecular sieving and surface diffusion. Molecular sieving membranes present considerable stability and fluxes, as the pore sizes are similar to H_2 molecule size, selectivity values between 4 and 20 have been reported, Kluiters, 2004. However, they are not yet commercially available at large scale.

Ceramic membranes

Ceramic membranes are made of a metal in the form of an oxide, nitride or carbide combined with non-metal species. They can be porous or dense. Porous ceramic membranes have two-layers: the separation layer made of alumina, zirconia, titania, or silica and a supporting layer more thicker and more porous. These ceramic membranes work at temperatures in the range of 200 to 600°C, but their stability is affected by the presence of steam. According to Delft et al., 2003 the stability may be increased by modifying silica with methyl groups. High H_2 fluxes through these membranes may be attained, leading to selectivities that can reach values around 140, depending on gaseous components. However, further developments are still needed before that may be used at large scale. Zeolites are another kind of porous ceramic membranes that operate as molecular sieves, due to the small channels inside their structure. The production of these zeolites to achieve the separation of small molecules is still difficult and further developments are still needed to overtake several problems, Kluiters, 2004.

Dense ceramic membranes are usually made of $SrCeO_{3-\delta}$ and $BaCeO_{3-\delta}$. They work at temperatures between 600 and 900°C. As H_2 is transported through the solid phase in the form of protons, only hydrogen ions can move through the membrane, thus, high selectivity can be achieved, especially at temperatures near 900°C. The main drawback of these materials is their stability in the presence of CO_2 and H_2S, which prevent their utilization for syngas. Again, further developments are still needed for this type of membranes.

Metallic membranes

In metallic membranes, H_2 is adsorbed on one side of the membrane, then it forms two atoms that move through the metal matrix and recombine at the other side, afterwards H_2 desorbs at the permeate side. Palladium and palladium alloys have shown good performance for producing pure hydrogen and usually work at temperatures between 300 and 600°C. If H_2 is locked inside the palladium lattice, which may happen at low temperatures, membrane may breakdown. To avoid this problem, some authors have added other elements such as silver or copper to palladium membranes. The main disadvantage of these membranes is the very high cost of palladium, thus, thin layers of this element deposited on a porous ceramic or metallic support have been used, which have also the advantage of improving the flux. Another drawback is the high sensitivity to sulfur and chlorine compounds that may poison the membrane and reduce H_2 fluxes from 20 to 100%. Johnson Matthey produces palladium-silver alloy membranes that are commercially available.

In Table 3 are presented the main characteristics of hydrogen selective membranes, Kluiters, 2004, which allows membranes comparison. Metallic membranes present high H_2 selectivity and flux and can operate at suitable temperatures for syngas. However, they may be poisoned by some of syngas components like H_2S, HCl, and CO, thus syngas needs to have very low contents of these components. The selection of the type of membrane to be used depends on gas flow characteristics, such as: composition, temperature and pressure, membrane performance, like selectivity and flux and also membrane cost.

Table 3. Main characteristics of most important types of H_2 selective membranes, Kluiters, 2004

	Dense Polymer	Metallic	Porous Ceramic	Porous Carbon	Dense Ceramic
Building Material	Polymer	Palladium alloys	alumina, zirconia, titania, silica	Carbon	$SrCeO_{3-\delta}$ $BaCeO_{3-\delta}$
Operating Temperature (°C)	< 100	300-600	200-600	500-900	600-900
H_2 Selectivity	Low	High (> 1000)	Medium	Medium	High (> 1000)
H_2 Flux	Low	High	High	High	Medium
Stability	Swelling, compaction, mechanical strength	Phase transition	Stability in H_2O	Brittleness, Oxidation	Stability in C_2O
Poisoning	HCl, SO_x, (CO_2)	H_2S, HCl, (CO)		Strong adsorbing vapours, organics	H_2S
Transport Mechanism	Solution/ Diffusion	Solution/ diffusion	Molecular sieving	Surface diffusion; molecular sieving	Solution/ diffusion (proton difusion)

After the choice of the membrane type, it is necessary to decide if the module in which the membrane is commercially available is adequate or needs to be changed. The module is the structure that contains the membrane area and the supporting units. Usually there are five types of modules: a) plate-and-frame and b) spiral-wound modules both for flat membranes, c) tubular, d) capillary and e) hollow-fibre all for tubular membranes. Kluiters, 2004 reviewed these different types of modules. In the choice of a module or before taking the decision of changing an existing one, different aspects should be considered: manufacturability, operability (temperature, pressure and fouling), efficiency, deterioration, maintainability and cost.

A membrane stage is a unit that contains one or more membrane modules. The simplest membrane processes only have one membrane stage, to increase membrane system performance, more than one stage are usually used. In Figure 8 configuration are used two membrane stages, In the first membrane unit is produced a permeated with the right purity and a retentate with a high amount of the wanted gas, thus it goes to another membrane to separated the undesired product that is rejected, while the permeate that do not has the required purity is recycled to the feed stream.

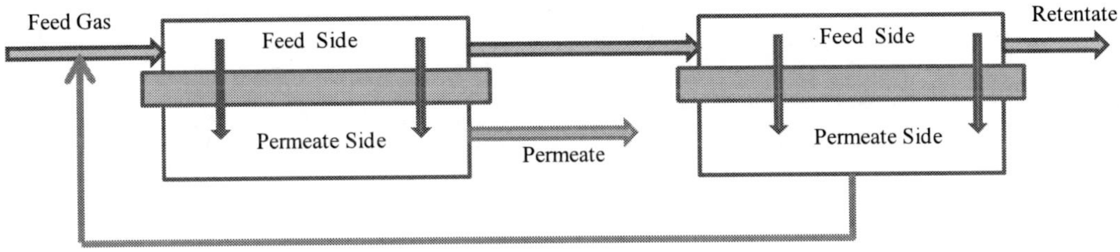

Figure 8. Membrane with two stages and recirculation of the second permeate cycle.

In Figure 9 scheme, the permeate does not have the required purity, so it goes to another membrane unit, while the retentate is rejected. In the second membrane unit, a permeate with the desired purity is obtained, whilst the retentate is recycled to the feed flow.

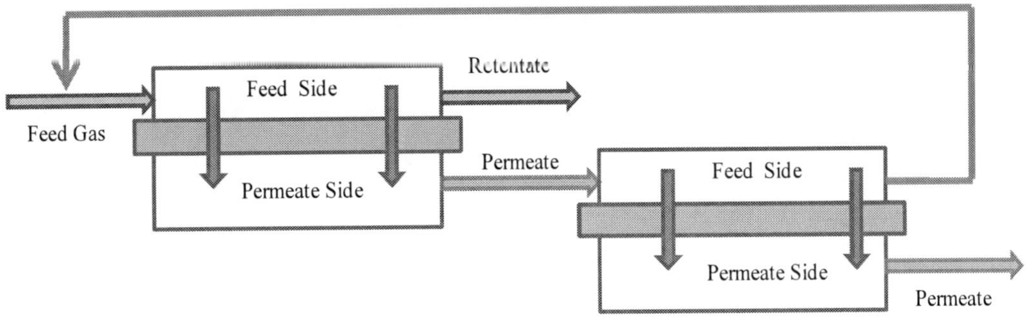

Figure 9. Membrane with two stages and recirculation of the second retentate cycle.

Multistage systems are often used, as the additional investment costs may be compensated by the increase in separation efficiency, Noble and Stern, 1995. However, as multistage systems include flow recycling, gas compression is needed, which increases process cost, thus the rise in overall process efficiency has to be enough to balance the high costs. With the rise of membrane modules used, increase the different ways of stream recycling, Kluiters, 2004 reviewed multistage membrane systems.

Table 4. Techniques for membrane preparation, Kluiters, 2004

Technique	Main features	Pore sizes
Sintering	Compression of powders and formation of solids by sintering at high temperatures.	0.1 - 10 μm
Stretching	Expansion of an extruded film or foil made from a partially crystalline polymeric material.	0.1 - 3 μm
Track-etching	A film or foil is subjected to perpendicular high energy particle radiation, next the film is immersed in an acid or alkaline bath to etch away polymeric material along the tracks obtained.	0.02 - 10 μm
Template leaching	Formation of a structure with a soluble and a non-soluble phase. The soluble phase is afterward leached out by an acid or base.	> 5 nm porous glass membranes
Phase inversion	A polymer in a liquid state is transformed into a solid state to produce a membrane	porous and non-porous
Coating	A number of coating procedures may be used: dip coating, plasma polymerization, interfacial polymerization and in-situ polymerization to create composite membranes. The membrane is supported by a porous sub layer.	
Sol-gel process	Transformation of an alkoxide precursor into a gel via: colloidal suspension route or polymeric gel route. Colloidal suspension route - the precursor is hydrolysed to form a sol (a colloidal dispersion of particles in a liquid), by altering the surface charge of the particles in the sol or by rising the concentration, thus particles tend to agglomerate and a gel (three-dimensional network structure) is obtained. Polymeric gel route - the precursor is hydrolysed slowly to originate an inorganic polymer and next a polymer network.	
Chemical vapour deposition (CVD)	A metal organic component is vaporized into the carrier gas to go into from one side of the membrane, while the other reactant go through the other side of the membrane (an inorganic polymeric gel layer on top of a support). Reactants diffuse into the pores and react there and product is deposited on the pores.	dense or micro-porous
Physical vapour deposition (PVD)	A solid material is evaporated in a vacuum system to form a thin to medium thickness film which is next condensed and deposited on a cooled substrate.	
Alloy casting and rolling	It is usually used for metallic plates and sheets manufacturing at large scale. This method is most used for metallic membranes preparation.	
Electroless plating	Controlled auto catalyzed decomposition or reduction of meta-stable metallic salt complexes on target surfaces. This method is most used for metallic membranes preparation.	
Electroplating	In a plating bath, the metal or an alloy is coated on the substrate, the substrate work as the cathode.	
Pyrolysis	Degradation of polymers coated on porous supports. Method to prepare silica and carbon molecular sieves, like silicone rubbers and thermosetting polymers.	

There are different techniques for manufacturing membranes, as reviewed by Kluiters, 2004 and Mulder, 2000. In Table 4 are summarized the main techniques for membrane preparation.

Many researchers and research institutions have put great effort in developing H_2 selective membranes and a large number of publications have appeared lately. Stiegel and Ramezan, 2006 reported that Oak Ridge National Laboratory (ORNL) developed a porous membrane, whose pore size and distribution can be precisely controlled to allow primarily hydrogen to diffuse through the pores, thus leading to a high separation performance. US Department of Energy (DOE) and ORNL have also developed dense ceramic solid electrolyte membranes, whose operating temperature are too high for applications to coal-based hydrogen production, and hydrogen fluxes have not achieved the level of commercial significance, Stiegel and Ramezan, 2006. Some research work has been dedicated to metallic membranes, most of which are based on Palladium (Pd). However, these membranes have been found to be susceptible to degradation from the presence of both sulphur and CO and more work needs to be done to improve its performance.

DOE has been investigating high temperature ceramic membranes that can affect the water–gas shift reaction in the synthesis gas stream while simultaneously separating the hydrogen, Stiegel and Maxwell 2001. After the removal of particulate in candle filters, the gasification gas is cooled down by steam addition in the reactor with high temperature membrane where the water–gas shift reaction occurs simultaneously with the separation of hydrogen. The equilibrium conversion of the water–gas shift reaction is favoured by the removal of hydrogen. The hydrogen is cooled and compressed to the desired pressure. The non-permeate stream from the membrane is treated to convert the remaining CO, H_2, and H_2S to CO_2, H_2O, and SO_2, Stiegel and Maxwell, 2001.

Dolan et al., 2006 reported that catalytic membrane reactors, composed of a suitable hydrogen-selective membrane and water-gas-shift catalyst, can be used to produce a pure hydrogen stream from coal-derived syngas with high yield. Amorphous alloy membranes can operate at 400 °C, therefore, they are suitable for use in a catalytic membrane reactors based on a metal-oxide catalyst. Dolan et al., 2006 present a critical review of amorphous alloys, their manufacture and properties, and compare them with other types of membrane reactors. According to these authors membranes that operate at 700 °C, and thus suitable for use with metallic catalysts, are still a technological challenge, because of to the tendency of amorphous alloys to crystallize after prolonged use at elevated temperatures.

Bounaceur et al., 2006 stated that membranes can provide significant improvements in relation to the best available technology, amine absorption. Membrane processes based on already existing materials can purify flue gas with about 0.5–1 GJ/tonne CO_2 recovered, while the energetic costs of amine absorption range from 4 to 6 GJ/tonne CO_2 recovered. Vasileiadis et al., 2004 analysed impermeable and membrane type reformers to treat syngas gas from coal gasification. Membranes were modified by adding nickel in the mesoporous multiplayer alumina tubes by the dry impregnation method using $NiNO_3$ solutions. However, successive (multiple) modifications of the membrane or modification with highly concentrated solutions resulted in almost complete blocking of the pores.

Inorganic membranes appear to be more suitable for gasification gas applications, they can be either porous or dense and the latter can be metallic or solid electrolytes (ceramic). Kaldis et al., 2004 studied two alternative separation options were studied: (a) low temperature separation, using polymer membranes, and (b) high temperature separation, by ceramic membranes, whose cost analysis show that CO_2 removal was technically feasible. Incorporation of shift reaction and membrane separators results in an energy penalty of 8–14%. However, this technology would permit reduction of CO_2 emissions exceeding 50%, compared to conventional IGCC.

Basu et al., 2004 reviewed gases separation by different membrane systems and the progression of membrane technology during the last years. H_2 may progress through the membrane by three different mechanisms: ionic (proton), atomic and molecular transport Ryi et al., 2007. Molecular transport membranes, like ceramics, act as molecular sieves and allow hydrogen separation from gas mixtures. In atomic transport membranes, usually called dense metallic, hydrogen is adsorbed onto the metal surface and then dissociated, atomic hydrogen diffuses through the bulk metal membrane, after hydrogen association, hydrogen is desorbed from the metal surface. In proton transport membranes, hydrogen is dissociated on one side of the membrane, protons and electrons are transported independently through the membrane and on the other side of the membrane hydrogen is reconstituted Ryi et al., 2007.

Metallic membranes based on palladium (Pd), nickel, platinum and the metals in groups 3-5 of the periodic table (previously known as groups IIIb–Vb), have been studied, due to their permeability to hydrogen. Pd and its alloys have proven to have high hydrogen permeability, good hydrogen selectivity and chemical compatibility with many hydrocarbons present in gas streams, Hatlevik et al., 2009. However, commercial application of pure Pd membranes has some limitations, as they have been found to be susceptible to degradation from the presence of both sulphur and CO, Pizzi et al., 2008. At temperatures higher than 723K, Pd membranes can be deactivated by carbon compounds. At temperatures below 573 K, H_2 diffusion through the membrane may cause brittleness, due to the formation of β-PdH, which leads to the formation of pinholes in the metal film and is responsible for membrane fast deactivation Gade et al., 2009. This problem may be solved by alloying Pd with metals such: silver (Ag), nickel (Ni), and copper (Cu), which improve membrane stability.

Some companies, like Bishop and Company (later became Johnson-Matthey) and Union Carbide developed Pd alloys with different compositions and metal contents with the aim of achieving good hydrogen separations from gaseous stream with several components. Pd alloys with Ag increased hydrogen permeability in relation to pure Pd, prevented hydrogen embrittlement, by preventing α to β phase transition, increased robustness during thermal cycling and improved resistance to fouling Pizzi et al., 2008.

On the contrary, alloys with Cu led to an initial reduction in hydrogen permeability, but when the alloy contained around 40% of Cu, hydrogen permeability was higher than that of pure Pd. Some Pd binary alloys with gold (Au) and Cu showed higher hydrogen permeability than that of pure Pd. These alloys were unaffected by thermal cycling and presented higher resistance to poisoning by H_2S. For Pd alloys with Ag and Au, the maximum hydrogen permeability was obtained for the alloy with 27 and 10% of Ag and Au, respectively Hatlevik et al., 2009. Gryaznov et al., 2000 tested several Pd alloys, both binary and ternary and reported the good results obtained with $Pd_{95.5}$ $Ru_{4.5}$, due to the high tensile strength and with Pd_{68} Ag_{30} Ru_2, because the hydrogen permeability was 2.2 times higher than that of pure Pd at 500°C.

Some authors stated that hydrogen permeability could be increased by reducing membrane thickness, which would also have economical advantages, due to the high cost of Pd, though mechanical stability had to be maintained, which obliged new construction techniques. The use of composite membranes, thin Pd membranes supported on porous materials like stainless steel or ceramics, has led to higher permeances. Mejdell et al., 2009 tested various Pd-Ag membranes with three different configurations: self-supported (LP → 0), microchannel-supported (LP = ~ 0.001), and tubular-supported (LP = ~ 1). The first configuration has the drawback of giving an undefined gas flow pattern, while microchannel-supported configuration led to a more defined gas flow pattern, easing modelling and interpretation of flux. Besides, thin membranes allowed higher differential pressures in comparison with self-supported membranes. Even higher-pressure differences could be achieved in tubular configuration and a negligible support resistance was obtained for supports with comparable pore size dimension to the membrane thickness. Mejdell et al., 2009 results showed that each configuration had its strengths and weaknesses, the membranes studied presented high mechanical strength, as relatively high differential pressures could be used, both in the microchannel configuration and in the porous support.

Several Pd alloy membranes have been fabricated on different porous metallic substrates including PdAg, PdCu, PdAu, PdNi and PdFe, using different techniques: electroless plating, physical vapor deposition, chemical vapor deposition and electroplating Hatlevik et al., 2009. A ceramic support may have some advantages, namely: chemical stability, small pore sizes and uniform pore size distribution, which are important features for achieving thin membranes without defects. Nevertheless, differences in thermal expansion in the Pd and the ceramic support may lead to brittleness. The achievement of good connection between the metallic elements and the support are one of the challenges of supported membranes. Thus, several substrates have been also studied, such as: porous glass, porous ceramics, porous metals and composite structures (porous ceramic layers on porous stainless steel) with the aim of producing a uniform layer with similar hydrogen selectivity and mechanical stability of that of traditional rolled tubes Hatlevik et al., 2009.

Ishitsuka et al., 2008 tested amorphous $Zr_{36}Ni_{64}$ membranes in substitution of Pd membranes, due to its high cost. The membrane had a coating of 100 nm Pd in each side to facilitate hydrogen dissociation on the surface. Two structures were studied a one-membrane module and a four-membrane module. The results obtained showed that wider membranes were more suitable for hydrogen separation and though the performance of four-membrane module was around four times higher than that of one module, the recoveries of a four-membrane module were much lower than that of a one-membrane module. The four-membrane module allowed achieving 82.7% recovery for a 31.7 mL (STP)/min permeation rate, being the permeated gas over 99.999% of pure hydrogen Ishitsuka et al., 2008.

There are some differently designed membrane modules: plate-and-flame modules, tubular modules, spiral wound modules and hollow-fiber modules. The first two types can be used for Pd-based alloy membrane. There are different methods for manufacturing Pd-based alloy membrane modules, Ryi et al., 2007 reviewed some of these methods. As most hydrogen purification modules have complicated structures its construction is difficult. These authors manufactured a plate-and-flame type hydrogen purification module with easy assembly from stainless-steel tubes and plates. The membrane module contained a unit cell with Pd–Cu–Ni ternary alloy membranes.

Xiong et al., 2010 manufactured a $Pd/Nb_{40}Ti_{30}Ni_{30}/Pd$/porous nickel support composite membrane, which showed very good permeation capability and acceptable mechanical properties. The H_2 flux through this membrane was several times higher than that of self-supported $Nb_{40}Ti_{30}Ni_{30}$ and Pd alloys, which make this membrane a possible substitute for Pd and PdAg alloys for hydrogen separation and purification.

Zahedi et al., 2009 prepared a Pd membrane electroless plated on the surface of a Porous Stainless Steel (PSS) disk, whose surface was modified with WO_3 to avoid metal penetration into palladium layer. Thus, mean thickness of the Pd membrane on the modified disk was 12 μm. PSS has several advantages, namely: corrosion resistance, mechanical strength, thermal stability and suitable thermal expansion coefficient. However, adverse PSS pore sizes have unfavorable consequences on the palladium layer thickness, this together with the difficulty of metal diffusion through PSS into the palladium layer at high temperatures are common problems. To solve these problems, authors modified the support surface with WO_3. The results obtained showed high permeability and selectivity for hydrogen and a stable membrane during long duration permeation tests.

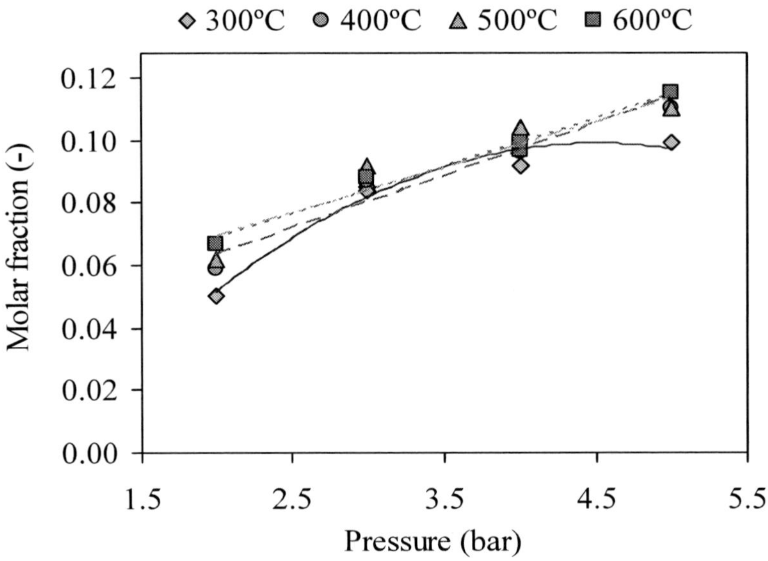

Figure 10. Effect of pressure on H_2 molar fraction in permeate flux when a Pd/Ag membrane is used to separate H_2 from a syngas mixture with 70% of H_2 and 30% of CO_2.

Membrane performance may be affected by the presence of undesirable compounds like, hydrocarbons, sulfur compounds and/or CO. In Figure 10 may be observed the good performance in H_2 separation obtained with a Pd/Ag membrane when the initial gas contained only 70% of H_2 and 30% of CO_2. In Figure 10 may be observed that H_2 molar fraction in permeate flux increases with the rise of pressure gap between the feed side and the permeate side of the membrane. On the other hand, the rise of temperature had a lower effect than pressure gap on H_2 molar fraction in permeate flux.

However, H_2 flux through the membrane decreased when a syngas mixture with 10% of CH_4 and the same amount of CO_2 was used, Figure 11. For the syngas mixture with 60% of H_2, 10% of CH_4 and 30% of CO_2 the rise of pressure gap between the feed side and the permeate side of the membrane seemed to have a milder effect on H_2 flux through the membrane than those observed when a syngas with only H_2 and CO_2 was tested. On the other hand, the rise of temperature from 300° to 500°C led to a greater increase in H_2 flux through the membrane then those observed for the syngas with only H_2 and CO_2. Both Figures 10 and 11 show that the presence of CH_4 in feed gas may affect H_2 flux through the membrane.

Keeping constant membrane base material, membrane performance may be improved by increasing membrane surface area, by decreasing membrane thickness and by increasing partial pressure difference across the membrane. The new challenges for membranes manufacturers are to achieve these goals without decreasing membrane stability. Another challenge is to guarantee membrane stability and performance even when syngas contains some impurities like, hydrocarbons, sulfur compounds and/or CO.

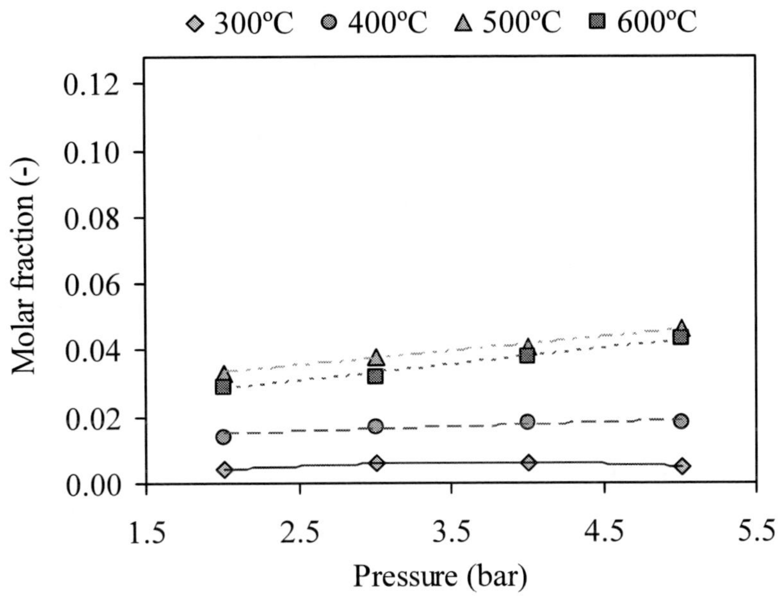

Figure 11. Effect of pressure on H_2 molar fraction in permeate flux when a Pd/Ag membrane is used to separate H_2 from a syngas mixture with 60% of H_2, 10% of CH_4 and 30% of CO_2.

6. Hybrid Systems

Some authors have reported the advantages of coupling WGS reaction with a membrane device, as the constant removal WGS reaction products promote the reaction pathway toward the product side, according to thermodynamic equilibrium. Thus, the removal of H_2 leads to an increase of CO conversion.

Hla et al., 2009 studied WGS reaction kinetics at 450°C in presence of two Fe-Cr commercial catalysts, named as HTC1 and HTC2. Both catalysts were copper promoted Fe-Cr based catalysts and due to the similarity in composition, both led to identical results, CO conversion was maximum at the beginning of the reaction when CO concentration was higher, as CO concentration decreased leading to the rise of CO_2 and H_2 concentrations, CO conversion decreased. H_2 removal from the reactional medium allowed increasing CO conversion. Thus, Hla et al., 2009 suggested that the use of a membrane reactor with H_2 separation was a good option to increase the conversion of WGS reaction without having to separate H_2 downstream.

A catalyst needs to be placed inside the membrane reactor and the all equipment is named as membrane reactor. A possible configuration for this type of reactors is shown in Figure 12.

These equipments are more difficult to design and to operate, as reactions that occur inside the device may lead to gaseous composition changes, alterations in gaseous components partial pressures and formation or reduction of contaminants, consequently membrane operation may be greatly affected. However, when a Pd membrane is used, a stream with high H_2 purity is obtained at a lower pressure than the pressure of the inlet gas stream, because the H_2 transported through the membrane is proportional to the pressure gap between the two sides of the membrane. On the other hand, CO_2 is retained at the inner side of the membrane at a high pressure, thus the power needed to recompress CO_2 is lower, which is another advantage of membrane reactors.

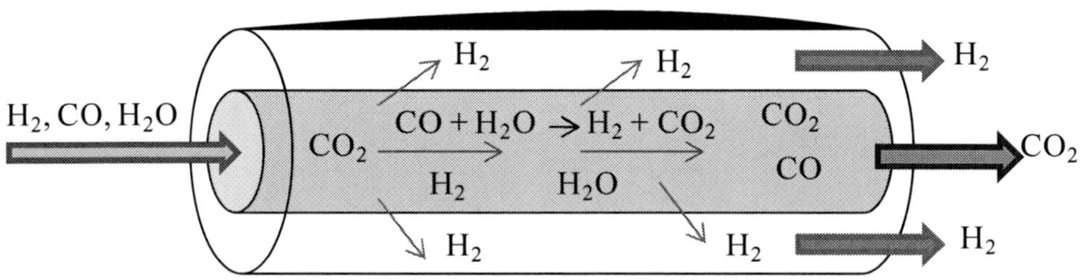

Figure 12. Possible configuration for a membrane reactor.

There are different options for placing the catalyst in a membrane reactor: a) catalyst placed inside the membrane, b) catalyst placed in a membrane top layer and c) catalyst placed inside the feed stream, Kluiters, 2004. The last option is the simplest one, as it is easy to organize, operate and replace the catalyst. On the other hand, the replacement of the catalysts for the two other options is complicated and usually involves replacing the complete membrane reactor.

Though membrane reactors may have several applications, according to Mulder, 2000 the commercial applications have been delayed due to different limitations such as: catalyst poisoning, low separation factors, outflows at higher temperatures and mass transfer restrictions. Therefore, several researchers have concentrated their work and effort in solving some of these problems and recently have appeared a large number of publications in this area. Pinacci et al., 2010 studied the WGS reaction in a reactor with a tubular palladium membrane inside of the reactor. A pulverized Fe/Cr commercially available catalyst was used at a temperature around 410ºC, leading to CO conversion up to the 85.0%, against the maximum of 37% reached with the traditional reactor without membrane. The use of the reactor membrane allowed to recover 82% of H_2 and to obtain a stream with H_2 purity higher than 97%. Bracht et al., 1997 used a catalytic membrane reactor, integrating a commercially available Fe-Cr catalyst for WGS reaction with a ceramic membrane selective for hydrogen. The main disadvantage of the Fe-Cr catalyst was the low activity at temperatures below 300°C. Thus, a Pt/ZrO_2 catalyst was also tested, though the disadvantage of a higher cost, Pt/ZrO_2 catalyst showed higher activity, better sulfur resistivity than the Fe-Cr catalyst and a great selectivity. Though, it was used a lower steam/CO ratio than usually employed for WGS reaction, the formation of adverse products like carbon or CH_4 was nor observed for either Fe-Cr or Pt/ZrO_2 catalysts.

Bi et al., 2009 also tested a membrane reactor, a $Pt/Ce_{0.6}Zr_{0.4}O_2$ catalyst was used for WGS reaction and a 1.4 mm thick high-flux Pd membrane was incorporated inside the reactor for H_2 recovery. CO conversions above thermodynamic equilibrium were reported, but H_2 recovery decreased from 84.8% to 48.7% at the highest space velocity, probably due to slow H_2 diffusion through the catalyst bed. Bi et al., 2009 suggested that this problem could be solved by decreasing the catalyst bed thickness, to reduce H_2 external mass flow resistance.

Another problem was identified by Battersby et al., 2009, it was pore collapse of the film matrix. These authors manufactured cobalt silica membranes for H_2 separation using the sol–gel technique. The membrane was tested in a membrane reactor at temperatures up to 300°C that used a commercial pelletised low temperature CuZnAl catalyst for WGS reaction. Fused alumina was used as catalyst diluting agent to ensure catalyst distribution along the membrane length. H_2 permeation purities of 89–95% were reported at high conversions, as the high GWS conversion provided greater membrane operational performance. The presence of cobalt helped to keep membrane structural microporosity, as H_2 selectivity was not influenced by steam presence. However, it was observed some flux decreased, due to pore collapse of the film matrix. Battersby et al., 2009 reported good operational selectivity for the membrane that provided a H_2 permeation purity of 90–95% at high conversions.

On important challenges of membrane reactors is to manufacture a membrane with high H_2 flux, by placing a thin metallic film on commercial ceramic membrane supports. These composite membranes were prepared by Tosti et al., 2000 using a thin film of Pd or Pd-alloy, which was placed on a porous support. Nevertheless, these membranes presented some defect formation, due to the appearance of pinholes on the membrane surfaces during thermal and hydrogenation-dehydrogenation cycling and because of that could not produce high purity H_2, Tosti et al., 2003. To overtake this problem, denser films of Pd or Pd-alloy could be tested.

Tang et al., 2010 modified a porous alumina tube supported MFI-type zeolite membrane by depositing molecular silica species in the zeolitic channels. Afterwards, the tube of the produced modified zeolite membrane was filled with a cerium doped ferrite catalyst to promote WGS reaction at the temperature that could vary from 400 to 550°C. The results obtained showed that higher CO conversions were obtained, together with H_2 separation with high selectivity and permeance. Though the modified membrane showed considerable stability in the WGS reaction temperature conditions, further membrane upgrading and operation at higher pressure could lead to higher H_2 selectivity and permeance.

Franz and Scherer, 2010 studied RGS reaction followed by H_2 selective polymeric membranes to separate H_2 from CO_2, using a developed detailed ASPEN model. A two stage concept was used, in the first one, the CO rich gas was first fed to a catalytic reactor, where it is heated and saturated with water to be converted into CO_2 and H_2. A single membrane module and a cascade of membranes were analyzed for CO_2 separation. For a single CO_2 selective membrane a CO_2/H_2 selectivity of 150 (state of the art 15.5 for polymeric membranes) would be needed to obtain power plant efficiency losses below 10% points with separation degrees above 85%. While for the option of cascade membranes, CO_2/H_2 selectivity around 60 would be necessary to achieve the same values. Thus, the additional electrical energy could be used to overcome the drawbacks of these membrane configurations. On the other hand, for H_2 selective membranes, selectivities around 50 (state of the art 5.9) would be needed to attain efficiency losses below 10 % point at separation degrees of 85%. Franz and Scherer, 2010 predicted results showed that high levels of CO_2 separation and high hydrogen recovery and purity were necessary, but difficult to obtain with existing membranes, thus H_2 selective membranes need to have very large areas, resulting in a decreased purity. The results obtained with ASPEN showed that the major efficiency loss does not take place in the membrane separation process itself, but was due to the water gas shift reaction. Therefore it is important to optimize this reaction to prevent efficiency losses.

7. Chemical Looping

To reduce climate change due to CO_2 emissions, which are predicted to be around 5.4Gt/year, Symonds et al., 2009, it is needed to control and reduce CO_2 emitted by cheap and widely available fossil fuels like coal and wastes. There are three possible ways to achieve effective CO_2 emissions reduction: a) to separate CO_2 from a flue gas (post-combustion capture) b) to separate CO_2 from a fuel gas (pre-combustion capture) and c) oxy-combustion.

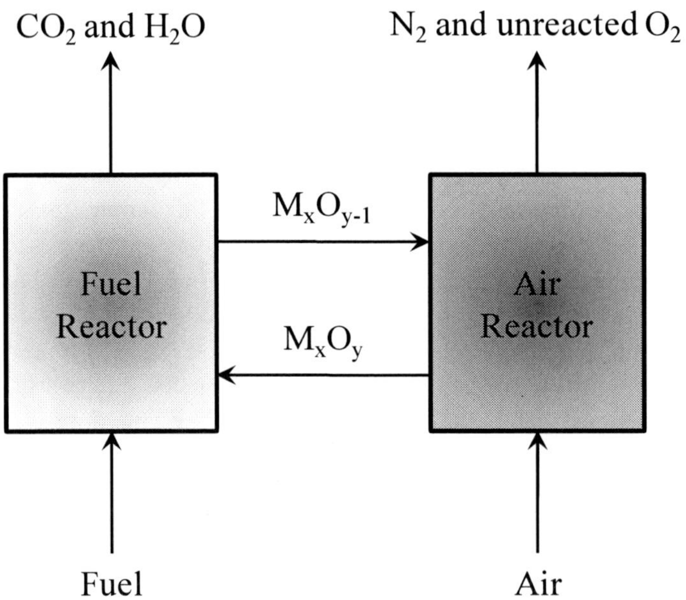

Figure 13. CO_2 capture from flue gas leaving a combustion process by chemical looping.

Nowadays, CO_2 capture is still one of the most expensive and energy consuming step in the CCS chain. Thus, more fuel needs to be consumed to achieve CO_2 capture. Hence, today´s challenge is the development of processes less energy demanding and more cost effective. One of such processes may be chemical looping. Chemical-looping combustion (CLC) was first developed for combustion systems that use an oxygen carrier, which transfers oxygen from air to the fuel without nitrogen and uses two reactors: an air reactor and a fuel reactor. CLC is a new indirect combustion process in which gas separation needed in post-combustion, oxy-combustion, and pre-combustion processes can be avoided. In CLC nitrogen from air is not mixed with flue gases and an almost pure CO_2 stream is produced, which avoids the expensive gas separation processes. In the fuel reactor, fuel is oxidized by a metal oxide (M_xO_y) that supplied the oxygen and CO_2 and H_2O are produced. The metal oxide (M_xO_y) is reduced to a metal (Me) or a reduced form (M_xO_{y-1}) and afterwards it is transferred to the air reactor, where it is oxidized with air and converted into the oxidized form (M_xO_y), which is again introduced into the fuel reactor to start a new cycle. As shown in Figure 13, the flue gas leaving the fuel reactor contains only CO_2 and H_2O, which after H_2O condensation is transformed into an almost pure CO_2 stream ready for storage. The exit gas from the air reactor contains N_2 and unreacted O_2.

The same concept may be applied to reforming, chemical-looping reforming (CLR), but the air to fuel ratio is kept low to allow only fuel partial combustion and to prevent its complete oxidation. This concept, shown in Figure 14, may be applied to natural gas or methane reforming. As the oxidation reaction of the metal oxide is exothermic and the reduction reactions are endothermic it is important to optimize the overall process by ensuring that the heat generated in the air reactor is enough to supply the heat needed for the fuel reactor to fulfill the heat balance in the system.

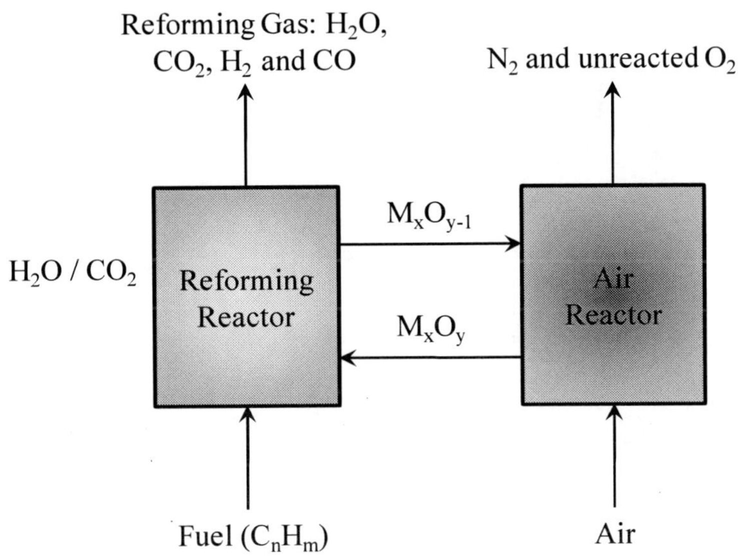

Figure 14. CO_2 capture by chemical looping applied to natural gas or methane reforming.

In a looping cycle, a carrier besides being used to supply oxygen to the fuel, it may also be used to remove CO_2 from the combustion or gasification gases, which can the released afterwards as a pure CO_2 stream, as shown in Figure 15. Either H_2O or CO_2 can be used, though H_2O is a better choice, as gasification reactions with CO_2 are slower than gasification with H_2O, Keller et al., 2010. In this concept it is fundamental the presence of a sorbent able of carbonation to remove CO_2 from flue gases produced by gasification, combustion or reforming by reaction (22). Afterwards the sorbent is regenerated by reaction (23).

Carbonation: $$M_xO + CO_2 \leftrightarrows M_xCO_3 \tag{22}$$

Regeneration: $$M_xCO_3 \leftrightarrows M_xO + CO_2 \tag{23}$$

$CaO/CaCO_3$ system has been widely studied, due to its low cost and availability. In the fuel reactor occurs the carbonation reaction and CaO reacts with CO_2 and is converted into $CaCO_3$, which in the regeneration reactor is again converted into CaO, releasing CO_2. After many carbonation/regeneration cycles occurs loss of activity and even of sorbent, thus a sorbent makeup flow is needed to compensate such loses. Anthony, 2008 reviewed solid carriers for looping cycles, both for oxygen supply and for CO_2 removal.

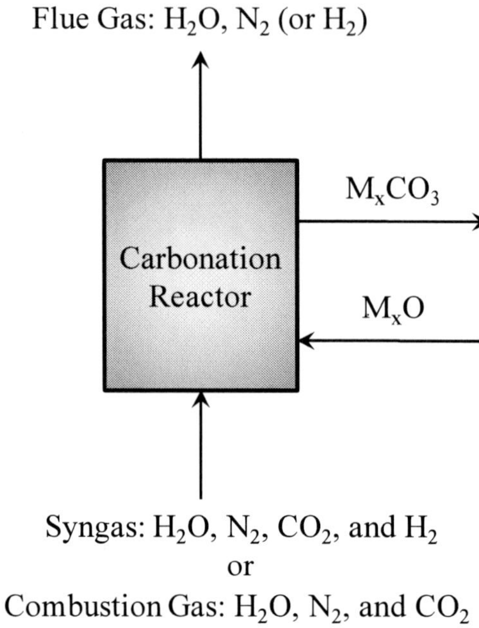

Figure 15. In a looping cycle, the carrier may supply oxygen to the fuel and remove CO_2 from the combustion or gasification gases.

Nowadays, it is considered that CLC concept may be applied to solid fuels mainly through three approaches: a) integration of gasification with oxygen carrier reduction in the fuel reactor b) submitting syngas from gasification to chemical looping, but it increases system complexity and decreases the efficiency and c) using gaseous oxygen from the decomposition of certain metal oxides at high temperatures. The last process is referred as Chemical Looping with Oxygen Uncoupling (CLOU) and improves coal gasification rate, Xiao et al., 2010.

In the first option a) integration of gasification with oxygen carrier reduction in the fuel reactor, after fuel pyrolysis, the residual char is gasified by CO_2 and H_2O through reactions (21), (24) and (25). Water gas shift reaction (WGS) (17) also occurs.

$$C + H_2O \leftrightarrows CO + H_2 \quad (24)$$
$$C + H_2O \leftrightarrows CO_2 + H_2 \quad (25)$$

In the fuel reactor also takes place the reduction of oxygen carrier by reactions (26) and (27).

$$Me_xO_y + H_2 \leftrightarrows Me_xO_{y-1} + H_2O \quad (26)$$
$$Me_xO_y + CO \leftrightarrows Me_xO_{y-1} + CO_2 \quad (27)$$

Thus, according to Xiao et al., 2010 in the fuel reactor occurs the overall chemical reaction (28).

$$CH_xO_y + (1.5 + 6 - 3y)\ Me_xO_y + z\ H_2O \leftrightarrows CO_2 + (x/2 + z)\ H_2O + (x + 4 - 2y)\ Me_xO_{y-1} \quad (28)$$

In the air reactor takes place the oxidation of the oxygen carrier by reaction (29):

$$Me_xO_{y-1} + \tfrac{1}{2}\ O_2 \leftrightarrows Me_xO_y \quad (29)$$

$CaO/CaCO_3$ system has been widely studied for chemical looping for CO_2 removal. However, $CaO/CaCO_3$ deactivates as the number of carbonation–calcinations cycles increases. Rodríguez et al., 2010 studied the deactivation of $CaO/CaCO_3$ system and found that incomplete carbonation and calcination conversions were obtained when large solid flows between reactors were used, due to particles limited residence times. Thus, to transfer the same amount of CO_2, particles will have to cycle higher number of times between carbonation and calcination reactors. Nevertheless, as the amount of CO_2 transported increased with higher solid circulation rates, both parameters needed to be optimized.

Another drawback of $CaO/CaCO_3$ is the possibility of melts formation with the calcium carrier at high temperatures and pressures. Paterson et al., 2001 showed that $CaCO_3/CaSO_4$ could produce eutectic mixtures with melting at about 1010 °C and 2 MPa. There is another concept for a calcium cycle, used by the HyPr-Ring process (hydrogen production by reaction integrated gasification). In this process the coal is steam gasified and $Ca(OH)_2$ is used to absorb CO_2, according to reaction (30). For the intermediate formation of $Ca(OH)_2$ temperature around 650 °C and pressure about 3 MPa are needed. After $CaCO_3$ calcination in a separate reactor, a pure CO_2 stream is released, Kuramoto et al., 2003 and Lin et al., 2005. But this option, also showed the problems of potential melts with the $CaO\text{-}Ca(OH)_2\text{-}CaCO_3$ system, besides the difficulties of separating the ash from the recycled CO_2 carrier.

$$C + CaO + 2H_2O \leftrightarrows 2H_2 + CaCO_3 \quad (30)$$

Anthony, 2008 and Abanades et al., 2004 reported that the tests done in a small pilot plant with calcium looping cycle showed CO_2 removal from the carbonator around 90%, thus CO_2 emissions remain below 2% as long as the carbonator temperature was kept above 650°C. At temperatures lower than 600°C, CO_2 emissions quickly increased. However, Abanades et al., 2004 stated that natural limestones loose their capture activity after 20 cycles and large makeup quantities are required to compensate the lack of activity. However, this drawback may be overtaken due to limestones low cost and to the possibility of using the exhausted limestones in the cement industry. Another advantage of limestone is their capacity to also remove SO_2, forming $CaSO_4$, which avoids the use of another sorbent for this purpose.

Silaban et al., 1996 studied the performance of dolomite in a TGA reactor and found that dolomite kept its reversibility better than did a calcitic sorbent, probably because of the extra pore volume created during the calcination of the magnesium existing in dolomite.

Natural sorbents are still studied by several authors due to its low cost, but may deactivate easily, due to sulfur poisoning, coal ash deposition, agglomeration and elutriation. But they have to be resistant to attrition, when such problems exists, pelletization of the fines escaping the primary reaction loop may be a possibility, but it will increase the costs of a calcium looping process.

Leion et al., 2008 b) and Mattisson et al., 2001 studied low-cost natural minerals such as ilmenite ($FeTiO_3/Fe_2TiO_5$) and iron. Leion et al., (2008a) reported that ilmenite demonstrated similar reactivity to synthetic $Fe_2O_3/MgAl_2O_4$ and that its reactivity increased with redox cycles, which showed the promising performance of iron-based oxygen carriers.

A wide range of sorbents have been used as oxygen carriers. Mattisson and Lyngfelt, 2004 used CuO/SiO_2 and NiO/SiO_2 as oxygen carriers and studied the thermodynamics and the heat balance of the fuel and air reactors and found that NiO/SiO_2 led to a higher selectivity towards H_2 production than CuO/SiO_2.

Dennis and Scott, 2010 a) tested a Cu-based oxygen carrier for gasification chemical looping to achieve CO2 separation. Cu was impregnated in θ-alumina and the support BET surface area decreased from $60m^2/g$, just after preparation, to $6m^2/g$ after 20 cycles. However, this drop did not appear to change the overall capacity of the oxygen carrier to react nor reactions kinetics. According to these authors surface area reduction could be achieved by avoiding large temperature rises during oxygen carrier oxidation. The carrier changed between Cu and Cu_2O, because CuO was only stable if the partial pressure of oxygen was higher than 0.03 bar at 1203 K (or 0.09 bar at 1250 K), Dennis and Scott, 2010 b). When oxygen concentration was reduced to zero, CuO decomposed rapidly to Cu_2O at the temperatures used.

Erlach et al., 2010 studied chemical looping process using NiO as oxygen carrier to oxidize a syngas with CO and H_2, by reactions (31) and (32). The reduced metal was afterwards oxidized with air in a second reactor to regenerate NiO by reaction (33).

$$NiO + CO \leftrightarrows Ni + CO_2 \quad (31)$$
$$NiO + H_2 \leftrightarrows Ni + H_2O \quad (32)$$
$$Ni + 1/2O_2 \leftrightarrows NiO \quad (33)$$

Oxygen carriers based on Ni- and Cu- were also studied by Zafar et al., 2005, who reported that NiO/SiO_2 showed high selectivity toward H_2. These authors also studied other oxygen carriers and found carbon deposition on Mn, Cu and Ni oxygen carriers during the reduction period. A significant deactivation as a function of the cycle was observed for the Ni, Fe, and Mn based carriers working at 950ºC was also reported by Zafar et al., 2005. These authors stated that Fe and Mn oxides supported on $MgAl_2O_4$ were a good option for CLR process, because they showed a high reactivity during reduction and oxidation. Ryden et al., 2006 also worked with an oxygen carrier made of NiO and $MgAl_2O_4$ and reported that natural gas was completely converted with high selectivity towards H_2 and CO. Besides, the addition of steam to natural gas considerable reduced or even eliminated the formation of solid carbon.

Diego et al., 2008 prepared and tested Ni-based oxygen carriers impregnated on alumina. Different types of alumina were used and these authors reported that though all oxygen carriers had a very high reactivity during oxidation, γ-Al_2O_3 showed the lowest reactivity during the reduction reaction, while, the Ni-based oxygen carrier impregnated on α-Al_2O_3 had the highest reactivity during the reduction reaction. The procedure used to prepare the oxygen carriers was also important, as γ-Al_2O_3 prepared by the deposition–precipitation process had higher reduction reactivity than the same support prepared by dry impregnation. On the other hand, α-Al_2O_3 prepared by both methods showed similar reactivities during reduction and oxidation processes. For all oxygen carriers prepared Diego et al., 2008 observed that an increase in the reaction temperature and/or in the H_2O/CH_4 ratio led to a decrease in the carbon deposition during reduction. These authors also reported that the tendency to increase carbon deposition was higher for oxygen carriers prepared by a deposition–precipitation method than for oxygen carriers prepared by dry impregnation. At 950°C and with H_2O/CH_4 molar ratios of 0.3–0.7 for the carriers NiO_{21}-γ-Al_2O_3, NiO_{16}-θ-Al_2O_3 and NiO_{11}-α-Al_2O_3, the carbon deposition only started when the oxygen carrier conversions were higher than 60–70%. Diego et al., 2009 reported high CH_4 conversion, higher than 98%, and that the NiO/CH_4 molar ratio was the most important parameter for gas product distribution. When NiO on α-Al_2O_3 was used a lower NiO/CH_4 molar ratio was needed to achieve the same gas product composition than when the NiO on γ-Al_2O_3 was tested.

Some more expensive sorbents need to be tested during thousands of sorption/ desorption cycles before can be chosen as viable options for CO_2 capture.

Many oxides have been tested as oxygen carries, however, the oxide chosen must have a exothermic behavior to balance the endothermic gasification reactions. It should also resist to gasification temperatures without suffering sintering or deactivation. Abanades et al., 2004 compared the performance and the cost of several sorbents commonly studied by several authors with that of monoethanolamine (MEA). These sorbents were divided into three groups: a) sorbents containing Li, Na, K, or Ca that react with CO_2 to form a carbonate b) and two generic substrates for adsorption-based systems (with or without those elements as active components) c) a group of metals with oxides suitable for supplying O_2. Some of these sorbents lose their capacity at different scales, due to attrition, escaping cyclones, filters and other solid capture devices, to chemical poisoning at sorbent active part and to surface deposits. The loss of activity oblige to using higher amounts of makeup sorbent, implying higher costs to retain the same amount of CO_2. Abanades et al., (2004) compared the cost of the amount of sorbent necessary to react (or absorb) 1 mol of CO_2 (or O_2) to the value presented by MEA and they also analyzed the characteristic number of cycles, $N_{1/2}$, that points to that 50% of the sorbent molecules or particles have completed the sorption-desorption cycle more than $N_{1/2}$ times. According to Abanades et al., 2004 analysis, the cheapest sorbent is Fe_2O_3, but it has performance limitations that decrease their applicability, which are not observed for Ni-based carriers. However, this carrier is expensive, thus $N_{1/2}$ value is around 17 000. Co (cobalt) had the highest $N_{1/2}$ value and Ni (nickel), hydrotalcites and Li_2CO_3 (lithium) also presented $N_{1/2}$ higher than MEA. However, Co also had the highest cost to react (or absorb) 1 mol of CO_2 (or O_2) and all the three previously mentioned sorbents had higher costs than MEA. Synthetic sorbents and catalysts are usually used in many industrial applications to retain or remove a minor contaminant in a large gas flow, thus the cost of the sorbent per 1 tonne of the impurity removed, or per the product produced may be high. However, for CO_2 removal, as the amount of CO_2 is very large, sorbent cost should be comparatively low, Abanades et al., 2004.

Manovic et al., 2009 studied a thermal pretreatment of three limestones with quite different geographical origins: Kelly Rock(Canada), La Blanca (Spain),and Katowice (Poland). These authors found that the pretreatment at high temperature in presence of CO_2 was more effective than that with N_2, as the sorbent showed good properties for CO_2 removal and proper particle sizes for fluidized bed systems. Kelly Rock and Katowice limestones were effectively self-reactivated, especially when CO_2 was used at 1000 °C during 24h. At high temperatures, re-crystallization of CaO occurred leading to the formation of a porous structure more resistant during CO_2 cycles. On the other hand, the performance of La Blanca limestone did not improve, because of its high Na content, which due to sintering led to low carbonation conversions that were even reduced along cycles.

A different approach of CLC is to have a single reactor where an immobile oxygen carriers is kept that have alternate periods of oxidation and reduction. Dennis and Scott, 2010 a) used a single reactor for gasification chemical looping. Lignite and its char in were gasified in situ using CO_2 in the presence of supported copper oxide. When the metal oxide became exhausted, fuel feeding is stopped. The supply of fuel is reduced by further gasification and then reactor content was re-oxidized by the introduction of air to the reactor, to begin the cycle again. When lignite was used and due to its high reactivity, char accumulation inside the bed was much lower than that observed with two bituminous coals, as due to its lower reactivity a steady accumulation of char in the bed with the number of cycles was observed, Dennis and Scott, 2010 b). Pavone, 2005 also suggested a single reactor alternately fed into with air or fuel gas through thin channels coated with a thin layer of oxygen carrier. Shimomura, 2003 proposed a rotary CLC reactor, with separate compartments of a rotating device, where fuel and air are fed into. Dahl et al., (2009) tested this configuration and reported CO_2 capture rates of 50%, which could be increased by improving reactor construction and by reducing internal gas mixing.

Nowadays challenge is the integration of CLC process with power generation systems to accomplish higher plant efficiencies and CO_2 recovery at high pressure. Thus, pressurized chemical looping combustion combined cycle (PCLC-CC) system has been proposed by some authors. The fuel reactor where the fuel is gasified by steam and/or CO_2 over the oxygen carrier is pressurized, together with the air reactor that would be pressurized moving bed reactor. Flue gases at high temperature and pressure would be used in gas turbines. Heat recovered from flue gases would be used to produce steam which would produce further electricity in a steam turbine. The use of high pressure would also allow increasing operating temperature. Temperatures higher than 900°C are usually forbidden, as the usually low melting points of oxygen carrier and coal ash limit the CLC operating temperature.

As mentioned by Xiao et al., 2010 there are few studies on pressurized CLC of solid fuels. The rise of pressure is expected to increase gasification reactions rate, as steam and/or CO_2 concentration increases. Xiao et al., 2010 studied pressurized CLC of coal with an iron ore-based oxygen carrier in a laboratory fixed bed reactor and reported that the rise of pressure led to higher CO_2 concentration and to lower CO, CH_4, and H_2 contents. The increase of pressure also enhanced combustion of residual char during oxidation and increased final carbon conversion during the reduction period. However, excessive steam partial pressure was disadvantageous for coal gasification and higher pressures also might lead to lower conversions of the oxygen carrier.

Xiao et al., 2010 stated that ore-based oxygen carriers showed good performance at high temperature and pressure, as the particles become porous, they maintained structure and size after several cycles and no particles agglomeration was detected. Though EDX analysis showed a small amount of coal ash on oxygen carrier surface, no appreciable changes were observed in crystalline phases.

ENEA (an Italian public agency operating in energy, environment and new technologies fields) has leading ZECOMIX project (Zero Emission COal MIXed technology), Romano and Lozza, 2009. This project includes four sections: (i) the chemical island that produces a hydrogen rich syngas, (ii) the power island, where high temperature steam is produced from syngas oxy-firing and a lower temperature heat recovery steam cycle, (iii) an air separation unit (ASU), where oxygen is produced in a double reboiler distillation process and (iv) the CO_2 compression and liquefaction section. Romano and Lozza, 2009 reports that recycled steam and hydrogen are used as gasifying agents to convert carbon into CH_4, which goes into two carbonator reactors, where CH_4 is mixed with steam and in presence of calcium oxide, to be converted into an H_2 rich syngas, whilst CO_2 is absorbed by the solid sorbent to originate $CaCO_3$. According to Romano and and Lozza, 2010 the predicted overall net efficiency is about 45 to 47%, including CO_2 compression and air separation consumptions and about 100% CO_2 is captured, as incondensable gases exiting the LP steam turbine (including pollutants from the combustion chamber) can be recycled to the calciner. The main challenge is the demonstration of this concept at large scale, as several technical problems will have to be solved.

Erlach et al., 2010 studied different configurations for CLC integrated with IGCC (CLC-IGCC), which were compared with conventional IGCC process with carbon capture by physical absorption. Process simulation was done using Aspen Plus and GateCycle. For the CLC-IGCC reference case with a reactor temperature of 1200°C the net plant efficiency was 37.7%, whilst the efficiency of the IGCC with de-carbonization in the base case was 34.9%. Different simulation cases were studied to study the effect of i) gas turbine inlet temperature on the IGCC and on CLC-IGCC, ii) addition of a CO_2 turbine to the CLC system, and iii) production of subcritical (127 bar) and supercritical (280 bar) steam in the CLC-IGCC. Process efficiency is very sensitive to changes in the gas turbine inlet temperature. A gas turbine working at 1430°C is possible for the conventional IGCC and lead to an efficiency increase of approximately 1.3 percentage points. However, oxygen carriers for CLC should not work at temperatures higher than 1200°C, as mentioned before. Another CO_2 turbine in the CLC system increases electricity production to more 160 MW, but it leads to lower steam temperatures and consequently to lower power output of steam turbines. Another drawback is that the CO_2 previously expanded has to be re-compressed before storage. According to Erlach et al., 2010 the additional increase in efficiency achieved by CO_2 turbine is not enough to compensate the rise in investment costs and in possible operational problems. The CLC-IGCC offers the advantages of higher plant efficiency and more CO_2 capture, as CLC-IGCC allows CO_2 capture around 100% conventional IGCC only reaches values about 86%. Thus, the relative cost effectiveness of both processes is quite dependent on price of carbon emissions. The development of low cost oxygen carrier with high thermal stability is also very important to the success of CLC processes.

CLC concept still needs to be demonstrated at large both scale technical and economically. CLC has to compete with existing processes that ensure total costs for CO_2 capture below \$40/tonne of CO_2 separated, Abanades et al., 2004. CO_2 capture cost includes several items: capital costs of capture plant, operating and maintenance costs, the cost of additional fuel to compensate for the efficiency penalty due to the capture plant and the cost of CO_2 compression. Thus, it is of most importance that that the carrier in chemical looping has a high activity and stability after many cycles and that it has a low cost.

Rezvani et al., 2009 studied several CO_2 capture processes in IGCC, using the simulation package "ECLIPSE": physical absorption, water gas shift reactor membranes and two chemical looping combustion cycles (CLC) single and doubled. Single stage CLC reached temperatures higher than 1 200°C, due to exothermic reactions, which increased the risk of sintering and the agglomeration of the oxygen carrier. Thus, to avoid high temperature zones in the fuel reactor, the oxygen carrier was cooled down, which was difficult to do, unless the air mass flow was increased. Alternatively, it was proposed a double stage CLC reactor concept by Naqvi and Bolland, 2007, which helps to control the temperatures between the two CLC cycles, without cyclone cooling and improves the combustion efficiencies.

Naqvi and Bolland, 2007 applied the CLC concept to natural gas for power generation with CO2 capture, using a CLC combined cycle, which included CLC-reactor system, an air turbine, a CO2-turbine and a steam cycle. These authors reported net plant efficiency of about 52% at an oxidation temperature of 1200°C. To keep this net plant efficiency at lower oxidation temperatures, the option of reheating gas turbine was studied by employing multi CLC-reactors. Different CLC configurations were studied. The base case was CLC-combined cycle (CLCCC) and three multi-stage CLC for combined cycles were analyzed: single reheat CLC-combined cycle (SR-CLCCC), double reheat CLC-combined cycle (DR-CLCCC) and single reheat CLC-combined cycle with exhaust recuperation (SRCLC-ER). Naqvi and Bolland, 2007 results showed that single reheat CLC-combined cycle led to net plant efficiency of above 51% at oxidation temperature of 1000°C, net plant efficiency could increase to 53%, but the oxidation temperature of 1200°C. The double reheat cycle only led to minor efficiency improvements. Nevertheless, these authors stated that the single reheat CLC combined cycle was a more attractive option, because it showed reasonable net plant efficiency at good oxidation temperature, without the complexity of using several reactors. The CLC-cycles were compared with conventional option using amine solution in post-combustion. All the CLC-cycles showed higher net plant efficiencies with very high CO2 capture close to 100%.

Rezvani et al., 2009 cost analysis showed that the electricity costs of the IGCC system with CLC are high enough to prevent the competitiveness of this option. On the other hand, CLC systems showed the highest CO_2 capture efficiencies and IGCC with a double CLC stage had the highest CO_2 avoidance rates, due to the lower CO_2 release. One of CLC drawbacks was its complexity, which required a good cooperation among multiple reactors, another one was the very high cost of this technology

Chiesa et al., 2008 proposed a process with iron oxide circulating among three reactors for chemical looping: fuel reactor, steam reactor and air reactor. In the fuel reactor, Fe_2O_3 is mostly reduced to FeO, while natural gas is oxidized by endothermic reaction (34). In the steam reactor, most of the reduced metal FeO reacts with steam to form Fe_3O_4 (magnetite) and hydrogen, by reaction (35). In the air reactor Fe_3O_4 with traces of unreacted FeO is completely oxidized to Fe_2O_3 by reaction (36), which sustains the thermal balance in the whole system.

$$4\ Fe_2O_3 + CH_4 \leftrightarrows 8\ FeO + 2\ H_2O + CO2 \quad (34)$$

$$8\ FeO + 8/3\ H_2O \leftrightarrows 8/3\ Fe_3O_4 + 8/3\ H_2 \quad (35)$$

$$8/3\ Fe_3O_4 + 2/3O_2 \leftrightarrows 4\ Fe_2O_3 \quad (36)$$

The overall reaction (37) corresponds to methane steam reforming with oxidization of CO into CO_2.

$$CH_4 + 2/3\ H_2O + 2/3\ O_2 \leftrightarrows 8/3\ H_2 + CO_2 \quad (37)$$

According to Chiesa et al., 2008 the three reactors concept is very interesting, because it completely removes CO_2 and produces pure and pressurized hydrogen, after condensation of the stream leaving the steam reactor with H_2 and steam. As only metal oxides are circulated, it is predictable to have less problems in solids handling and to have a higher flexibility towards operating temperature of the reactors. This process probably would need lower investment costs, as it does not need gas treatments (reforming, shift), separation processes nor air separation unit (ASU). The main drawbacks are achieving good circulation conditions of solids through the reactors and development of carriers with larger lifetime and stability.

CONCLUSION

Gasification may play an important role in CO_2 capture, thus it is considered a pre-combustion capture process. Through gasification, coal and/or different types of wastes may be converted into a H_2 rich fuel, almost free of carbon and a CO_2 stream ready for storage is obtained. Although the existing technologies to achieve these goals are most promising, R&D activities are still needed, especially to demonstrate the existing processes at industrial scale and to improve the economic viability of these processes, mainly syngas cleaning and upgrading, syngas conversion into H_2 and CO_2 and effective H_2 separation.

The cost of power generation with CO_2 capture is higher than those of existing processes without CCS. But, as CO_2 emissions are predicted to continue to grow, especially in emerging economies, but also all over the world, putting in danger life in earth, it is fundamental to carry on with R&D activities to develop processes less energy demanding and with lower costs. To attain these targets it is of most importance to have the support of governmental authorities, through right legislation measures and by launching financial supporting programs for R&D.

REFERENCES

Abanades J.C., Rubin E. S., Anthony E. J. (2004) Ind. Eng. Chem. Res. 43, 3462–3466.

Alauddin Z.A.B.Z., Lahijani P., Mohammadi M., Mohamed A.R. (2010) Renewable and Sustainable Energy Reviews 14, 2852–2862.

Ahmed I., Gupta A.K. (2009) Applied Energy 86(12), 2626–2634

Anthony E. J. (2008) Ind. Eng. Chem. Res. 47, 1747–1754.

Atakul H., Wakker J. P. (1995) Fuel 74, 187–191.

Aznar M.P., Caballero M.A., Corella J., Molina G., Toledo J.M. (2006) Energy & Fuels20, 1305–1309.

Basinska A., Klimkiewicz R., Teterycz H. (2004) React. Kinet. Catal., Lett. 82, 271–277.

Basu A., Akhtar J., Rahman M.H., Islam M.R. (2004) Petroleum Science and Technology 22(9&10), 1343–1368.

Battersby S., Smart S., Ladewig B., Liu S., Duke M.C., Rudolph V., Costa J.C.D. (2009) Separation and Purification Technology 66, 299–305.

Bi Y., Xu H., Li W., Goldbach A. (2009) International Journal of Hydrogen Energy 34, 2965–2971.

Bounaceur R., Lape N., Roizard D., Vallieres C., Favre E. (2006) Energy 31, 2556–2570.

Bracht M., Alderliesten P.T., Kloster R., Pruschek R., Haupt G., Xue E., Ross J.R.H., Koukou M.K., Papayannakos N. (1997) Energy Convers. Management 38, Suppl., SI59–S164.

Brown D., Gassner M., Fuchino T., Maréchal F. (2009) Applied Thermal Engineering 29, 2137–2152.

Butterman H.C., Castaldi M.J. (2009) Environ. Sci. Technol. 43(23), 9030–9037.

Butterman H.C., Castaldi M.J. (2007) Ind. Eng. Chem. Res. 46, 8875–8886.

Cameron D., Holliday R., Thompson D. (2003) J. Power Sources 118, 298–303.

Campoy M., Gómez-Barea A., Vidal F.B., Ollero P. (2009) Fuel Processing Technology 90(5), 677–685.

Cao Y., Wang Y., Riley J.T. Pan W.P. (2006) Fuel Processing Technology 87(4), 343–353.

Carbo M.C., Boon J., Jansen D., van Dijk H.A.J., Dijkstra J.W., van den Brink R.W., Verkooijen A.H.M. (2009) International Journal of Greenhouse Gas Control 3, 712–719.

Chen W.-H., Jheng J.-G. (2007) Journal of Power Sources 172, 368–375.

Chen C.-S., Lin J.-H., Lai T.-W., Li B.-H. (2009) Journal of Catalysis 263, 155–166.

Chiesa P., Lozza G., Malandrino A., Romano M., Piccolo V. (2008) International Journal of Hydrogen Energy 33, 2233–2245.

Choudhary V.R., Rajput A.M (1996) Ind. Eng. Chem. Res. 35, 11, 3934—3939.

Cormos C., Starr F., Tzimas E., Peteves S. (2008) International Journal of Hydrogen Energ 33(4), 1286–1294.

Dahl I., Bakken E., Larring Y., Spjelkavik A., Hakonsen S., Blom R. (2009) Energy Procedia 1, 1513–1519.

Delft Y.C. van, Hugill J.A., Dorst E.M., Veen, H.M. (2003) Hydrogen Preparation in Process Industry with Inorganic Membranes, ECN-CX-03-110, ECN, November.

Devi L., Craje M., Thüne P., Ptasinski K., J., Janssen F.J.J.G. (2005) Applied Catalysis A 294, 1, 68–79.

Dennis J.S., Scott S.A. (2010a) Fuel 89, 1623–1640.

Dennis J.S., Müller C.R., Scott S. A. (2010b) Fuel 89, 2353–2364.

Devi L., Ptasinski K.J., Janssen F.J.J.G. (2003) Biomass & Bioenergy 24, 125–40.

Diego L.F., Ortiz M., García-Labiano F., Adánez J., Abad A., Gayán P. (2009) Journal of Power Sources 192, 27–34.

Diego L.F., Ortiz M., Adánez J., García-Labiano F., Abad A., Gayán P. (2008) Chemical Engineering Journal 144, 289–298.

Dolan M.D., Dave N.C., Ilyushechkin A.Y., Morpeth L.D., McLennan K.G. (2006) Journal of Membrane Science 285, 30–55.

El-Moemen A.A., Karpenko A., Denkwitz Y., Behm R.J. (2009) Journal of Power Sources 190(1), 64–75.

El-Moemen A.A., Kucˇerova G., Behm R.J. (2010) Applied Catalysis B: Environmental 95, 57–70.

Erlach B., Schmidt M., Tsatsaronis G. (2010) Energy 1-12, doi:10.1016/j.energy.2010.08.038.

Evan J.G., O'Brien T. (2005) Fuel Processing Technology 86(14–15), 1423–1434.

Favre E., Bounaceur R., Roizard, D. (2009) Journal of Membrane Science 328, 11–14.

Fajín J.L.C., Cordeiro M.N.D.S., Illas F., Gomes J.R.B., (2009) Journal of Catalysis 268, 131–141.

Felice L., Courson C., Janda N., Galluccia K., Foscoloa P.U., Kiennemann A. (2009) Chemical Engineering Journal 154(1–3), 373–383.

Frank, P. (2003) Sulphur Tolerant Shift Catalyst - Dealing With the Bottom of the Barrel Problem. Synetix, Johnson Matthey Group, Billingham.

Franz J., Scherer V. (2010) Journal of Membrane Science 359, 173–183.

Gade S.K., Keeling M.K., Davidson A.P., Hatlevik O., Way J.D. (2009) International Journal of Hydrogen Energy 34, 6484–6491.

Gallucci F., Comite A., Capannelli G., Basile A. (2006) Ind. Eng. Chem. Res. 45(9), 2994–3000.

Galvita V., Schroder T., Mundera B., Sundmacher K. (2008) International Journal of Hydrogen Energy 33, 1354–1360.

Gangwal S.K., Gupta R.P., Portzer J.W., Turk B.S. (1998) Simultaneous Removal of H_2S and NH_3 Using Metal Oxide Sorbents, Research Triangle Institute K. Jothimurugesan, Department of Chemical Engineering Hampton University, Advanced Coal-Based Power & Environmental Systems '98 Conference, Conference Proceedings.

Gangwal, S.K., Gupta R., Krishnan G.N., Hung S.L. (1993) NH_3/H_2S Advances, Proceedings of the Coal-Fired Power Systems '93 - Advances in IGCC and PFBC Review Meeting, p. 399, DOE/METC - 93/6131 (DE93000289).

Gil J., Aznar M.P., Caballero M.A., Francés E., Corella J. (1997) Energy & Fuels 11(6), 1109–1118.

Gnanapragasam N.V., Reddy B.V., Rosen M. A. (2010 a) International Journal of Hydrogen Energy 35, 4933–4943.

Gnanapragasam N. V., Reddy B.V., Rosen M. A. (2010 b) International Journal of Hydrogen Energy 35, 4788–4807.

Gryaznov V.M., Sep. Purif. Methods 29 (2000) 171–187.

Gunawardana P.V.D.S., Lee H.C., Kim D.H. (2009) International Journal of Hydrogen Energy 34, 1336 —1341.

Guo P., Chen L., Yang Q., Qiao M., Li H., Li H., Xu H., Fan K. (2009) International Journal of Hydrogen Energy 34, 2361–2368.

Hatlevik Ø., Gade S.K., Keeling M.K., Thoen P.M., Davidson A.P., Way J.D. (2009) Separation and Purification Technology 73(1), 59-64 .

Hla S.S., Parka D., Duffy G.J., Edwards J.H., Roberts D.G., Ilyushechkin A., Morpetha L.D., Nguyen T. (2009 a) Chemical Engineering Journal 146, 148–154.

Hla S.S.E., Duffy G.J., Morpeth L.D., Cousins A., Roberts D.G., Edwards J.H. (2009 b) Catalysis Communications 10, 967–970.

Hou K., Hughes R. (2001) Chemical Engineering Journal 82, 311–328.

Huang T.-J., Yu T.-C., Jhao S.-Y. (2006) Ind. Eng. Chem. Res. 45(1) 150–156

Hu X., Huang Y., Shu S., Fan Y., Xu N. (2008) Journal of Power Sources 181, 135–139.

IEA, 2009, World Energy Outlook, International Energy Agency, OECD/IEA, Paris.

IEA, 2008, Energy Technology Perspectives, OECD/IEA, Paris.

Iggland M., Leion H., Mattisson T., Lyngfelt A. (2010) Chemical Engineering Science 65, 5841–5851.

Ishitsuka M., Haraa S., Mukaida M., Haraya K., Kita K., Kato K. (2008) Desalination 234, 293–299.

Kalamaras C.M., Panagiotopoulou P., Kondarides D.I., Efstathiou A.M., (2009) Journal of Catalysis 264, 117–129.

Kaldis S.P., Skodras G., Sakellaropoulos G.P. (2004) Fuel Processing Technology 85, 337–346.

Ke Z., Huiyuan G., Zebao R., Yuesheng L., Yongdan L. (2007) Chin. J. Chem. Eng. 15, 643–647.

Keller M., Leion H., Mattisson T., Lyngfelt A. (2010) Combustion and Flame, doi:10.1016/j.combustflame.2010.09.009.

Kim Y.T., Park E.D., Lee H.C., Lee D., Lee K.H. (2009) Applied Catalysis B: Environmental 90, 45–54.

Kim Y.J., Lee J.M., Kim S.D. (2000) Fuel 79(1), 69–77.

Kluiters S.C.A. (2004) Status review on membrane systems for hydrogen separation, Intermediate report EU project MIGREYD NNE5-2001-670, ECN-C-04-102.

Koppatz S., Pfeifer C., Rauch R., Hofbauer H., Marquard-Moellenstedt T., Specht M. (2009) Fuel Processing Technology 90 (7-8), 914–921.

Kuramoto K., Fujimoto S., Morita A., Shibano S., Suzuki Y., Hatano H., Shi-Ying L., Harada M., Takarada T. (2003) Ind. Eng. Chem. Res. 42(5), 975–981.

Leion H., Mattisson T., Lyngfelt A. (2008a) Int. J. Greenhouse Gas Control 2(2), 180–193.

Leion H., Lyngfelt A., Johansson M., Jerndal E., Mattisson T. (2008b) Chem. Eng. Res. Des. 86, 1017–1026.

Leion H., Mattisson T., Lyngfelt A. (2007) Fuel 86, 1947–1958.

Leppälahti J., Koljonen T., Hupa, M., Kilpinen, P. (1997) Energy & Fuels 11(1), 39–45.

Li B., Maruyama K., Nurunnabi M., Kunimori K., Tomishige K. (2005) Ind. Eng. Chem. Res. 44, 485–494.

Li F., Zeng L., Fan L. (2010) Ind. Eng. Chem. Res. 49 (21), 11018–11028

Li F., Zeng L., Fan L.-S. (2007) Fuel 89, 3773–3784.

Lian Y., Wang H., Fang W., Yang Y. (2010) Journal of Natural Gas Chemistry 19, 61–66.

Lian Y., Wang H., Zheng Q., Fang W., Yang Y. (2009) Journal of Natural Gas Chemistry 18, 161–166.

Liang H., Zhang Y., Liu Y. (2009) Journal of Natural Gas Chemistry 18, 436–440.

Lin S., Harada M., Suzuki Y., Hatano H. (2005) Energy Convers. Manage. 46(6), 869–880.

Liu P., J. Rodriguez A., Takahashi Y., Nakamura K. (2009) Journal of Catalysis 262, 294–303.

Mahishi M. R., Goswami D. Y. (2007) International Journal of Hydrogen Energy, 32(14), 2803–2808.

Maluf S.S., Assaf E.M. (2009) Fuel 88, 1547–155.

Manovic V., Anthony E. J., Loncarevic D. (2009) Chemical Engineering Science 64, 3236–3245.

Mattisson T., Lyngfelt A., Cho P. (2001) Fuel 80, 1953–1962.

Mattisson T., Zafar Q., Lyngfelt A., Gevert B. (2004) Integrated hydrogen and power production fromnatural gas with CO2 capture, in: 15thWorld Hydrogen Energy Conference, Yokohama, Japan.

McKee B. (2002) Solutions for the 21st century, Zero emissions technologies for fossil fuels. International Energy Agency, Technology Status Report.

Mejdell A.L., Peters T.A., Stange M., Venvik H.J., Bredesen R. (2009) Journal of the Taiwan Institute of Chemical Engineers 40, 253–259.

Morf P., Haslerb P., Nussbaumer T. (2002) Fuel 81, 843–853.

Mulder M. (2000) Basic Principles of Membrane Technology, Kluwer Academic Publishers, Dordrecht.

Naqvi R., Bolland O. (2007) Int. J. Greenhouse Gas Control, 19–30.

Noble R.D., Stern S.A. (ed.) (1995) Membrane Separations Technology - Principles and Applications, Elsevier Science B.V., Amsterdam.

Oliveira E. L.G., Grande C.A., Rodrigues A. E. (2010) Chem. Engin. Science 65(5), 1539–1550.

Pavone D. (2005) CO2 capture by means of chemical looping combustion. In: Proceedings of COMSOL Multiphysics User's Conference, Paris, France, 15 Nov.

Paterson N., Elphick S., Dugwell D.R., Kandiyoti R. (2001) Energy & Fuels 15(4), 894–902.

Pecho J., Schildhauer T.J., Sturzenegger M., Biollaz S., Wokaun A. (2008) Chemical Engineering Science 63(9), 2465–2476.

Pfeifer C., Puchner B., Hofbauer H. (2009) Chemical Engineering Science 64 (23), 5073–5083.

Pinacci P., Broglia M., Valli C., Capannelli G., Comite A. (2010) Catal. Today doi:10.1016/j.cattod.2010.02.034.

Pinto F., André R.N., Lopes H., Franco C., Carolino C., Costa R., Gulyurtlu I. (2010) Fuel 89, 3340–3351.

Pinto F., André R.N., Gulyurtlu I., (2009 a), Innovation on biomass wastes utilization through gasification and co-gasification. Stage of deployment and needs for further R&D In: "*Biomass Gasification: Chemistry, Processes and Applications*", Editors: Jean-Pierre Badeau and Albrecht Levi (Ed.), New York: Nova Science Publishers, Inc. 2009, ISBN: 978-1-60741-461-2.

Pinto F., André R.N., Gulyurtlu I. (2009 b) Stage of deployment of syngas cleaning technologies and RD&D needs to accomplish the new challenges of syngas utilization, In: "*Syngas: Production Methods, Post Treatment and Economics*", Editors: Adorjan Kurucz and Izsak Bencik (Ed.), New York: Nova Science Publishers, Inc. 2009, ISBN: 978-1-60741-841-2.

Pinto F., André R.N., Lopes H., Franco C., Gulyurtlu I., Cabrita I. (2009 c) Fuel 88, 2392–2402

Pinto F., Lopes H., André R.N., Gulyurtlu I., Cabrita I. (2008) Fuel, 87, 7, *1050-1062.*

Pizzi D., Wortha R., Baschetti M.G., Sarti G.C., Noda K. (2008) Journal of Membrane Science 325, 446–453.

Rakass S., Oudghiri-Hassani H., Rowntree P., Abatzoglou N. (2006) Journal Power Source 158(1), 485-496.

Rapagnà S., Provendier H., Petit C., Kiennemann A., Foscolo P.U. (2002) Biomass & Bioenergy 22, 377–388.

Rezvani S., Huang Y., McIlveen-Wright D., Hewitt N., Mondol J.D. (2009) Fuel 88, 2463–2472

Rhodes C., Williams B.P., King F., Hutchings G.J. (2002) Catal. Commun. 3, 381–384.

Rodríguez N., Alonso M., Abanades J.C. (2010) Chemical Engineering Journal 156, 388–394

Rodríguez J.A., Liu P., Hrbek J., Pérez M., Evans J. (2008) Journal of Molecular Catalysis A: Chemical 281, 59–65.

Romano M., Lozza G. (2009) Energy Procedia 1, 1473–1480.

Romano M., Lozza G. (2010) International Journal of Greenhouse Gas Control 4, 459–468.

Ryden M., Lyngfelt A., Mattisson T. (2006) Fuel 85, 1631–1641.

Ryi S.-K., Park J.-S., Kim S.-H., Cho S.-H., Hwang K.-R., Kim D.-W., Kim H.-K., (2007) Journal of Membrane Science 297, 217–225.

Ryi S.-K., Park J.-S., Kim S.-H., Kim D.-W., Kim H.-K. (2009) Journal of Membrane Science 326, 589–594.

Shangguan J., Liang L., Fan H., Shen F. (2007) Journal of Natural Gas Chemistry 16, 53–59.

Shim H. M, Chun W. G., Kim H.T. (2010) International Journal Energy Research 34(5), 412–421.

Shimomura Y. (2003) Modern Power Systems 23(1), 15-7.

Silaban A., Narcida M., Harrison D.P. (1996) Chem. Eng. Commun. 146, 149–162.

Stevens R.W. Jr., Shamsi A., Carpenter S., Siriwardane R. (2010) Fuel 89, 1280–1286.

Stiegel G.J., Maxwell, R.C. (2001) Fuel Processing Technology 71, 79–97.

Stiegel G.J., Ramezan M. (2006) International Journal of Coal Geology 65, 173–190.

Souza T.R.O., Brito S.M.O., Andrade H.M.C. (1999) Appl. Catal., A 178, 7–15.

Tang Z., Kim S.-J., Reddy G.K., Dong J., Smirniotis P. S. (2010) Journal of Membrane Science 354, 114–122.

Tars Protocol guidelines (2005) Sampling and analysis of tars and particles in biomass producer gases – Technical report, CEN BT/TF 143, July.

Tibiletti D., Meunier F.C., Goguet A., Reid D., Burch R., Boaro M., Vicario M., Trovarelli A. (2006) J. Catal., 244, 183–191.

Tosti S., Basile A., Chiappetta G., Rizzello C., Violante V. (2003) Chemical Engineering Journal 93(1), 23–30.

Tosti S., Violante V., Basile A., Chiappetta G., Castelli S., Francesco M., Scaglione S., Sarto F. (2000) Fusion Engineering and Design 49–50, 953–958.

Valero A., Usón S. (2006) Energy, 31 (10–11) 1643–1655.

Vasileiadis S., Ziaka-Vasileiadou Z. (2004) Separation and PurificationTechnology 34, 213–225.

Wahlund B., Yan J., Westermark M. (2004) Biomass and Bioenergy, 26 (6), 531–544.

Wang L., Weller C.L., Jones D.D., Hanna M.A. (2008) Biomass and Bioenergy, 32, 573–581.

Wang W., Padban N., Ye Z., Andersson A., Bjerle I. (1999) Ind. Eng. Chem. Res., 38, 4175–4182.

Watanabe K., Miyao T., Higashiyama K., Yamashita H., Watanabe M. (2009) Catalysis Communications 10, 1952–1955.

Wu P., Li X., S. Ji, Lang B., Habimana F., Li C. (2009) Catalysis Today 146, 82–86.

Xiao R., Song Q., Song M., Lu Z., Zhang S., Shen L. (2010) Combustion and Flame 157 1140–1153.

Xiong L., Liu S., Rong L. (2010) International Journal of Hydrogen Energy 35, 1643–1649.

Yang H., Xu Z., Fan M., Gupta R., Slimane R.B., Bland A.E., Wright I. (2008) Journal of Environmental Sciences 20(1), 14–27.

Yang Z., Zhang Y., Wang X., Zhang Y., Lu X., Ding W. (2010) Energy Fuels 24, 785–788.

Zafar Q., Mattisson T., Gevert B. (2005) Ind. Eng. Chem. Res. 44 3485–3498.

Zahedia M., Afra B., Dehghani-Mobarake M., Bahmani M. (2009) Journal of Membrane Science 333, 45–49.

Zevenhoven R., Kilpinen P. (2004). Control of pollutants in flue gases and fuel gases (Sulfur chapter), The Nordic Energy Research Programme, Helsinki University of Technology, February 2004.

Zhu Z.H., Finnerty J., Lu G.Q., Yang R.T. (2002) Energy & Fuels 16 (6), 1359–1368.

In: Gasification: Chemistry, Processes and Applications
Editor: Michael D. Baker
ISBN: 978-1-61209-681-0

Chapter 2

AN OVERVIEW ABOUT CURRENT AND FUTURE EXPERIMENTAL ACTIVITIES IN A FLEXIBLE GASIFICATION PILOT PLANT

Alberto Pettinau*, Francesca Ferrara and Carlo Amorino
Sotacarbo S.p.A. c/o Grande Miniera di Serbariu, Carbonia, Italy

ABSTRACT

In the field of hydrogen production through coal gasification for distributed power generation, Sotacarbo, an Italian limited company which operates in the field of clean coal technologies, is developing different research and development projects for the tuning of a coal-to-hydrogen process configuration. Toward this goal, a flexible pilot platform was built in 2007-2008, and it is currently in operation. In particular, the platform includes demonstrative and pilot air-blown fixed-bed gasifiers, the latter equipped with a flexible syngas treatment line for a combined power generation and CO_2-free hydrogen production.

This chapter presents a detailed description of all the experimental equipment and the main results obtained during the first 700 hours of tests in the pilot unit. In particular, the optimum operating conditions with a low sulphur South African coal have been defined, with a syngas production of about 43 Nm^3/h (from the gasification of 8 kg/h of coal), characterized by a lower heating value of 4.5 MJ/kg, and a high cold gas efficiency, of about 97%.

A series of experimental tests has also been carried out in order to assess the plant performance under different operating conditions and the effects of the main parameters, such as air/fuel and steam/fuel mass ratios and oxidant composition (air, eventually enriched in oxygen, or mixtures of oxygen and carbon dioxide). Moreover, a series of preliminary tests has been carried out in order to evaluate the gasification performance with a portion of high sulphur Sulcis coal or biomass (wood pellets) in the fuel blend.

The experimental results obtained in syngas treatment processes show a very efficient removal of pollutants (in particular, a final concentration of sulphur compounds lower than 10 ppm, in volume, can be obtained, even with high sulphur coal in the fuel

* Corresponding author: apettinau@sotacarbo.it

blend) and the possibility to reduce carbon dioxide emissions of about 90% by using a single-stage amine-based carbon capture system, which assures a CO_2 absorption efficiency of about 99%.

Finally, the guidelines of the future research activities are described in this chapter, with particular reference to the further optimization of both gasification and syngas treatment processes in different operating conditions.

1. INTRODUCTION

In a long-term view for the future, a strong development of all the technologies based on renewable and nuclear energy sources can be expected due to the need of increasing energy production and to mitigate greenhouse gas emissions [1-2]. In the meantime, sustainable energy production needs further significant effort toward the reduction of energy consumption and toward the development of high efficiency fossil fuels-based power generation plants, equipped with carbon capture and storage (CCS) systems [3]. Moreover, due to its great price stability[1], coal represents a secure and feasible primary fuel for the production of electrical energy [4-5] and different energy carriers, such as hydrogen.

Among clean coal technologies, coal gasification could represent a competitive option for power generation. According to a recent study published by the European Commission [6], power generation through integrated gasification combined cycles (IGCC) currently involves a production cost of electricity (CoE) slightly higher than those corresponding to a conventional pulverized coal combustion (PCC) plant[2]. This difference tends to be reduced according to the projection for 2020, and IGCC becomes convenient with respect to PCC if a CCS configuration is considered[3]. As a matter of fact, the possibility to operate carbon capture from syngas before the combustion process (in general by using cold gas chemical absorption processes with alkanolamine solutions [7-9]), allows for operation with a relatively low gas flow and a relatively high CO_2 partial pressure, with subsequent advantages in terms of removal efficiency and equipment size [3].

Moreover, compared to combustion technologies, coal gasification allows the production of chemicals or clean fuels, with particular reference to hydrogen [10-12], universally considered one of the most important energy carriers [13-15] and characterized by a worldwide production (18% from coal) greater than one billion of cubic meters per day [16-18].

In this context, Sotacarbo[4] is engaged in a series of research and development projects in order to develop and optimize a coal gasification process and an integrated syngas treatment

---

[1] Compared to oil and natural gas, whose main fields are confined to a relatively small area, coal is widely available in the world and distributed more uniformly than other fossil fuels; this allows a great price stability and represents a secure source from a strategic point of view.

[2] The evaluation, referred to year 2007, indicates a CoE of 45-55 €/MWh for IGCC and CFBC (circulating fluidized bed combustion) plants and of 40-50 €/MWh for conventional PCC plants.

[3] The projection for the year 2020 indicates a CoE of 75-90 €/MWh for IGCC and 80-105 €/MWh for PCC, both equipped with a carbon capture and storage system.

[4] Sotacarbo S.p.A. is a limited company controlled by the Sardinian Regional Administration and by ENEA, the Italian National Agency for Energy and Environment; established in 1987, it operates research and development in the field of clean coal technologies. Currently Sotacarbo, with the aid of its research center and

line for hydrogen production from coal. Toward this goal, a flexible pilot platform has been built at the Sotacarbo Research Centre, in Carbonia (South-West Sardinia, Italy).

The choice of the configuration of the pilot platform is a compromise between the need to develop a coal gasification process for medium and small scale industrial applications and the interest in the development of coal-to-hydrogen integrated processes to be applied in large-scale power plants.

In particular, the target of the gasification process (medium and small scale, up to 10-15 MW_{th}) led the choice of a fixed-bed, air-blown gasifier. As a matter of fact, these processes can be simply managed [19-20], and it is well known that the counter current fluid dynamics assure a higher efficiency with respect to other gasification technologies [21-22]. Moreover, the selected gasification technology allows a great flexibility in terms of primary fuel (different kinds of coal and also biomass and wastes).

A typical large-scale coal-to-hydrogen plant configuration could include, for example, an oxygen-blown entrained-flow gasification process, operating at high pressure; the scale down of such a process to pilot scale involves a series of technical problems. Therefore, the need to analyze a hydrogen production line from coal syngas suggested use of the same fixed-bed gasifier as syngas producer, even if the raw syngas properties are quite different with respect to those produced in a typical large scale plant. This choice allowed a significant reduction of the research costs and strongly simplified the management of the plant.

This chapter reports a detailed description of the whole pilot platform, together with an overview of the main experimental results obtained during about 700 hours of experimental tests. Moreover, a description of the potential research activities which could be carried out in the pilot platform is here presented.

2. The Sotacarbo Pilot Platform

The Sotacarbo pilot platform was built in 2007-2008 (in the field of a research and development project[5] partially funded by the Italian Ministry of Education and Research) to test different plant solutions and different operating conditions; therefore, a very flexible and simple layout was selected during the design and construction phases.

Currently, the layout of the Sotacarbo pilot platform (figure 1) includes two fixed-bed up-draft and air-blown gasifiers: a 700 kg/h (corresponding to about 5 MW_{th}) demonstrative unit and a 8-10 kg/h (about 80 kW_{th}) pilot unit. The choice of this kind of gasification process is a consequence of the particular commercial interest in the field of medium and small scale industrial applications.

Whereas the main goal of the experimental tests on the demonstrative unit is the optimization of the gasifier (in different operating conditions) for its commercialization, the pilot plant is mainly used to study and develop the gasification and syngas treatment

its laboratories, is mainly engaged in research activities on coal gasification, hydrogen production and near zero emissions power generation.

[5] The project, called COHYGEN (coal-to-hydrogen generation), started in 2003 and it has been concluded in early 2009, with a total cost of about 11 M€; partner of the project were Ansaldo Energia (former Ansaldo Ricerche), ENEA and the Department of Mechanical Engineering of the University of Cagliari.

processes for a combined production of hydrogen and electrical energy. Therefore, the pilot plant is equipped with a complete and flexible syngas treatment process.

Figure 1. The Sotacarbo pilot platform.

2.1. Demonstrative unit

The demonstrative unit is based on a 5 MW_{th} gasifier (figure 2), equipped with a manual coal charging system, a wet scrubber for dust and tar (topping atmospheric residue) removal and a flare for syngas combustion.

For the feed of the gasifier, coal (characterized by a granulometry between 5 and 60 mm, according to the design conditions) is bought in big bags for the experimental tests; every bag, containing about 1 ton of coal, is drawn out from the storage area by a heaver and, through a tackle, charged in a proper hopper in order to empty the bag itself. Then, coal is drawn out from this hopper and sent to the gasifier through four different injection points in order to distribute the fuel as uniformly as possible and to optimize the gasification process.

The reactor is similar to Wellman-Galusha gasifiers and operates at about atmospheric pressure (0.11-0.14 MPa). Being a fixed-bed reactor, the fuel bed is characterized by different operating zones, where the coal drying, devolatilization, pyrolisis, gasification and combustion processes take place. As coal flows downwards, it is heated by the hot raw gas that moves upwards, coming from the gasification and combustion zones [23-24]. The gasification agents (air and steam) are introduced into the reactor near the bottom, below the fuel grate, so that they are pre-heated by the cooling bottom ashes, which are removed through the grate itself.

In order to distribute the fuel as uniformly as possible, the reactor is equipped with a stirrer (internally cooled, in order to keep a low metal temperature), which is characterized by two degrees of freedom: an axial rotation and a vertical translation [25]. Furthermore, the gasifier is equipped with a cooling water jacket, in order to operate an accurate temperature control (figure 3).

Figure 2. Demonstrative gasifier.

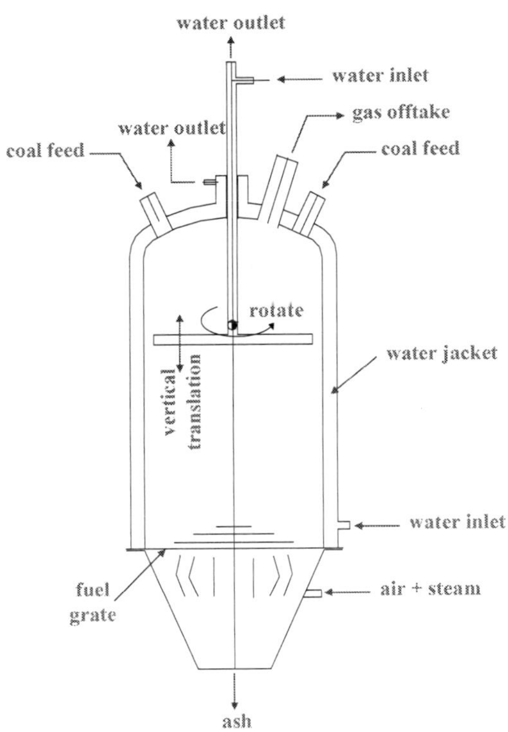

Figure 3. Simplified scheme of the demonstrative gasifier.

The start-up of the gasifier is carried out by using a series of three ceramic lamps, located near the bottom of the fuel bed, which heat the fuel (initially wood pellets) in an inert atmosphere. When the temperature reaches about 800-850 °C, air is injected into the reactor and the fuel combustion takes place. Finally, steam is also injected into the reactor (whereas, in the meantime, air injection is reduced to sub-stoichiometric conditions) and coal can be fed as gasification fuel. As for the temperature control, the gasifier is equipped by a series of 36 thermocouples (6 sensors, disposed around the circular section of the reactor, in 6 different height levels) in order to have a detailed temperature profile of the reactor.

The pilot plant does not include a syngas desulphurization section (raw syngas is only sent to a wet scrubber, for tar and dust removal, and then it is directly burned in a flare); therefore, it can be only fed with fuels characterized by a sulphur content lower than 0.5-0.6 % (in weight).

The experimental tests on the demonstrative plant have not been started as of when this chapter was prepared. In any case, a global syngas production of about 2500 Nm^3/h can be estimated (through the use of a mathematical model of the process). The expected raw syngas lower heating value is about 6.5 MJ/kg, with a hydrogen concentration of about 21%; it corresponds to a cold gas efficiency of 93-94%.

2.2. Pilot unit

The pilot unit, in which the experimental tests reported in this chapter have been carried out, were developed in order to study and optimize the gasification process and to produce a syngas flow to feed the experimental syngas treatment process for combined production of hydrogen and electrical energy.

Gasification process

Due to its dimension, the pilot gasifier (figure 4) is quite different from the demonstrative unit. In particular, whereas the 5 MW_{th} unit is cooled through a water jacket, the walls of the pilot gasifier are covered with a refractory material; moreover, coal is charged through a single inlet point and the reactor is not equipped with the intercooled stirrer. The fuel bed is held up by a grate which includes two series of metallic bars: the upper one is fixed and directly in contact with fuel, whereas in the lower one each bar is equipped with two blades and can rotate around its axis, thus increasing or reducing the openings. The temperature profile of the reactor can be determined through a series of 11 thermocouples disposed over a metallic probe (which can operate up to 1200 °C) inserted through the top of the gasifier and located near the reactor vertical axis. Each temperature sensor is about 150 or 200 mm (depending on the position into the reactor) far from each other.

The manually controlled fuel feeding system is identical with respect to the demonstrative plant and an extractor fan, which serves both pilot and demonstrative gasifiers, allows the extraction of dust during the coal handling and charging phases. As for the plant start-up, a series of three ceramic lamps allows the heat of the reactor, initially fed with wood pellets (with the addition of a small amount of paraffinic material, in order to promote the ignition) and subsequently charged with coke (only when the gasification reactions are running the coal is fed to the reactor).

Figure 4. Pilot gasifier.

Dust and tar removal system

As can be seen in the pilot plant simplified scheme (figure 5), raw syngas from the gasification process is sent to an integrated component which includes a wet scrubber, a first cold gas desulphurization stage and an electrostatic precipitator (ESP).

In particular, the wet scrubber reduces syngas temperature from 150-300 °C (depending on the particular operating conditions) to about 50 °C and operates a primary dust and tar separation. Sometimes, when fine particles and tar contents in raw syngas are very high[6], the wet scrubber is not able to operate a fine removal of these components; in this case, the electrostatic precipitator can be used in order to strongly reduce particulate and tar content in syngas.

Finally, the need to use coal with a very high sulphur content and to protect the electrostatic precipitator from the effects of an acidic atmosphere, prompted the insertion, between wet scrubber and ESP, of a first cold gas desulphurization stage, which generally uses sodium hydroxide (40% in volume, diluted in water), as solvent for H_2S removal.

Downstream the ESP, syngas can be sent to the power generation line; moreover, depending on the goals of each experimental test, a portion of syngas (20-25 Nm^3/h) can be sent to the hydrogen production line.

[6] Their concentration strongly depends on primary fuel and operating conditions. For example, the gasification of biomass or high volatiles coals involve a very high tar content in raw syngas.

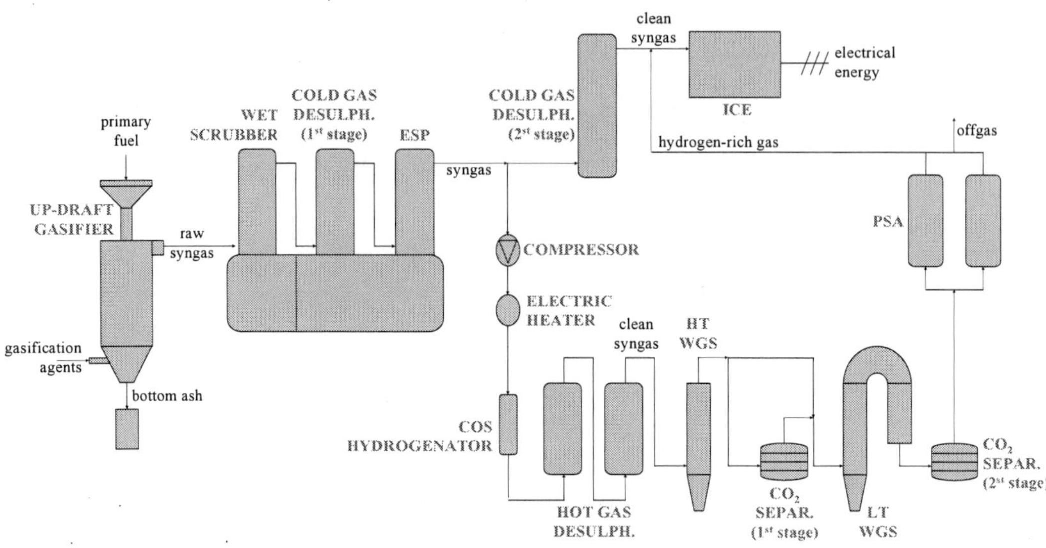

Figure 5. Pilot plant simplified scheme.

Power generation line

The power generation line is constituted by the second cold gas desulphurization stage directly followed by an internal combustion engine (ICE), characterized by a nominal power output of about 24 kW, fed with clean syngas, eventually enriched in hydrogen.

In particular, the second cold gas desulphurization stage (figure 6) is a packed column, made of polypropylene (and filled with Rashig rings, made of the same material, with a diameter of 25 mm and disposed randomly), in which hydrogen sulphide is chemically absorbed through a proper solvent. According to the design conditions, syngas is washed with an aqueous solution of sodium hydroxide and hypochlorite (typically 40% and 13% in volume, respectively), according to the well-known following reactions [26]:

$$2NaOH + H_2S \Leftrightarrow Na_2S + 2H_2O \tag{1}$$

$$Na_2S + 4NaOCl \Leftrightarrow Na_2SO_4 + 4NaCl \tag{2}$$

During some experimental tests, the packed column has been used with methildiethanolamine (MDEA) as solvent for H_2S absorption[7].

Hydrogen production line

The hydrogen production line includes a compressor followed by an electric heater, a dry hot gas desulphurization process, an integrated water-gas shift (WGS) and CO_2 absorption system and a hydrogen purification section.

In particular, the hot gas desulphurization process (figure 7) operates at about 300-500 °C and includes three main components: a catalytic filter for COS conversion and two H_2S adsorbers.

---

[7] The packed column designed as second cold gas desulphurization stage has been also used, during a series of experimental tests, as CO_2 absorber with different amine-based solutions.

Figure 6. Second cold gas desulphurization stage.

In the catalytic filter, the small amount of carbonyl sulphide contained in syngas reacts with hydrogen to be converted to H_2S, according with the hydrogenation reaction [27]:

$$COS + H_2 \Leftrightarrow H_2S + CO \tag{3}$$

promoted by $Ni\text{-}MoO_3/Al_2O_3$ catalyst.

On the other hand, the two hydrogen sulphide absorbers are disposed in lead-leg configuration and filled with a zinc oxide-based sorbent [28-30]: zinc oxide (ZnO) reacts with H_2S producing zinc sulphide and steam, according with the following reaction [31-33]:

$$ZnO + H_2S \Leftrightarrow ZnS + H_2O \tag{4}$$

The integrated water-gas shift and CO_2 separation system includes both high and low temperature shift reactors with an intermediate and a final CO_2 absorber.

In particular, the WGS process takes place in two reactors (operating at 300-450 °C and about 250 °C, respectively) which have been installed in order to test different catalysts, including conventional $Fe_3O_4/Cr_2O_3/CuO$ based catalysts for high temperature (HT) stage and $CuO/ZnO/Al_2O_3$ based catalysts for low temperature (LT) stage [34-36]; toward this goal, the reactors have been designed with a maximum bed volume of 9.6 and 17.5 dm^3, respectively.

Figure 7. Hot gas desulphurization system.

Conventional catalysts require lengthy in situ pre-reduction; they are sensitive to air (pyrophoric) and condensed water, and are poisoned by sulphur [37]. In order to avoid these kinds of problems, as of now the water-gas shift section has been tested with advanced non pyrophoric commercial catalysts. In particular, the high temperature catalyst is composed of cylindrical pellets with particles of platinum deposed in a matrix of zirconium doped ceria; on the other hand, the low temperature catalyst is composed of spherical particles with a matrix of alumina with zirconium, cerium and lanthanum and an external surface composed of the same elements with the addition of carbon and platinum [37-38].

Carbon dioxide absorption takes place in two identical bubbling reactors (figure 8), made of AISI 316L, in which syngas is injected through 40 diffusers based on ceramic membranes and reacts, at about 30 °C and atmospheric pressure, with amine-based solvents.

Figure 8. CO_2 absorbers.

In particular, different solvents, such as monoethanolamine (MEA) in different concentrations and a solution of methyldiethanolamine (MDEA) and piperazine (PZ), have been used in this process in some experimental tests. In particular, according to the design conditions, a 5M (about 30% in weight) aqueous solution of monoethanolamine reacts with

CO_2 in accordance with the following well-known mechanism [39-44] where R in equation (9), (10) and (11) represents CH_2-CH_2OH:

$$2H_2O \Leftrightarrow OH^- + H_3O^+ \quad \text{(ionization of water)} \quad (5)$$

$$CO_2 + 2H_2O \Leftrightarrow HCO_3^- + H_3O^+ \quad \text{(}CO_2\text{ hydration)} \quad (6)$$

$$HCO_3^- + H_2O \Leftrightarrow CO_3^{2-} + H_3O^+ \quad \text{(dissociation of bicarbonate)} \quad (7)$$

$$CO_2 + OH^- \Leftrightarrow HCO_3^- \quad \text{(formation of bicarbonate)} \quad (8)$$

$$CO_2 + 2RNH_2 \Leftrightarrow RNHCOO^- + RNH_3^{\cdot} \quad \text{(reaction of }CO_2\text{ with MEA)} \quad (9)$$

$$RNH_3^+ + H_2O \Leftrightarrow RNH_2 + H_3O^+ \quad \text{(dissociation of protonated MEA)} \quad (10)$$

$$RNHCOO^- + H_2O \Leftrightarrow RNH_2 + HCO_3^- \quad \text{(carbamate to bicarbonate reversion)} \quad (11)$$

Finally, the hydrogen purification section is based on the pressure swing adsorption (PSA) technology, which is widely common in the industrial applications due to its low costs [10]. In particular, PSA is composed of a simple double-stage process based on carbon molecular sieves.

The size of the secondary syngas treatment line, even if much smaller than the size of commercial scale plants, has been chosen in order to give reliable experimental data for the scale-up of future plants.

Moreover, with the goal being to ensure a full plant flexibility, as well as to simplify the management of the experimental pilot plant, the different cooling and heating devices are not fully integrated.

2.3. Control system and data collection

In order to support the experimental tests, the pilot plant is equipped with a control and sampling system which allows monitoring of the process performance. In particular, process operation can be managed mainly through the control system, which automatically collects the main process parameters (such as pressures, temperatures, volume flows and so on).

As for syngas composition and properties, the plant is equipped with a system for the real-time measurement of oxygen concentration in raw syngas (this measure plays a double role of safety control, to avoid the formation of explosive atmosphere, and performance indicator of the gasification process). Moreover, upstream and downstream of each plant component, a sampling outlet has been placed in order to operate syngas analysis through a gas chromatograph and to evaluate the concentration of the main chemical compounds (CO_2, H_2, O_2, CO, CH_4, N_2, H_2S, COS, C_2H_6, C_3H_8) in the selected stream.

Finally, the plant has been recently equipped with a tar sampling and analysis system, in order to evaluate, through a proper gas chromatograph, the content and the composition of the main hydrocarbon components in raw syngas.

3. Coal Gasification Performance in the Pilot Unit

Globally, the pilot gasifier has been tested for about 700 hours with different fuel blends. In particular, during the first experimental campaign (between June 2008 and March 2009) the gasifier was tested with operating parameters very far from the design conditions due to some technical problems, which were resolved in late 2009 through a series of plant modifications[8]. The last experimental campaign was mainly carried out in order to optimize the gasification process in its conventional operating condition. In particular, this chapter presents the main results obtained during this new campaign, with the experimental data collected during more than 400 hours of experimental tests carried out between December 2009 and September 2010.

3.1. "Standard" gasification performance

A great number of the experimental tests considered here have been operated using a low sulphur South African coal; the proximate, ultimate and thermal analyses, determined in the Sotacarbo laboratories (according with the international standard procedures), are shown in table 1. The same table also reports the characterization of a high sulphur Sulcis coal (extracted from the only Italian coal mine located in South-West Sardinia, near the Sotacarbo plant), which has been used (in mixture with South African coal) in some experimental tests.

During the first phase of the last experimental campaign, the plant operating conditions had been optimized in order to increase syngas production and lower heating value (LHV). In particular, a series of experimental tests has been carried out with different air/coal and steam/coal mass ratios, in order to analyze the effects of both these parameters on temperature profile and on syngas composition. The main result of all these tests is the definition of the "standard" operating conditions, which represent the best performance obtained by the gasification of only South African coal [45]. These results, shown in table 2, have been averaged during more than six hours of steady-state operation of the reactor.

Currently, the best performance of the pilot gasifier has been obtained with the injection of 36.8 kg/h of air and 6.0 kg/h of steam. In this way, the air-blown gasification of 8.0 kg/h of coal (corresponding to a thermal input of about 60 kW) involves the production of 43 Nm^3/h of raw syngas, which is characterized by a lower heating value of 4.50 MJ/kg. In particular, coal consumption is significantly influenced by a series of parameters such as fuel reactivity, fuel resident time [46] and steam/coal and air/coal mass ratios.

Raw syngas is composed of a high nitrogen content (51%, in volume), typical of air-blown gasification process [19, 47], and a relatively high hydrogen content (about 19%).

Cold gas efficiency is defined as the ratio between the chemical power associated with raw syngas and those associated with coal:

$$\eta_{CG} = \frac{m_G \cdot LHV_G}{m_C \cdot LHV_C} \tag{12}$$

---

[8] The main plant modification is the replacement of the old grate of the gasifier (which gave some problems during the ash discharging phases) with a new one, directly designed by Sotacarbo.

where m_G and m_C are the mass flows of syngas and coal and LHV_G and LHV_C are the lower heating values of the same fuels, respectively. This value is very consistent with the corresponding typical efficiencies of the industrial air-blown fixed-bed gasification plants.

Table 1. South African and Sulcis coal characterization

	South African coal	Sulcis coal
Proximate analysis (% in weight)		
Fixed carbon	72.58 %	40.65 %
Moisture	3.64 %	7.45 %
Volatiles	8.81 %	40.45 %
Ash	14.97 %	11.45 %
Ultimate analysis (% in weight)		
Total carbon	75.56 %	66.49 %
Hydrogen	3.86 %	6.18 %
Nitrogen	1.40 %	1.41 %
Sulphur	0.57 %	7.02 %
Moisture	3.64 %	7.45 %
Ash	14.97 %	11.45 %
Thermal analysis (MJ/kg)		
Higher heating value	28.10	22.59
Lower heating value	27.18	21.07

Table 2. Synthesis of "standard" gasification conditions

"Standard" operating parameters			
Primary fuel consumption (kg/h)			8.0
Air mass flow (kg/h)			36.8
Steam mass flow (kg/h)			6.0
Dry syngas composition (molar ratios)			
CO	0.1807	CH_4	0.0151
CO_2	0.0947	H_2S	0.0003
H_2	0.1889	COS	0.0001
N_2	0.5128	O_2	0.0074
Gasifier performance			
Syngas mass flow (kg/h)			46.83
Syngas volume flow (Nm^3/h)			42.90
Syngas lower heating value (MJ/kg)			4.50
Specific heat (kJ/kg K)			1.23
Syngas outlet pressure (MPa)			0.14
Syngas outlet temperature (°C)			126
Maximum reaction temperature (°C)			1034
Cold gas efficiency			96.93%
Hot gas efficiency			92.90%
Gasifier yield (Nm^3/kg)			5.36

Hot gas efficiency represents a more detailed performance indicator of the gasification section, taking in account sensible heats, steam production and, eventually, mechanical work; in this case, it can be defined through the following simplified equation:

$$\eta_{HG} = \frac{m_G \cdot LHV_G + Q_G}{m_C \cdot LHV_C + Q_S} \tag{13}$$

where Q_G and Q_S are the thermal energy associated to produced syngas (sensible heat) and injected steam (the latter takes into account the inefficiencies of the steam production and injection system), respectively. This equation does not consider the steam production through syngas cooling and the mechanical work: both these terms are null in this gasification process.

Finally, gasifier yield is defined as the ratio between the volume flow of produced syngas, expressed in normal cubic meters per hour, and the mass flow of primary fuel, in kilograms per hour.

Raw syngas composition is strongly influenced by the primary fuel properties and the gasification parameters. In particular, raw syngas from South African coal (in "standard" conditions) presents a very low content of sulphur compounds (274.3 ppm of H_2S and 100.9 ppm of COS, both expressed in volume). This result can be compared with the sulphur content in raw syngas from the gasification of only Sulcis coal (characterized by a sulphur content of about 7% in weight), which typically amounts to about 12000 ppm.

Thermal profiles in the reactor are significantly variable during the process (even during the steady-state phases, when steam and air mass flow are about constant), due to being influenced by a series of parameters such as the succession of coal charging and ash extraction, both of which are discontinuous. Figure 9 shows the typical temperature profiles (measured by the previously described temperature probe located near the vertical axis of the reactor) in the pilot gasifier.

The three dotted lines correspond to three different steps of the gasifier start-up process, measured 30, 60 and 90 minutes after the start of air injection, respectively. On the other hand, the continuous line represents the temperature profile during the steady-state operation (reached about 120-180 minutes after the start of air injection), averaged during the same period that has been assumed to define the "standard" conditions.

In this figure it is possible to distinguish the different operating zones of the gasifier: the freeboard (between the top of the reactor and about 1200 mm from the bottom), in which temperature remains near uniform at 125-130 °C; the coal heating, drying, devolatilization and pyrolisis zone (800-1200 mm); the combustion and gasification zone (100-800 mm, in which the maximum temperature of about 1000-1050 °C is reached) and the ash cooling area.

When fresh coal is charged into the reactor, a remarkable temperature decrease can be observed in the upper part of the fuel bed, whereas there are not appreciable variations of the thermal profile in the combustion and gasification zone. On the other hand, ash discharge involves a slight vertical translation of the thermal profile; in this case, the high temperature zone slightly moves down near the grate.

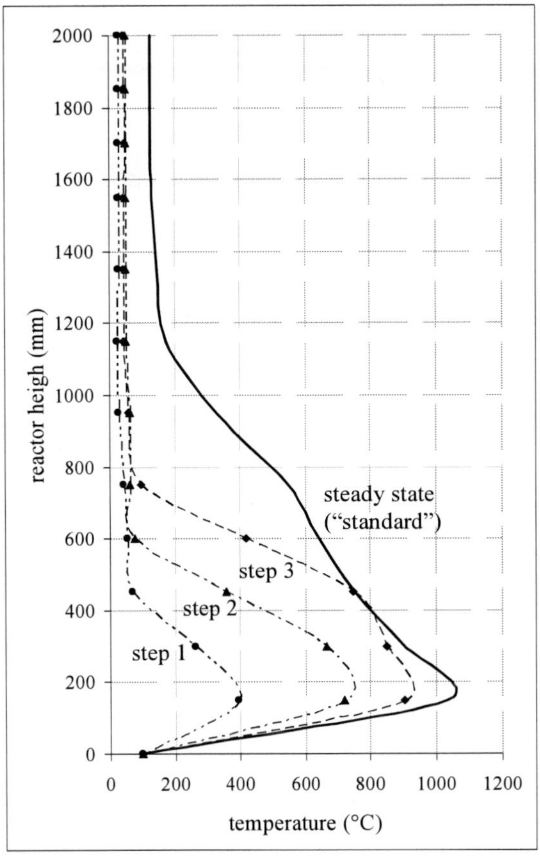

Figure 9. Temperature profiles into the gasifier.

3.2. Effects of air/coal mass ratio

Theoretically, an increasing of the air/coal mass ratio (with respect to the "standard" conditions) involves a reduction of hydrogen and carbon monoxide concentrations [19]. This is mainly due to the combination of three different phenomena: the rising temperature in the reactor, which reduces the effects of water-gas shift conversion; the increasing nitrogen content, which dilutes raw syngas; and, mainly, the effects of combustion reactions, promoted by the higher oxygen supplying.

On the other hand, the subsequent raising of steam injection[9] promotes both the gasification and water-gas shift reactions (increasing the reactants concentration and reducing the operating temperature in the reactor) and involves a raising of hydrogen concentration and a decreasing of CO content; carbon dioxide concentration remains almost constant because the increase of CO_2 content due to the water-gas shift reaction is offset by the syngas dilution by steam [5].

---

[9] In order to maintain the same temperature profile, in particular in the combustion and gasification zone, near the bottom of the gasifier, steam injection has been increased together with the raise of air/coal mass ratio.

Table 3 shows the effects of the raising air/coal mass ratio through a comparison between the gasification results in "standard" operating conditions and those obtained during the test run in which the maximum South African coal consumption has been obtained. Both tests were operated at the same maximum temperature (1000-1050 °C), controlled through a variation of steam injection.

Table 3. Effects of air/coal mass ratio

	"Standard"	Max. cons.
Operating parameters		
Primary fuel consumption (kg/h)	8.0	10.3
Air mass flow (kg/h)	36.8	45.0
Steam mass flow (kg/h)	6.0	10.0
Dry syngas composition (molar ratios)		
CO	0.1807	0.1533
CO_2	0.0947	0.1076
H_2	0.1889	0.1893
N_2	0.5128	0.5147
CH_4	0.0151	0.0109
H_2S	0.0003	0.0001
COS	0.0001	0.0001
O_2	0.0074	0.0240
Gasifier performance		
Syngas mass flow (kg/h)	46.83	57.78
Syngas volume flow (Nm^3/h)	42.90	52.26
Syngas lower heating value (MJ/kg)	4.50	3.98
Specific heat (kJ/kg K)	1.23	1.22
Syngas outlet pressure (MPa)	0.14	0.14
Syngas outlet temperature (°C)	126	275
Maximum reaction temperature (°C)	1034	1000
Cold gas efficiency	96.93%	82.14%
Hot gas efficiency	92.90%	80.96%
Gasifier yield (Nm^3/kg)	5.36	5.07

As shown in table 3, an increase of air flow involves a higher coal consumption and a significant reduction of syngas quality and cold gas efficiency. In these conditions, a higher raw syngas production has been obtained with respect to the "standard" conditions (52.26 Nm^3/h, against 42.90 Nm^3/h), with a significant decrease of lower heating value (3.98 MJ/kg, against 4.50 MJ/kg in "standard" conditions). Globally, the cold gas efficiency decreases from 96.93% to 82.14%.

As for the main chemical compounds, the expected significant decrease of CO concentration can be noticed, with a corresponding increase of CO_2 concentration. Hydrogen content remains almost constant: as a matter of fact, the effect of a high air injection (and the subsequent reduction of hydrogen content) is completely offset by the significant increasing of steam injection.

3.3. Gasification of high sulphur Sulcis coal

Being the only coal reserve in Italy, Sulcis coal basin plays a very important role from a strategic point of view [48]. Moreover, the closeness between the Sulcis coal mine and the Sotacarbo pilot platform makes the possibility of testing the gasification of Sulcis coal (in mixture with low sulphur coal) particularly interesting[10].

As reported above, a series of experimental tests were carried out by using high sulphur Sulcis coal, and the proximate, ultimate and thermal analyses were previously shown in table 1. In particular, during the last experimental campaign, different runs were carried out by mixing 20% (in weight) of Sulcis coal with South African coal. Due to its volatile content and its high reactivity [49-50], the addition of a small portion of Sulcis coal in the fuel blend involves a significant improvement of the gasification performance. The main results of this kind of test are shown in table 4, in comparison with "standard" operating conditions.

Table 4. Gasification performance with 20% of Sulcis coal

	"Standard" (S.A. coal)	20% of Sulcis coal
Operating parameters		
Primary fuel consumption (kg/h)	8.0	10.5
Fuel lower heating value (MJ/kg)	27.18	25.96
Air mass flow (kg/h)	36.8	41.2
Steam mass flow (kg/h)	6.0	7.8
Dry syngas composition (molar ratios)		
CO	0.1807	0.1772
CO_2	0.0947	0.0969
H_2	0.1889	0.2149
N_2	0.5128	0.4780
CH_4	0.0151	0.0151
H_2S	0.0003	0.0006
COS	0.0001	0.0001
O_2	0.0074	0.0172
Gasifier performance		
Syngas mass flow (kg/h)	46.83	54.84
Syngas volume flow (Nm^3/h)	42.90	51.49
Syngas lower heating value (MJ/kg)	4.50	4.83
Specific heat (kJ/kg K)	1.23	1.27
Syngas outlet pressure (MPa)	0.14	0.14
Syngas outlet temperature (°C)	126	170
Maximum reaction temperature (°C)	1034	1050
Cold gas efficiency	96.93%	97.20
Hot gas efficiency	92.90%	93.93
Gasifier yield (Nm^3/kg)	5.36	4.90

[10] One of the main goals of the research carried out by Sotacarbo is the relaunch of the industrial activities related with the use of Sulcis coal.

In particular, with respect to the "standard" operating conditions, the experimental tests with 20% of Sulcis coal show a significant increase of coal consumption (10.5 kg/h, higher than the maximum value obtained with only South African coal) as a consequence of a small increasing of air flow (41.2 kg/h, with a steam flow of 7.8 kg/h). Syngas production raises up to 51.49 Nm^3/h, with a lower heating value of 4.83 MJ/kg (to be compared with 42.90 Nm^3/h and 4.50 MJ/kg in the "standard" conditions). Obviously, the sulphur compounds' content increases up to 634.4 ppm (in volume) of H_2S and 126.3 ppm of COS (to be compared with 274.3 and 100.9 ppm, respectively, obtained in the "standard" conditions). Globally, a very high cold gas efficiency (97.20%) has been calculated in these particular operating conditions.

3.4. Gasification with mixtures of CO_2 and O_2

A series of preliminary experimental tests (about 100 hours) were carried out, with South African coal, in order to verify the possibility of operating the gasification process with a mixture of oxygen and carbon dioxide (instead of air) as gasification agent.

In the scientific literature, different studies (mainly based on experimental tests in bench-scale plants) report the influence of CO_2 injection in catalytic fluidized-bed biomass gasifiers or pyrolisers [51-52] or in high pressure entrained flow coal gasifiers [53]. In these cases, a CO_2-rich syngas characterized by a low H_2/CO molar ratio has been obtained by laboratory-scale experimental tests. The results of these studies suggest an assessment of the effect of CO_2 on an autothermal coal gasification process in a fixed-bed low-pressure reactor, as the Sotacarbo pilot gasifier.

Toward this goal, a series of experimental tests were planned in order to investigate the effects of CO_2 injection in the coal gasification process and in the syngas treatment line and some preliminary runs have been already carried out. In particular, during this kind of test, the pilot gasifier was started up according to the conventional procedures; after the reach of the air-blown steady-state, air injection is interrupted and the mixture of oxygen and carbon dioxide is injected into the reactor. These preliminary tests have been conducted significantly far from the "standard" operating conditions, due to the difficulty during this phase of having complete control over the process. In particular, a low oxygen concentration (about 15% in volume) in the oxidant mixture involves a series of problems related to the temperature control (figure 10); as a matter of fact, the low oxygen content reduces the contribution of combustion reactions, and the endothermic gasification reactions are strongly limited. As a consequence, a very low fuel consumption (about 5 kg/h) takes place, with the production of a raw syngas characterized by a very low lower heating value (2.78 MJ/kg) and by a hydrogen content of about 13% (in volume). Globally, cold gas efficiency is very low (59.94%, to be compared with 96.93% in "standard" air-blown conditions).

As shown in figure 10, a rise of oxygen concentration (up to 20% in volume) and oxidant mass flow involves an increase of the temperatures in the reactor and, at the same time, the improvement of the syngas quality (the lower heating value rises up to 3.41 MJ/kg, with a hydrogen content of about 18% in volume). On the other hand, a further reduction of the cold gas efficiency (48.53%) takes place. Figure 11 shows the concentrations (in terms of molar fractions) of the main chemical species (CO, CO_2, H_2 and N_2) before and after the injection of the mixture of carbon dioxide and oxygen (with O_2 concentration of 20% in volume); in particular, the last 25 minutes of conventional steady-state air-blown gasification have been reported, together with the transient phase when the oxidant mixture was injected instead of air. During this phase, nitrogen concentration strongly reduces to an average value of about 0.3% (a small nitrogen injection is still used for fuel pressurization during the coal charging phases), whereas carbon dioxide concentration strongly rises up to about 58%.

$$C + CO_2 \Leftrightarrow 2CO \tag{14}$$

This high CO_2 concentration promotes the Boudouard reaction [54-55] and involves an increase of CO concentration in raw syngas; finally, the hydrogen concentration remains almost constant before and after CO_2 injection (except for a short period of instability during the transient phase).

It is important to note that, due to the preliminary nature of the past experimental tests, the process has not yet been optimized in this particular operating condition. The optimization of the CO_2/O_2-blown gasification process is one of the main goals for the future experimental campaigns [45].

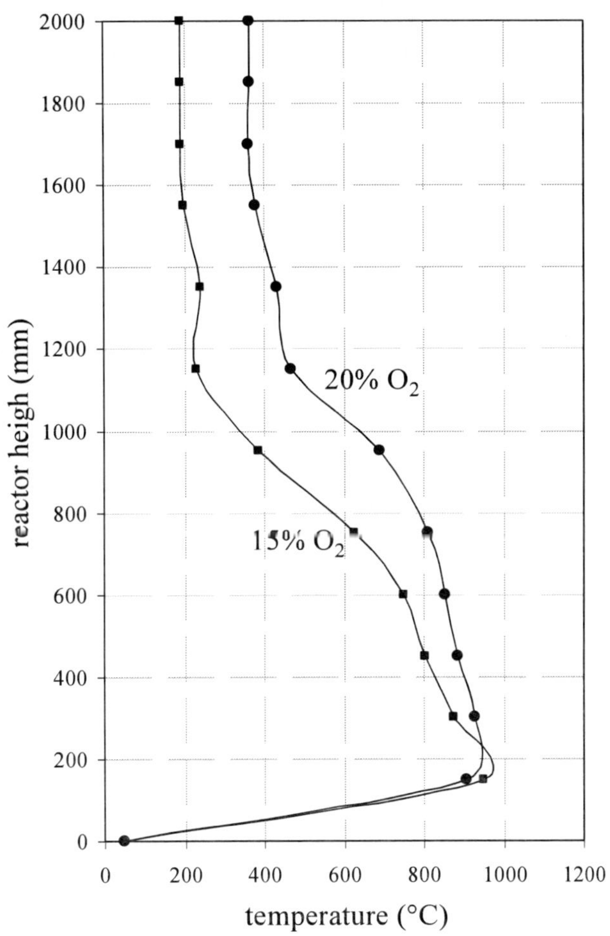

Figure 10. Typical temperature profiles during CO_2/O_2 gasification tests.

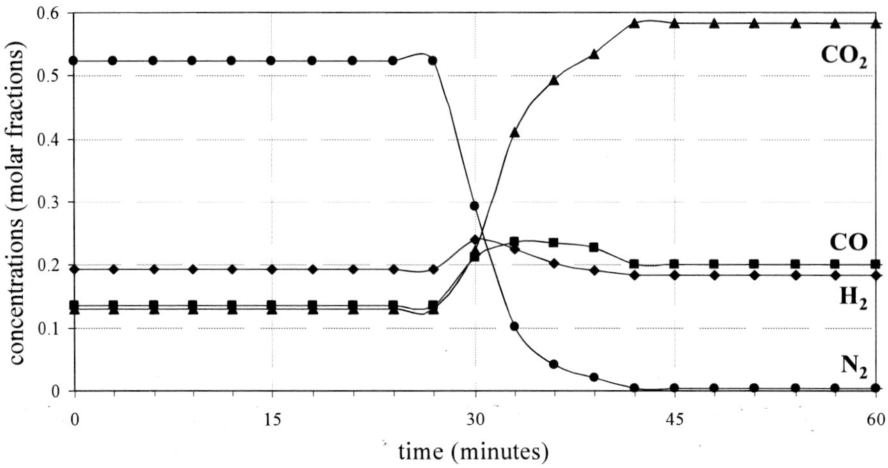

Figure 11. Syngas composition before and after CO_2/O_2 injection.

4. Global Pilot Plant Performance

A global mass balance of the Sotacarbo pilot plant has been carried out in order to evaluate the overall plant performance and both hydrogen and carbon balances. The results reported here come from a detailed analysis made up on the basis of the experimental data collected for every plant section. Through the processing of these data, a global mass balance of the plant has been assessed and, for every section, the performance and properties of each flow have been determined with good accuracy.

This evaluation considers a hypothetical commercial near zero emissions configuration, which includes the gasification section and the CO_2-free hydrogen production line. Therefore, the balance has been carried out, on the basis of the experimental data, with the theoretical hypothesis that all syngas from the depulverization system[11] is heated, compressed and sent to the hot gas treatment line.

4.1. Global plant performance in "standard" gasification condition

Figure 12 and table 5 show the main results of this evaluation, with reference to the above described "standard" steady-state gasification of low sulphur South African coal.

In particular, figure 12 shows a simplified scheme of the Sotacarbo pilot plant with the main results of a global balance of the whole of the experimental equipment, mainly expressed in terms of hydrogen and carbon content. For every stream, mass flow (m, in kilograms per hour) and lower heating value (LHV, in megajoule per kilogram) have been reported, together with the molar flows (in moles per hour) of equivalent hydrogen (H_{eq}), of effective molecular hydrogen (F_{H2}), of atomic carbon (C_{eq}) and of molecular CO_2 (F_{CO2}). In

[11] It includes wet scrubber, first cold gas desulphurization (CGD) stage and ESP.

particular, the molar flow of equivalent hydrogen is the global amount of hydrogen contained in the chemical species which can potentially contribute to the overall hydrogen production:

$$H_{eq} = F_{H_2} + \frac{F_{CH_4}}{2} + F_{H_2S} + F_{H_2O} \tag{15}$$

where F_i is the molar flows of each chemical species (hydrogen, methane, hydrogen sulphide and steam, respectively). According to this definition, the global balance of H_2 is influenced by the steam injection and condensation, by the hydrogen content in coal, and, slightly, by the air moisture.

In parallel, the molar flow of atomic carbon represents the global amount of carbon contained in the chemical species which could contribute to CO_2 production when syngas is burned:

$$C_{eq} = F_{CO} + F_{CO_2} + F_{CH_4} + F_{COS} \tag{16}$$

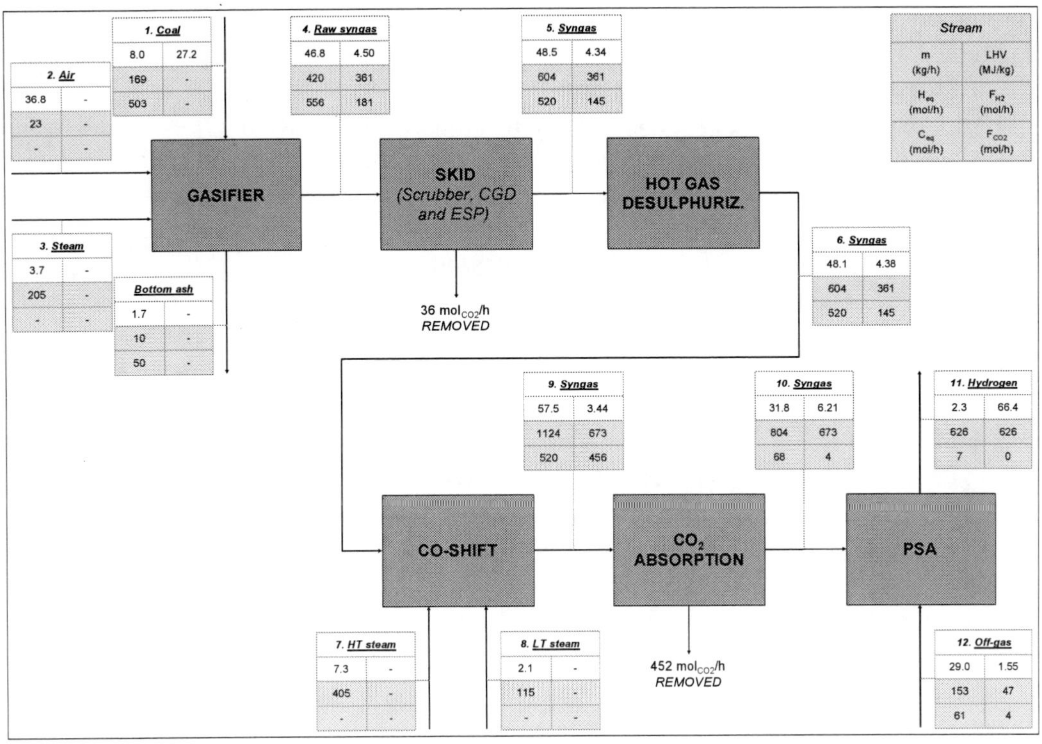

Figure 12. Synthesis of the global plant balance.

The global balance of the gasification section is directly evaluated by the experimental data; this justifies the apparent incongruence in the hydrogen and carbon balances. As a matter of fact, the reported data for coal, gasification agents and produced syngas are averaged in a time window in which the process can be considered in steady-state. On the contrary, it is currently impossible to evaluate the direct effects of the gasification process on the ash composition, being that the discharge temporization time is comparable with the duration of the considered time window. Therefore, ash composition is significantly conditioned by a series of phenomena which take place before the reach of steady-state phase.

Table 5 shows, stream by stream, the mass and volume flows, pressure, temperature, lower heating value (LHV) and specific heat, together with the gaseous streams composition, in terms of molar fractions.

In both figure 12 and table 5, raw syngas composition (stream 4) is expressed in dry basis, the data directly provided by the gas chromatograph. Downwards of the wet scrubber, syngas is saturated by water injection; therefore, in the reported data, the moisture content has been considered.

4.2. Global hydrogen balance

As for the gasification section, the material balance in "standard" conditions show that hydrogen production is mainly influenced by steam injection (52%) and secondarily by the hydrogen content in coal (43%, considering the primary fuel with its humidity), whereas the contribution of air moisture (5%) is very low. Obviously, this distribution is strongly conditioned by the gasification parameters and a variation of these percentages can be expected for different operating conditions.

Hydrogen content remains almost constant through syngas cleaning processes (washing section and hot gas desulphurization processes), while it significantly changes through the water-gas shift section, in which CO reacts with steam and H_2 content is strongly increased. In particular, the experimental results in this section show that the process operates very close to the chemical equilibrium: CO conversion in the first stage is about 90%, whereas in the second stage a conversion of about 85% takes place. Globally, a final CO concentration lower than 2% (in volume) can be achieved by the double-stage process.

Globally, the gasification of 8.0 kg/h of low sulphur South African coal allows for production of about 2.3 kg/h of hydrogen. About 1.86 Nm^3 of hydrogen can be produced for every kilogram of South African coal or, in other words, for every kilowatt of gasified coal a production of 0.70 kW of hydrogen can be obtained. In general, based on results from some preliminary tests, a slight decrease of hydrogen production can be expected when only high sulphur Sulcis coal is gasified, due to the different operating conditions of the gasification process.

Even if hydrogen purity (94-97%) is significantly lower with respect to the typical values (99.999%) that can be obtained with the modern PSA systems [10, 56], these values are fully acceptable considering that, in the pilot plant, produced hydrogen is only used to enrich clean syngas which feeds the internal combustion engine

Considering the whole process, as shown in figure 13, about 58% of produced syngas comes from steam injected into the water-gas shift section, while 22% comes from the gasification steam, 18% comes from coal and only 2% comes from the gasification air (considered with its humidity).

Table 5. Stream properties and composition

	1 *Coal*	*2* *Air*	*3* *Steam*	*4* *Raw gas**	*5* *Syngas*	*6* *Syngas*
Mass flow (kg/h)	8.00	36.80	3.70	46.83	48.54	48.09
Volume flow (Nm^3/h)	-	28.68	4.60	42.90	46.21	45.89
Pressure (MPa)	0.14	0.14	0.14	0.14	0.13	0.14
Temperature (°C)	25	25	110	126	50	350
LHV (MJ/kg)	27.18	0.00	0.00	4.50	4.34	4.38
Specific heat (kJ/kg K)	-	1.014	1.67	1.233	1.276	1.280
Stream composition (molar fraction)						
CO	-	-	-	0.1807	0.1677	0.1678
CO_2	-	-	-	0.0947	0.0703	0.0703
H_2	-	-	-	0.1889	0.1753	0.1753
N_2	-	0.7742	-	0.5128	0.4760	0.4760
CH_4	-	-	-	0.0151	0.0140	0.0140
H_2S	-	-	-	0.0003	0.0000	0.0000
COS	-	-	-	0.0001	0.0001	0.0000
O_2	-	0.2076	-	0.0074	0.0069	0.0069
H_2O	-	0.0182	1.0000	-	0.0897	0.0897
	7 *HT steam*	*8* *LT steam*	*9* *Syngas*	*10* *Syngas*	*11* *Hydrogen*	*12* *Off-gas*
Mass flow (kg/h)	7.29	2.08	57.46	31.82	2.30	29.05
Volume flow (Nm^3/h)	9.08	2.59	57.56	40.26	14.85	24.84
Pressure (MPa)	0.24	0.24	0.13	0.13	0.13	0.13
Temperature (°C)	350	250	250	30	45	45
LHV (MJ/kg)	-	-	3.44	6.21	66.38	1.55
Specific heat (kJ/kg K)	1.674	1.674	1.388	1.672	8.602	0.124
Stream composition (molar fraction)						
CO	-	-	0.0134	0.0191	0.0104	0.0247
CO_2	-	-	0.1767	0.0025	0.0007	0.0037
H_2	-	-	0.2605	0.3715	0.9444	0.0419
N_2	-	-	0.3801	0.5420	0.0444	0.8479
CH_4	-	-	0.0012	0.0160	0.0000	0.0257
H_2S	-	-	0.0000	0.0000	0.0000	0.0000
COS	-	-	0.0000	0.0000	0.0000	0.0000
O_2	-	-	0.0055	0.0078	0.0000	0.0126
H_2O	1.0000	1.0000	0.1526	0.0411	0.0000	0.0434

* Dry basis

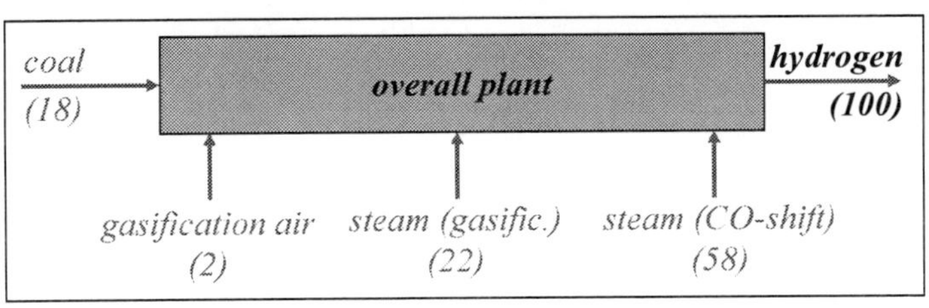

Figure 13. Global hydrogen balance.

4.3. Global carbon balance

Hypothetically, the combustion of 8.0 kg/h of South African coal (characterized by a total carbon content of 75.56%, as shown in table 1) involves the production of about 22 kg/h of

CO_2. Through the gasification process, carbon reacts with gasification agents and it is converted in a mixture of CO, CO_2, CH_4 and a small amount of COS, whereas a portion (currently about 8-10% of total carbon amount in coal[12], as results from the experimental tests) remains unreacted in the bottom ash.

Carbon content in raw syngas remains almost constant through the wet scrubber and electrostatic precipitator (carbon amount removed as tar can be neglected), whereas a little variation takes place in the first cold gas desulphurization stage. In particular, during the experimental tests with low sulphur coal, a pH value of about 9.5-10.0 was used for the desulphurization solvent (an aqueous solution of sodium hydroxide), with a subsequent absorption of about 20% of CO_2.

Through the two-stage water-gas shift section (in the reported analysis, the intermediate CO_2 removal system was not used, in order to verify the performance of a single-stage carbon capture system), carbon monoxide reacts with steam producing hydrogen and carbon dioxide. Actually, syngas is enriched in hydrogen and CO_2 but the global carbon content remains constant.

As mentioned, carbon dioxide absorption takes place in a bubbling reactor which operates at about 30 °C through different amine-based solvents; in particular, in this analysis a 5M monoethanolamine (MEA) solution was considered, according to the plant design conditions. In this way, the process (which has been significantly improved during the recent experimental campaign) operates with a global efficiency of about 99%. The remaining carbon content in syngas is separated through PSA and can be found in off-gas, which is sent to the flare, except for a small amount which remains in hydrogen-rich stream.

Globally, the gasification of 8.0 kg/h of South African coal, with the hypothesis that all syngas should be sent to the hydrogen production line, involves the emission of about 2 kg/h of CO_2, with a global removal efficiency of about 90%. In other words, as shown in figure 14, being 100 the atomic carbon which enters in the overall process through coal, about 8-10% is removed through the bottom ash, 6% is separated in the first cold gas desulphurization stage and 75% is separated as CO_2 in the carbon dioxide absorber, while the remaining 10-11% can be found in the process products (hydrogen-rich gas and off-gas).

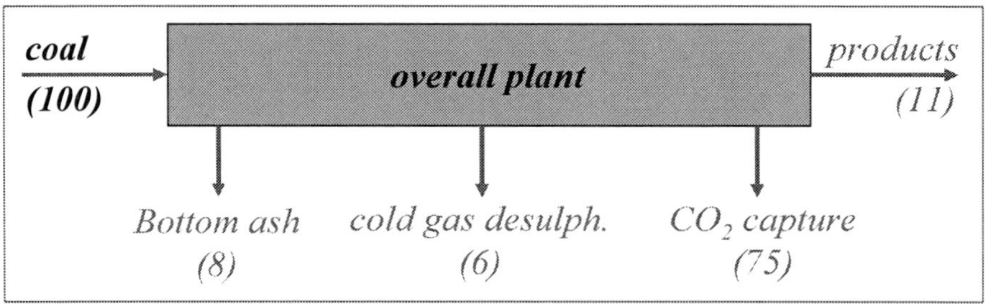

Figure 14. Global carbon balance.

[12] This high value is not referred to the "standard" operating conditions, but it is the typical value measured during the last experimental campaign. As specified later, the reduction of the unburnt content is one of the main goal of the future gasification tests.

4.4. Pollutant emissions

As for pollutant emissions, the wet scrubber allows for fine dust and tar removal, in particular for South African coal. When Sulcis coal is used, tar content in raw syngas becomes very high, even when the syngas outlet temperature is higher with respect to the gasification of South African coal; in this case, depending on the percentage of Sulcis coal in the fuel blend, the wet scrubber could be unable to assure an adequate removal. Even in this case, fine tar removal can be achieved by using the electrostatic precipitator.

The hot gas desulphurization system typically operates at about 350-400 °C by using commercial zinc oxides-based sorbents. It should be able to assure an efficient H_2S separation even when high sulphur coal is used, but, in order to protect the ESP from an acid atmosphere, in general it operates together with the first cold gas desulphurization (CGD) stage. As mentioned above, the latter allows a first significant reduction of sulphur compounds operating the absorption of H_2S in a metallic packed column, in which syngas can be washed by an aqueous solution of sodium hydroxide.

Otherwise hot gas desulphurization, it is possible to treat syngas in the second sulphur compounds absorber stage. Both these units (first and second cold gas desulphurization systems) operate a separation of hydrogen sulphide (its efficiency strongly depends on the pH of the solvent solution), but the COS concentration remains about constant.

Both cold and hot gas desulphurization systems (considered independently of each other) allow for obtainment of a final H_2S concentration lower than 10 ppm (in volume), compatible with the use of clean syngas to feed the internal combustion engine. In many cases, downwards of the hot gas desulphurization system, a final H_2S and COS concentration lower than 1 ppm has been measured, even using high sulphur coal. This concentration is generally compatible with some technologies for distributed power generation, like internal combustion engines, micro gas turbines and different kinds of fuel cells [57-59].

5. Future Developments of the Gasification Process

The performance of the gasification process was significantly improved during the last experimental campaign; in particular, as reported above, the best operating conditions have been defined for the gasification of South African coal, assumed as the reference fuel.

The interest in optimizing the technology for different commercial applications involves the need to further improve the performance of the process with the reference fuel (in particular reducing the carbon content in the bottom ash) and to assure a great flexibility in terms of primary fuel, oxidant agents and so on.

5.1. Coal gasification tests

As for the optimization of the coal gasification process, the effects of fuel granulometry and reactivity needs to be systematically analyzed. Currently, coal gasification tests have been carried out using a constant coal granulometry (between 5 and 10 mm), obtained through a crushing of the raw fuel and its riddling, with the separation of fines and powders. A reduction of coal granulometry could involves an increase of reaction rate[13], a subsequent rise of fuel load and syngas production, the reduction of the eventual formation of preferential paths into the fuel bed and a more efficient fuel utilization with a lower carbon content in the bottom ash[14] [60-62]. On the other hand, a rise in pressure drops can be expected with the subsequent need to increase the injection pressure of the gasification agents. The goal of the planned experimental tests is to define the minimum fuel granulometry compatible with the operation of the grate and the pressure profile in the reactor.

The pilot gasifier has been recently equipped with a system for sampling and the characterization of tar; after the setting of this system, a more detailed characterization of produced syngas and of the tar separation system will be assessed in order to analyze the effects of the operating conditions (in particular fuel composition and thermal profiles) on tar formation and properties.

Some experimental tests will also be carried out in order to evaluate the performance of the gasifier with air enriched in oxygen [63-64] and to optimize the gasification with mixtures of oxygen and carbon dioxide as oxidant agent.

In particular, as for air enrichment, preliminary tests show that the gasification process can be operated with an oxygen concentration in the oxidant agent up to 40% (in volume); oxygen concentration in the gasification air has been gradually increased, thus slightly reducing the actual oxygen/coal mass ratio; as a matter of fact, the reduction of nitrogen dilution requires a low heat provided by the combustion reactions. As a consequence, an increase of the hydrogen concentration (up to the average value of about 50%, with picks of 55-56%) has been obtained in raw syngas, with a carbon monoxide concentration between 20 and 25% and a reduction of CO_2 concentration (5-12%). It is important to notice that these results are preliminary, and the process needs to be optimized.

5.2. Biomass or coal/biomass gasification tests

Biomass is an abundant carbon-neutral source of renewable energy, and gasification provides a competitive way to convert biomass to syngas containing hydrogen, carbon monoxide, methane and carbon dioxide [19]. Syngas, obtained from biomass gasification, can be used to generate heat and power like natural gas [65-66], to synthesize other chemicals and liquid fuels [67-68] or to produce hydrogen [69-71]. Significant advances have been made in the technology of biomass gasification and syngas utilization [19]; however, more research is needed to improve syngas quality for its commercial uses.

In the co-gasification process, with the blending of biomass and coal, it is possible to obtain a syngas which has great flexibility to be used in several ways. Co-gasification of biomass with coal can be considered as a bridge between the energy production systems based on fossil fuels and those based on renewable energy sources. The high reactivity of

---

[13] This is mainly due to the increasing of the contact surface between solid fuel particles and gaseous gasification agents (air and steam).

[14] According with the unreacted-core shrinking model, presented by Wen in 1968 [60], the process is characterized by three resistances: external film diffusion, diffusion through the ash layer and the reaction at the surface of the unreacted core. In large coal particles, the formation of an ash layer gradually reduces the reaction rate, those generating an inert ash layer, an unreacted core and as a consequence, a high global carbon content in the discharged ash.

biomass and its highly volatile content suggest that some synergetic effects might occur in simultaneous thermochemical treatment of coal and biomass, depending on the gasification conditions [72].

To this aim, a series of gasification tests have been planned in order to evaluate the performance of the gasification process feed with wood-based biomass, mixtures of biomass and coal [73] and, eventually, waste [74-75]. The effect of the main variables, such as temperature profile, feedstock composition and air/fuel mass ratio, will be studied on gas composition, lower heating value, cold gas efficiency, gasifier yield and tar content.

Preliminary tests of co-gasification of wood-based biomass and coal (with air as gasification agent), have been carried out in the pilot up-draft fixed-bed gasifier. The effect of the feedstock composition with a varying content of wood-based biomass on the co-gasification behaviour has been preliminarily studied.

In particular, table 7 shows a comparison between the "standard" operating conditions and the main results of a gasification test in which 20% (in weight) of wood-based biomass (the same wood pellet generally used as start-up fuel, and for which the characterization is shown in table 6) has been mixed with South African coal. It is important to underline that these co-gasification conditions come from a preliminary test, and the process has not been optimized yet.

Table 6. Wood pellet characterization

	100% wood pellets	20% wood p. 80% S.A. coal
Proximate analysis (% in weight)		
Fixed carbon	17.70 %	61.60 %
Moisture	8.57 %	4.63 %
Volatiles	73.32 %	21.71 %
Ash	0.42 %	12.06 %
Ultimate analysis (% in weight)		
Total carbon	79.51 %	76.35 %
Hydrogen	11.18 %	5.32 %
Nitrogen	0.32 %	1.18 %
Sulphur	0.00 %	0.46 %
Moisture	8.57 %	4.63 %
Ash	0.42 %	12.06 %
Thermal analysis (MJ/kg)		
Higher heating value	18.45	26.17
Lower heating value	16.85	25.11

In order to compare the results and extrapolate the effect of the addition of biomass in the feeding fuel, the considered co-gasification tests have been carried out with the same operating conditions assumed as "standard" (except for primary fuel composition): air injection of 35-37 kg/h, steam injection of 6-7 kg/h and maximum temperature between 1000 and 1050 °C.

As expected, biomass being more reactive than South African coal [76-78], its introduction involves, with respect to "standard" operating conditions, an increase of fuel consumption (from 8.0 to 10.2 kg/h) and a subsequent rise of syngas production (from 43 to 45 Nm^3/h).

Syngas composition and properties remain almost constant[15], particularly hydrogen concentration (about 18-19%) and lower heating value (4.5 MJ/kg), while the cold gas efficiency strongly decreases.

Table 7. Gasification performance with 20% of wood pellets

	"Standard"	Co-gasific.
Operating parameters		
Primary fuel consumption (kg/h)	8.0	10.2
Air mass flow (kg/h)	36.8	35.0
Steam mass flow (kg/h)	6.0	6.7
Dry syngas composition (molar ratios)		
CO	0.1807	0.1681
CO_2	0.0947	0.0883
H_2	0.1889	0.1888
N_2	0.5128	0.5046
CH_4	0.0151	0.0173
H_2S	0.0003	0.0000
COS	0.0001	0.0001
O_2	0.0074	0.0328
Gasifier performance		
Syngas mass flow (kg/h)	46.83	49.23
Syngas volume flow (Nm^3/h)	42.90	45.14
Syngas lower heating value (MJ/kg)	4.50	4.44
Specific heat (kJ/kg K)	1.23	1.23
Syngas outlet pressure (MPa)	0.14	0.14
Syngas outlet temperature (°C)	126	142
Maximum reaction temperature (°C)	1034	1047
Cold gas efficiency	96.93%	85.30%
Hot gas efficiency	92.90%	82.47%
Gasifier yield (Nm^3/kg)	5.36	4.43

5.3. Preliminary tests on biochar production

Biomass is a very versatile renewable energy resource which can also be used to produce CHP (combined heat and power) and both biochar and bio-oil, each of which has potential economic value for various uses. In particular, soil application of biochar may enhance both soil quality and can be an effective means of sequestering large amounts of carbon, thereby helping to mitigate global climate change through carbon sequestration [79-81] and enhanced plant growth, which will also enhance the sustainability of biomass as a renewable energy resource. Moreover, biochar is highly absorbent and therefore increases the soil's ability to retain water, nutrients and agricultural chemicals, preventing water contamination and soil erosion. Another potential use for biochar is metal adsorption [82].

Biochar can be produced by pyrolysis or gasification systems. To this aim, the possibility of carrying out a series of gasification and biochar demonstration and characterization runs, in

[15] Except for oxygen concentration in raw syngas, which is still high (being that the process is not optimized) in the syngas from co-gasification tests.

order to optimize the biochar production in the pilot up-draft fixed-bed gasifier, is currently under evaluation in strict cooperation with Hamilton Maurer International, Inc. (Texas, United States).

In particular, after about six hours of steady-state biomass gasification, the run will be stopped using a nitrogen purge to "freeze" the gasifier fuel-bed for biochar recovery and characterization. In order to optimize the biochar production and recovery as a co-product, the characterization of the biochar within the gasifier can be important as a function of the bed level. For all the demonstration and characterization runs, both the biomass feedstock and biochar as a function of bed depth will be characterized: proximate and ultimate analysis, heat of combustion, surface area and XRF (X-ray fluorescence) mineral analysis will be done in Sotacarbo Laboratories. Biochar will also be characterized by the University of Florida both in the laboratory and applied to different soils to characterize the impact of biochar on plant growth and soils absorption of plant nutrients.

6. Future Developments of the Syngas Treatment Processes

The flexibility of plant configuration will allow a long series of experimental tests in order to optimize each syngas treatment process in different operating conditions and to assess the performance of the whole plant on the basis of different specific applications.

In particular, a series of experimental tests has been planned in order to evaluate the performance of the hot gas desulphurization process by varying the operating conditions (temperature and sorbent composition), to optimize the water-gas shift process with different catalysts (and with different steam/CO molar ratios) and to evaluate the effects of solvent composition and operating parameters on H_2S and CO_2 cold gas absorption processes.

Moreover, a series of bench-scale tests are currently in planning in order to preliminarily evaluate the possibility of applying the coal-to-liquids (CtL) technology in the Sulcis coal basin.

6.1. Hot syngas desulphurization process

In order to support the hot gas desulphurization experimental tests in the pilot platform, a bench-scale plant (called MOSCA, Metal Oxides for Sulphur Compounds Adsorption) has been developed and tested with three different commercial zinc oxides-based sorbents and in different operating conditions [83].

The plant was conceived to reproduce the operative conditions of the hot gas desulphurisation process in the Sotacarbo pilot plant. It consists of a feed gas apparatus, a fixed-bed desulphurisation reactor, a washing bottle for the off-gas cleanup and a gas sampling and analysis equipment.

A bi-component mixture of H_2S and N_2 (with a hydrogen sulphide concentration of 1.5% in volume) has been selected in order to reproduce the typical H_2S concentration in a coal syngas obtained through the gasification of Sulcis coal in an air-blown fixed-bed gasifier. In this way the operating conditions of the Sotacarbo pilot plant have been obtained as for H_2S concentration.

Gas stream depuration takes places in a quartz fixed-bed reactor (figure 14), characterized by an internal diameter of 50 mm and filled with a known volume of sorbent. A fixed quartz grate and an inert disc made of ceramic fiber (completely inert for H_2S adsorption, as experimentally verified) allow the stability of the sorbent bed. A quiet zone has been located upwards of the sorbent bed in order to assure a uniform distribution of the gaseous flow and to allow its heating.

Figure 15. The MOSCA reactor.

The reactor is placed horizontally in a tubular furnace in which a constant temperature can be maintained by a PID (proportional-integral-derivate) controller. The experimental system operates in an open circuit with a constant pressure of about 0.14 MPa. Downstream of the reactor, treated gas is sent to a washing bottle, in which it bubbles into an aqueous 5 M NaOH solution, which allows the chemical absorption of the remaining traces of hydrogen sulphide.

Until now, the experimental tests only allowed for the sorbent characterization during H_2S adsorption on the bi-component mixture, in order to focus the attention on the reaction between hydrogen sulphide and zinc oxide. A series of sulphidation/regeneration cycles is currently in progress in order to evaluate the behaviour of the process in its whole development and the effects of oxygen[16] or steam in the regeneration phase [84]. These tests should give the information needed to predict the performance and to design the regeneration system for the pilot plant.

Moreover, a series of experimental tests has been planned in order to verify the effects of steam and of a reducing ambient [85-86] on the sulphidation process; to this goal, the experimental apparatus will soon be equipped with a gas mixer and a peristaltic pump in order to create a gas mixture as similar as possible to the coal syngas.

At the moment, a thermogravimetric study is in progress in order to evaluate the amount of sulphur released by the sorbents during the regeneration process. In this way, interesting indications of the kinetic behaviour of the process will be obtained [87]. The thermogravimetric analysis is operated on the spherical particles and, in parallel, on the powdered sorbent, in order to separate the effects of diffusion on the overall kinetic.

[16] Currently, regeneration process is carried out by using oxygen (3% in volume) diluted in nitrogen, provided in bottles; a gas mixture will be soon introduced in order to test the process with different oxygen concentrations.

6.2. Optimization of CO_2 absorption process

Before the optimization of the pilot CO_2 capture system, a set-up of the double-stage water-gas shift process is need in order to stabilize the syngas output properties and to verify the effects of operating temperature [88] and steam/CO molar ratio [88-89] in both high and low temperature stages. Moreover, a series of experimental tests will be carried out in order to test the process with conventional catalysts [90-91] instead of non pyrophoric platinum-based catalysts.

As for CO_2 absorption, a series of tests has been planned in order to evaluate the effect of the main operating parameters (such as temperature, CO_2 partial pressure, CO_2 loading, gas flow rate and solvent composition) on the performance of the carbon dioxide capture system.

The temperature influences the chemical and physical equilibrium of the process, along with all the chemical-physical parameters such as viscosity, density and so on [92]. Therefore, the need to evaluate the effects of a temperature variation on the absorption system suggests carrying out a series of experimental tests with different operating temperatures[17].

CO_2 partial pressure can be indirectly modified by varying steam injection in the water-gas shift process or through a direct carbon dioxide injection upwards of the carbon capture section. This parameter influences mass transfer and CO_2 reaction into the liquid phase, but only a little variation of global removal efficiency is expected [93].

An increase of gas flow rate (which can be controlled through the compressor that sends syngas to the hydrogen production line, as shown in figure 5) involves a rise of the effective interfacial area between liquid and gas phases; therefore a raise in the whole absorption efficiency, up to a maximum value, can be expected.

With a relatively simple plant modification, it will also be possible to test the CO_2 absorption system at different pressures. As a matter of fact, with the operating pressure of PSA 0.3 MPa, it is possible to compress syngas before the second CO_2 absorption reactor, using the same compressor for both processes.

The choice of the absorption solvents can be determined by a number of factors, such as absorption efficiency, energy consumption, corrosion and solvent degradation. Most of the experimental tests have been made, in the Sotacarbo pilot plant, using MEA in different concentrations due to their high reactivity and low cost [94-95]. Obviously, solvent concentration influences the process up to a maximum value, depending to the kind of reactor [93, 96]. The increasing MEA concentration implies a higher amount of active solvent available to react with carbon dioxide; moreover, this increase involves a rise of solvent viscosity, which effects will be evaluated in the experimental tests.

Moreover, there is a growing interest in using mixed amine solvents: these solutions often provide greater absorption performance or meet unique needs when compared to the solutions of single alkanolamines [97-101]. To this aim, the possibility of carrying out a series of experimental tests, in the Sotacarbo pilot plant, with different solvent blends, with particular reference to an aqueous solution of piperazine (PZ) activated MDEA solution appears interesting. As a matter of fact, the regeneration of MEA-based solvents involves a heat request significantly higher than those needed for the regeneration of MDEA, which has a higher maximum loading capacity and is is less corrosive than MEA [102-104]. However

[17] These tests require some simple modifications to the plant equipment, in order to allow a rigorous control of the operating temperature.

MDEA, being a tertiary amine, does not react with CO_2 directly and it is necessary to use a promoter which can enhance the reaction rate without diminishing the MDEA advantages. Piperazine has been shown to be an effective promoter due to its rapid formation of carbamates with CO_2 [105-107]. During the preliminary tests, PZ concentration in the activated aqueous MDEA solution was 2% (in weight) while the total amine concentration in the solution was 30% in weight. When CO_2 is absorbed into an aqueous mixed amine solution of MDEA ($RRCH_3N$, where R represents CH_2-CH_2OH) and PZ, the following reaction may take place in the liquid phase [108]:

$$CO_2 + RRCH_3N + H_2O \Leftrightarrow RRCH_3NH^+ + HCO_3^- \quad \text{(reaction } CO_2\text{/MDEA)} \quad (17)$$

$$CO_2 + PZ + H_2O \Leftrightarrow PZCOO^- + H_3O^+ \quad \text{(formation of monocarbamate)} \quad (18)$$

$$CO_2 + PZCOO^- + H_2O \Leftrightarrow PZ(COO^-)_2 + H_3O^+ \quad \text{(formation of dicarbamate)} \quad (19)$$

$$CO_2 + RRCH_3N + PZ \Leftrightarrow RRCH_3NH^+ + PZCOO^- \quad \text{(form. of monocarbamate by PZ/MDEA)} \quad (20)$$

$$CO_2 + RRCH_3N + PZCOO^- \Leftrightarrow RRCH_3NH^+ + PZ(COO^-)_2 \quad \text{(formation of dicarbamate by PZCOO}^-\text{/MDEA)} \quad (21)$$

$$CO_2 + OH^- \Leftrightarrow HCO_3^- \quad \text{(formation of bicarbonate)} \quad (22)$$

$$HCO_3^- + H_2O \Leftrightarrow CO_3^{2-} + H_3O^+ \quad \text{(dissociation of bicarbonate)} \quad (23)$$

$$PZ + H_3O^+ \Leftrightarrow PZH^+ + H_2O \quad \text{(protonation of PZ)} \quad (24)$$

$$PZCOO^- + H_3O^+ \Leftrightarrow PZH^+COO^- + H_2O \quad \text{(protonation of monocarbamate)} \quad (25)$$

$$RRCH_3N + H_3O^+ \Leftrightarrow RRCH_3NH^+ + H_2O \quad \text{(protonation of MDEA)} \quad (26)$$

$$2H_2O \Leftrightarrow OH^- + H_3O^+ \quad \text{(ionization of water)} \quad (27)$$

According to the preliminary results, MDEA/PZ solutions appear to be attractive new blended solvents for CO_2 removal. By varying the relative concentrations of the amines, an optimum absorption system can be obtained for the specific application.

The design of a solvent regeneration system is currently under development. The introduction of this system will allow for the possibility to test the integrated absorption and regeneration process and to strongly reduce the operating costs of the experimental tests. This system will be optimized for the regeneration of MEA-based solvents but, due to its flexibility, it will allow operation with different solvent solutions.

In order to reduce the experimental tests in the pilot carbon capture unit, the above mentioned evaluations will be supported through a series of experimental tests which will be carried out in a bench-scale CO_2 absorption unit (called GAIA, greenhouse-gas absorption in amine-based solvents), which has been recently installed (in cooperation with the Department of Chemical Engineering of the University of Cagliari) in the Sotacarbo laboratories (figure 16). This unit includes a glass packed column characterized by an internal diameter of 80 mm and a bed height of 1000 mm. The reactor is randomly filled with Rashig rings (with a diameter of about 8 mm) made of glass. Syngas is simulated through a flow of CO_2 diluted in nitrogen (both these gases are supplied in bottles) and solvent (up to 225 Ndm^3/h, with a pressure up to 1.0 MPa) is sent to the reactor through a proper pump. The bench-scale plant includes a regeneration section in which exhaust solvent is heated with a steam flow and releases captured CO_2. The whole system is equipped with different pressure and temperature probes, and gas composition is analyzed through the same gas chromatograph used in the pilot platform.

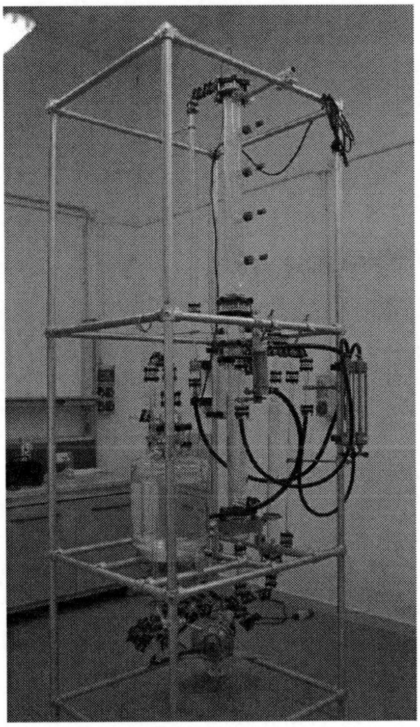

Figure 16. The GAIA bench-scale CO_2 absorption column.

6.3. Coal-to-liquids experimental tests

Besides hydrogen and electrical energy, coal syngas can be used for the production of a series of chemical products such as substitute natural gas, methanol (MeOH, CH_3OH), dimethylether (DME, CH_3OCH_3), ammonia, gasoline and so on. This kind of production could involve a series of potential benefits such as the reducing of dependence of oil and natural gas import, low pollutant emissions, a high feedstock flexibility and high efficiency of poligeneration processes [109-111].

The technologies for chemical synthesis of liquid fuels from coal syngas are well known and established, but the increasing interest in this kind of process needs a scientific and technical effort in order to optimize the plant configuration and reduce the production costs [112].

In this field, Sotacarbo is starting a study for the optimization of a coal-to-liquids (CtL) process for the production of methanol, DME and (in a second phase) Fisher-Tropsch (FT) liquids from high sulphur Sulcis coal.

In general, methanol synthesis from coal syngas takes place according to a well known mechanism which can be described through the two following reactions [21]:

$$CO + 2H_2 \Leftrightarrow CH_3OH \tag{28}$$

$$CO_2 + 3H_2 \Leftrightarrow CH_3OH + H_2O \tag{29}$$

Both these reactions, being exothermic[18] and taking place with reduction of volume, are promoted by low temperature[19] and high pressure. They historically took place in gas-phase fixed-bed reactors filled with copper catalysts, which mainly promote reaction 28. Therefore, a stoichiometric number (SN) is defined as:

$$SN = \frac{x_{H_2} - x_{CO_2}}{x_{CO} + x_{CO_2}} \tag{30}$$

where x_i represents the molar fraction of species. The best conversion efficiency can be achieved when SN is about 2.02-2.04 [21], with a CO_2 concentration of about 2.5-3.5% (in volume) [113].

Recently the performance of methanol synthesis from coal syngas has been significantly improved through the development of liquid-phase slurry bubble column reactors[20] (SBCR). In these processes, solid copper ($Cu/ZnO/Al_2O_3$) catalysts [114-117] are in powder form and it is slurried in an inert mineral oil, which operates as a temperature moderator and a heat removal medium, transferring the heat of reaction from the catalyst surface via the liquid slurry to boiling water in an internal tubular heat exchanger [111, 118].

As for dimethylether synthesis, among the methanol synthesis reactions, a further reaction (dehydration of methanol) takes place in the meantime and in the same reactor [119]:

$$2CH_3OH \Leftrightarrow CH_3OCH_3 + H_2O \qquad \text{(dehydration of methanol)} \tag{31}$$

This reaction is promoted by the same catalysts used for methanol synthesis with the addition of some solid-acid catalysts (such as ferrierite, γ-alumina, tungsten-zirconia and sulphated-zirconia) [120-123].

In this context, the first step of the Sotacarbo research involves the development of a flexible bench-scale slurry bubble column reactor for methanol and DME synthesis from coal syngas, which will be simulated through a gas mixer. The data collected during the experimental tests could allow (in a second phase of the project) the design of a small scale

---

18 The reactions 26 and 27, strongly exothermic, involve the production of 90.84 and 49.57 kJ/mol, respectively.

19 Methanol synthesis processes take place at about 250-270 °C. As a matter of fact, low temperature shifts the equilibrium of the reaction to the products, but the minimum operating value is defined by the activation of catalysts.

20 This liquid phase process has been initially developed in United States since 1981 by Air Products and Chemicals, Inc., and called LPMeOH™ (Liquid Phase Methanol).

pilot unit[21] in which the whole syngas production and treatment processes could be tested and optimized.

CONCLUSION

At present, the 700 hours of experimental tests carried out in the Sotacarbo coal-to-hydrogen pilot plant allowed for optimization of the gasification process fed with South African coal, evaluation of the global effects of the main operating parameters and, in general, definition of the performance, in design conditions, of every syngas treatment section.

As obtained by the experimental tests, the best gasification performance with a low sulphur South African coal takes place with an air flow of about 37 kg/h and a steam flow of about 6 kg/h. As a consequence of these operating conditions, fuel consumption amounts to 8 kg/h, and produced syngas (about 43 Nm^3/h) is characterized by a lower heating value of 4.5 MJ/kg, with a hydrogen content of about 19% (in volume).

The gasification cold gas efficiency (which is strongly influenced by the operating parameters) has been considered as the main performance indicator for the comparison of different operating conditions, representing the ratio between the chemical power associated to produced syngas with respect to those associated to the primary fuel. In "standard" operation, it amounts to 96.93%. In general, as expected, the experimental results show that an increase of air injections leads to a rise of coal consumption and syngas production, but, on the other hand, a significant decrease of syngas lower heating value and of the cold gas efficiency takes place.

In general, a slight increase of the fuel reactivity (corresponding to the use of a small percentage of a more reactive fuel, such as Sulcis coal or biomass) involves a significant rise in coal consumption and syngas production. In particular, the additon of 20% of Sulcis coal in the fuel blend involves an increase of syngas production (about 20%, from 42.90 to 51.49 Nm^3/h) and a rise in cold gas efficiency (from 96.93% in "standard" condition to 97.20% with Sulcis coal). On the other hand, the addition of 20% of wood pellets still involves a slight increase in syngas production (5%, from 42.90 to 45.14 Nm^3/h), but in the meantime a significant reduction of cold gas efficiency (from 96.93% to 85.30%) takes place; this is mainly due to the preliminary nature of the tests, with the process not optimized yet.

As for the whole coal-to-hydrogen process, with the theoretical hypothesis that all produced syngas is treated in the CO_2-free hydrogen production line, the potential specific hydrogen production is about 1.86 Nm^3 per kilogram of coal. Obviously, in the Sotacarbo pilot plant, hydrogen production has not been optimized, due to the experimental aim of the plant. This justifies the relatively high energy content in the off-gas, which is currently sent to the flare. With a view towards the application of this technology to an industrial coal-to-hydrogen plant, it is possible both to maximize hydrogen production (thus reducing energy content in the offgas) or to reduce hydrogen production itself and use the rich off-gas for the co-production of electrical energy.

---

[21] Due to the properties of the Sotacarbo pilot platform (air-blown gasification process, atmospheric pressures and so on), the CtL pilot unit will be totally independent from the current plant, except for some auxiliaries equipments.

Global carbon dioxide emissions can be strongly reduced (up to about 90%) through a one-stage CO_2 capture plant, which is currently characterized by an absorption efficiency of about 99% by using a MEA solution as solvent. The global CO_2 emission can be further reduced by using a two-stage capture system, with a first intermediate absorption stage between high and low temperature water-gas shift reactors. With reference to an industrial application of the technology, a two-stage CO_2 capture system equipped with a solvent regeneration section and a carbon sequestration plant should allow separation and storage of more than 90-95% of the global carbon content, with some economical advantages related with the Emissions Trading System.

As for pollutant emissions, the integration of a wet scrubber and electrostatic precipitator allows for obtainment of a negligible dust emission, whereas tar content is strongly reduced, even when Sulcis coal is gasified. The hot gas desulphurization process allows for obtainment of a final H_2S concentration lower than 10 ppm (in volume, even with high sulphur Sulcis coal) and, in many cases, lower than 1 ppm.

The studies and experimentations carried out on the Sotacarbo pilot platform represent only the first phase of a large series of experimental campaigns (by using the pilot platform with the support of the Sotacarbo Laboratories) which are in progress or in planning in order to optimize the gasification process and syngas treatment line for different applications in the field of distributed power generation through the use of hydrogen as energy carrier.

ACKNOWLEDGMENT

The authors are very grateful to Rolf E. Maurer of Hamilton Maurer International, Inc. for his useful suggestions, and all the colleagues in Sotacarbo (especially Caterina Frau, Monica Caboni and Alessandro Orsini) who supported this work.

NOMENCLATURE

<u>*Symbols*</u>

C_{eq}	molar flow of atomic carbon (mol/h)
F_i	molar flow of species i (mol/h)
H_{eq}	molar flow of equivalent hydrogen (mol/h)
LHV_G	raw syngas lower heating value (MJ/kg)
LHV_C	coal lower heating value (MJ/kg)
m	stream mass flow (kg/h)
m_C	coal mass flow (kg/h)
m_G	raw syngas mass flow (kg/h)
Q_G	raw syngas sensible heat (MJ/h)
Q_S	thermal energy required for steam production (MJ/h)
x_i	molar fraction of species i (dimensionless)
η_{CG}	cold gas efficiency (dimensionless)
η_{HG}	hot gas efficiency (dimensionless)

Acronyms

CCS	carbon capture and storage
CGD	cold gas desulphurization
CHP	combined heat and power
CoE	cost of electricity
COHYGEN	coal-to-hydrogen generation research and development project
CtE	coal-to-liquids
DME	dymethylether (CH_3-O-CH_3)
ENEA	Italian National Agency for Energy and Environment
ESP	electrostatic precipitator
FT	Fisher-Tropsch
GAIA	greenhouse-gas absorption in amine-based solvents bench-scale plant
HT	high temperature
ICE	internal combustion engine
IGCC	integrated gasification combined cycle
LHV	lower heating value (MJ/kg)
LT	low temperature
MDEA	methyldiethanolamine ($CH_3N(C_2H_4OH)_2$)
MEA	monoethanolamine ($(CH_2CH_2OH)NH_2$)
MeOH	methanol (CH_3OH)
MOSCA	metal oxides for sulphur compounds adsorption bench-scale plant
PCC	pulverized coal combustion power plant
PID	proportional-integral-derivate controller
PSA	pressure swing adsorption
PZ	piperazine ($C_4H_{10}N_2$)
SBCR	slurry bubble column reactor
SN	stoichiometric number
tar	topping atmospheric residue
WGS	water-gas shift conversion
XRF	X-ray fluorescence

REFERENCES

[1] Koljonen T.; Flyktman M.; Lehtilä A.; Pahkala K.; Peltola E.; Savolainen I. The role of CCS and renewables in tackling climate change. Energy Procedia 2009, vol 1, pp 4323-4330.

[2] Kothari R.; Tyagi V. V.; Pathak A. Waste-to-energy: A way from renewable energy sources to sustainable development. Renewable and Sustainable Energy Reviews 2010, vol 14, pp 3164-3170.

[3] M. Genova, R. Casirati, A. Calabrò, P. Chiesa, D. Cocco, A. Pettinau. PR.A.T.O.: a project to optimize pre-combustion technology for CO_2 capture, Proceedings of Coal-Gen Europe 2008, Warsaw, Poland, July 1-3, 2008.

[4] U. S. Department of Energy (2008). Annual Energy Outlook 2008. Available at http://www.eia.doe.gov.

[5] Pettinau A., Orsini A., Calì G., Ferrara F. The Sotacarbo coal gasification experimental plant for a CO_2-free hydrogen production. International Journal of Hydrogen Energy 2010, vol 35, pp 9836-9844.

[6] Commission of the European Community (2008). Energy sources, production costs and performance of technologies for power generation, heating and transport. Available at http://eur-lex.europa.eu.

[7] Aaron D.; Tsouris C. Separation of CO_2 from flue gas: a review. Separation Science and Technology 2005, vol 40, pp 321-348.

[8] Chen-Hung L.; Meng-Hui L. Kinetics of absorption of carbon dioxide into aqueous solutions of monoethanolamine + N-methyldiethanolamine. Chemical Engineering Science 2002, vol 57, pp 4569–4582.

[9] Issacs E. E.; Otto F. D.; Mather A. E. Solubility of mixtures of H_2S and CO_2 in a monoethanolamine solution at low partial pressures. Journal of Chemical and Engineering Data 1980, vol 25, pp 118–120.

[10] Chiesa P.; Consonni S.; Kreutz T.; Williams R. Co-production of hydrogen, electricity and CO_2 from coal with commercially ready technology. Part A: Performance and emissions. International Journal of Hydrogen Energy 2005, vol 30, pp 747-767.

[11] Kreutz T.; Williams R.; Consonni S.; Chiesa P. Co-production of hydrogen, electricity and CO_2 from coal with commercially ready technology. Part B: Economic analysis. International Journal of Hydrogen Energy 2005, vol 30, pp 769-784.

[12] De Micco G.; Fouga G. G.; Bohé A. E. Coal gasification studies applied to H_2 production. International Journal of Hydrogen Energy 2010, vol 35, pp 6012-6018.

[13] Balat M. Potential importance of hydrogen as future solution to environmental and transportation problems. International Journal of Hydrogen Energy 2008, vol 33, pp 4013-4029.

[14] Cormos C. C.; Starr F.; Tzimas E.; Peteves S. Innovative concepts for hydrogen production processes based on coal gasification with CO_2 capture. International Journal of Hydrogen Energy 2008, vol 33, pp 1286-1294.

[15] Shoko E.; McLellan B.; Dicks A. L.; Diniz da Costa J. C. Hydrogen from coal: production and utilization technologies. International Journal of Coal Geology 2006, vol 65, pp 213-222.

[16] Collot A. G. Prospects for hydrogen from coal. IEA Clean Coal Centre, Report No. CCC/78, London, UK, 2003.

[17] Perna A. Combined power and hydrogen production from coal. Part A – Analysis of IGHP plants. International Journal of Hydrogen Energy 2008, vol 33, pp 2957-2964.

[18] Cicconardi S. P.; Perna A.; Spazzafumo G. Combined power and hydrogen production from coal. Part B: Comparison between the IGHP and CPH systems. International Journal of Hydrogen Energy 2008, vol 33, pp 4397-4404.

[19] Wang L.; Weller C. L.; Jones D. D.; Hanna M. A. Contemporary issues in thermal gasification of biomass and its application to electricity and fuel production. Biomass and Bioenergy 2008, vol 32, pp 573 – 581.

[20] Basu P. Combustion and gasification in fluidized. CRC Press, Boca Raton, FL, 2006.

[21] Supp E. How to produce methanol from coal. Springer-Verlag, Berlin, DE, 1990.

[22] Rajvanshi A. K. Biomass gasification. In: Alternative Energy in Agriculture, volume II, edited by D. Y. Goswami, CRC press, 1986, pp 83-102.

[23] Hobbs M. L.; Radulovic P. T.; Smoot L. D. Modeling fixed-bed coal gasifier. A.I.Ch.E. Journal 1992, vol 38, pp 681-702.

[24] Hahn O. H.; Wesley P. D.; Swisshelm B. A.; Maples S.; Withrow J. A mass and energy balance of a Wellman-Galusha gasifier. Fuel Processing Technology 1979, vol 2, pp 332-334.

[25] Ferrara F.; Orsini A.; Caboni M.; Pettinau A. Coal gasification in the Sotacarbo coal-to-hydrogen experimental plant. Proceedings of the International Conference HYPOTHESIS VIII, Lisbon, Portugal, April 1-3, 2009.

[26] Chen L.; Huang J.; Yang C. Absorption of H_2S in NaOCl Caustic Aqueous Solution. Environmental Progress 2001, vol 20, pp 175-181.

[27] Yao X.; Li Y. Density functional theory study on the hydrodesulfurization reactions of COS and CS_2 with Mo_3S_9 model catalyst. Journal of Molecular Structure: THEOCHEM 2009, vol 899, pp 32-41.

[28] Thambimuthu K. Hot gas clean-up of sulphur, nitrogen, minor and trace elements. IEA Coal Research, Report No. IEACR/12, London, UK, 1998.

[29] Sánchez-Hervás J. M.; Otelo J.; Ruiz E. A study on sulphidation and regeneration of Z-sorb III sorbent for H_2S removal from simulated ECOLGAS IGCC syngas. Chemical Engineering Science 2005, vol 60, 2005, pp 2977-2989.

[30] Li L.; King D. L. H_2S removal with ZnO during fuel processing for PEM fuel cell applications. Catalysis Today 2006, vol 116, pp 537-541

[31] Sun J.; Modi S.; Liu K.; Lesieu R.; Buglass J. Kinetics of zinc oxide sulfidation for packed-bed desulfurizer modeling. Energy & Fuels 2007, vol 21, pp 1863-1871.

[32] Rosso I.; Galletti C.; Bizzi M.; Saracco G.; Specchia V. Zinc oxide sorbents for the removal of hydrogen sulphide from syngas. Industrial & Engineering Chemistry Research 2003, vol 42, pp 1688-1697.

[33] Tamhankar S. S.; Bagajewicz M.; Gavalas G. R.; Sharma P. K.; Flytzani-Stephanopoulos M. Mixed oxide sorbents for high-temperature removal of hydrogen sulphide. Industrial Engineering Chemistry Process Design and Development 1986, vol 25, pp 429-437.

[34] Dufour J.; Martos C. ; Ruiz A. Promotion of HT-WGS Fe_2O_4/MoO_3 based catalysts. Proceedings of the International Conference HYPOTHESIS VIII, Lisbon, Portugal, April 1-3, 2009.

[35] Edwards M. A.; Whittle D. M.; Rhodes C.; Ward A. M.; Rohan D.; Shannon M. D.; Hutchings G. J.; Kiely C. J. Microstructural studies of the copper promoted iron oxide/chromia water-gas shift catalyst. Physical Chemistry Chemical Physics 2002, vol 4, pp 3902-3908.

[36] Rhodes C., Hutchings G. J., Ward A. M. Water-gas shift reaction: finding the mechanistic boundary. Catalysis Today 1995, vol 23, pp 43-58.

[37] Cau G., Mura G., Murgia S. Experimental investigation and numerical simulation of CO to CO_2 conversion and hydrogen production from the water gas shift reaction. Proceedings of the International Conference HYPOTHESIS VIII, Lisbon, Portugal, April 1-3, 2009.

[38] Haryanto A.; Fernando S.; Adhikari S. Ultrahigh temperature water gas shift catalyst to increase hydrogen yield from biomass gasification. Catalysis Today 2007, vol 129, pp 355-364.

[39] Hikita H.; Asai S.; Ishikawa H.; Honda M. The kinetics of reactions of carbon dioxide with monoethanolamine, diethanolamine and triethanolamine by a rapid mixing method. Chemical Engineering Journal 1977, vol 13, pp 7-12.

[40] Hikita H.; Asai S.; Katsu Y.; Ikuno S. Absorption of carbon dioxide into aqueous monoethanolamine solutions. A.I.Ch.E Journal 1979, vol 25, pp 793-800.

[41] Aboudheir A.; Tontiwachwuthikul P.; Chacma A.; Idem R. Kinetics of the reactive absorption of carbon dioxide in high CO_2-loaded, concentrated aqueous monoethanolamine solutions. Chemical Engineering Science 2003, vol. 58, pp 5195-5210.

[42] Danckwerts P. V. The reaction of CO_2 with ethanolamine. Chemical Engineering Science 1979, vol 34, pp 443-446.

[43] Blauwhoff P. M.; Versteeg G. F.; van Swaaij W. M. A study of the reaction between CO_2 and alkanolamines in aqueous solutions. Chemical Engineering Science 1984, vol 39, pp 207-225.

[44] Versteeg G. F.; van Swaaij W. M. On the kinetics between CO_2 and alkanolamines both in aqueous and non-aqueous solutions – I. Primary and secondary amines. Chemical Engineering Science 1988, vol 43, pp 573-585.

[45] Pettinau A.; Frau C.; Ferrara F. An updating of coal gasification experimental tests in a pilot CO_2-free coal-to-hydrogen plant. Proceedings of the 27th International Pittsburgh Coal Conference, Istanbul, Turkey, October 11-14, 2010.

[46] HM Associates Inc.; Princeton Energy Resources International LLC; TFB Consulting. Assessment of the commercial potential for small gasification combined cycle and fuel cell systems. Final report prepared for U.S. Department of Energy, March 2003.

[47] Gordillo G.; Annamalai K.; Carlin N. Adiabatic fixed-bed gasification of coal, dairy biomass, and feedlot biomass using an air–steam mixture as an oxidizing agent. Renewable Energy 2009, vol 34, pp 2789-2797.

[48] Domazetis G.; Barilla P.; James B. D. Lower emission plant using processed low-rank coals. Fuel Processing Technology 2010, vol 91, pp 255-265.

[49] D'Alessio A.; Vergamini P.; Benedetti E. FT-IR investigation of the structural changes of Sulcis and South Africa coals under progressive heating in vacuum. Fuel 2000, vol 79, pp 1215-1220.

[50] Calemma V.; Radovic L. R. On the gasification reactivity of Italian Sulcis coal. Fuel 1991, vol 70, pp 1027-1030.

[51] Garcia L.; Salvador M. L.; Arauzo J.; Bilbao R. CO_2 as gasifying agent for gas production from pine sawdust at low temperatures using Ni/Al coprecipitated catalyst. Fuel Processing Technology 2001, vol 69, pp 157-174.

[52] Arauzo J.; Radlein D.; Piskors J.; Scott D. S. A New Catalyst for the Catalytic Gasification of Biomass. Energy & Fuels 1994, vol 8, pp 1192-1196.

[53] Kajitani S.; Suzuki N.; Ashizawa M.; Hara S. CO_2 gasification rate analysis of coal char in entrained flow coal gasifier. Proceedings of the 21st International Pittsburgh Coal Conference, Osaka, Japan, September 13-17, 2004.

[54] Basu P. Biomass Gasification Design Handbook 2010, Academic Press, London, United Kingdom, 2010.

[55] Yergey A. L.; Lampe F. W. Carbon gasification in the Boudouard reaction. Fuel 1974, vol 53, pp 280-281.

[56] Ribeiro A. M.; Grande C. A.; Lopes F. V. S.; Loureiro J. M.; Rodrigues A. E. A parametric study of layered bed PSA for hydrogen purification. Chemical Engineering Science 2008, vol 63, pp 5258-5273.

[57] Gould B. D.; Baturina O. A.; Swider-Lyons K. E. Deactivation of Pt/VC proton exchange membrane fuel cell cathodes by SO_2, H_2S and COS. Journal of Power Sources 2009, vol 188, pp 89-95.

[58] Trembly J. P.; Marquez A. I.; Ohrn T. R.; Bayless D. J. Effects of coal syngas and H_2S on the performance of solid oxide fuel cells: Single-cell tests. Journal of Power Sources 2006, vol 158, pp 263-273.

[59] Marquez A. I.; Ohrn T. R.; Trembly J. P.; Ingram D. C.; Bayless D. J. Effects of coal syngas and H_2S on the performance of solid oxide fuel cells: Part 2. Stack tests. Journal of Power Sources 2007, vol 164, pp 659-667.

[60] Wen C. Y. Noncatalytic heterogeneous solid-fluid reaction models. Industrial & Engineering Chemistry Research 1968, vol 60, pp 34-54.

[61] Wen C. Y.; Chaung T. Z. Entrainment Coal Gasification Modeling. Industrial and Engineering Chemistry Process Design and Development 1979, vol 18, pp 684-695.

[62] Murgia S.; Vascellari M.; Cau G. Two-dimensional CFD model o fan Air-Blown Coal Updraft Gasifier. Proceedings of the 4th International Conference on Clean Coal Technologies, Dresden, Germany, May 18-21, 2009.

[63] Campoy M.; Gómez-Barea A.; Vidal F. B.; Ollero P. Air–steam gasification of biomass in a fluidised bed: Process optimisation by enriched air. Fuel Processing Technology 2009, vol 90, pp 677-685.

[64] Mathieu P.; Dubuisson R. Performance analysis of a biomass gasifier. Energy Conversion and Management 2002, vol 43, pp 1291-1299.

[65] Rodrigues M.; Walter A.; Faaij A. Co-firing of natural gas and biomass gas in biomass integrated gasification/combined cycle systems. Energy 2003, vol 28, pp 1115–1131.

[66] Kinoshita C. M.; Turn S. Q.; Overend R. P.; Bain R. L. Power generation potential of biomass gasification systems. Journal of Energy Engineering 1997, vol 23, pp 88–99.

[67] Tijmensen M. J. A.; Faaij A. P. C.; Hamelinck C. N.; van Hardeveld M. R. M. Exploration of the possibilities for production of Fischer Tropsch liquids and power via biomass gasification. Biomass & Bioenergy 2002, vol 23, pp 129–152.

[68] Ng K. L.; Chadwick D.; Toseland B. A. Kinetics and modeling of dimethyl ether synthesis from synthesis gas. Chemical Engineering Science 1999, vol 54, pp 3587–3592.

[69] Watanabe M.; Inomata H.; Arai K. Catalytic hydrogen generation from biomass (glucose and cellulose) with ZrO_2 in supercritical water. Biomass & Bioenergy 2002, vol 22, pp 405–410.

[70] Rapagna S.; Provendier H.; Petit C.; Kiennemann A.; Foscolo P.U. Development of catalysts suitable for hydrogen or syn-gas production from biomass gasification. Biomass & Bioenergy 2002, vol 22, pp 377–378.

[71] Hanaoka T.; Yoshida T.; Fujimoto S.; Kamei K.; Harada M.; Suzuki Y.; et al. Hydrogen production from woody biomass by steam gasification using a CO_2 sorbent. Biomass & Bioenergy 2005, vol 28, pp 63–68.

[73] McLendon T. R.; Lui A.P.; Pineault R.L.; Beer S. K.; Richardson S.W. High-pressure co-gasification of coal and biomass in a fluidized bed. Biomass and Bioenergy 2004, vol 26, pp 377-388.

[73] Amorino C.; Pettinau A.; Maurer R. E.; Hughes E.; Larceri F.; Repetto F.; Wellhausen P.; Lange P. Clean and economical gasification of combined coal and biomass pelletized fuels by industries worldwide. Proceedings of the 3rd International Conference on Clean Coal Technologies, Cagliari, Italy, May 15-17, 2007.

[74] He M.; Hu Z.; Xiao B.; Li J.; Guo X.; Luo S.; Yang F.; Feng Y.; Yang G.; Liu S. Hydrogen-rich gas from catalytic steam gasification of municipal solid waste (MSW): Influence of catalyst and temperature on yield and product composition. International Journal of Hydrogen Energy 2009, vol 34, pp 195-203.

[75] Straka P.; Bučko Z. Co-gasification of a lignite/waste-tyre mixture in a moving bed. Fuel Processing Technology 2009, vol 90, pp 1202-1206.

[76] Kastanaki E.; Vamvuka D. A comparative reactivity and kinetic study on the combustion of coal–biomass char blends. Fuel 2006, vol 85, pp 1186-1193.

[77] Cetin E.; Moghtaderi B.; Gupta R.; Wall T. F. Influence of pyrolysis conditions on the structure and gasification reactivity of biomass chars. Fuel 2004, vol 83, pp 2139-2150.

[78] Collot A. G.; Zhuo Y.; Dugwell D. R.; Kandiyoti R. Co-pyrolysis and co-gasification of coal and biomass in bench-scale fixed-bed and fluidised bed reactors. Fuel 1999, vol 78, pp 667-679.

[79] Hansen J.; Sato M.; Kharecha P.; Beerling D.; Berner R.; Masson-Delmotte V.; Pagani M.; Raymo M.; Royer D. L.; Zachos J. C. Target atmospheric CO_2: where should humanity aim?. Open Atmospheric Science Journal 2008, vol 2, pp 217–231.

[80] Laird D. A. The charcoal vision: a win-win-win scenario for simultaneously producing bioenergy, permanently sequestering carbon, while improving soil and water quality. Agronomy Journal 2008, vol 100, pp 178–181.

[81] Woolf D.; Amonette J. E.; Street-Perrott F. A.; Lehmann J.; Joseph S. Sustainable biochar to mitigate global climate change. Nature Communications 2010, vol 1, pp 1-9.

[82] Mullen C. A.; Boateng A. A.; Goldberg N. M.; Lima I. M.; Laird D. A.; Hicks K. B. Bio-oil and bio-char production from corn cobs and stover by fast pyrolysis. Biomass and Bioenergy 2010, vol 34, pp 67-74.

[83] Frau C.; Ferrara F.; Orsini A.; Pettinau A. Bench-scale testing of ZnO based sorbents for hot gas cleanup. Proceedings of 65th ATI National Confenrence, Domus de Maria, Italy, September 13-17, 2010.

[84] Hassan K. H. Regeneration and Activity Test of Spent Zinc Oxide Hydrogen Sulfide Removal Catalyst. European Journal of Scientific Research 2010, vol 39, pp 289-295.

[85] García E.; Cilleruelo C.; Ibarra J. V.; Pineda M.; Palacios J. M. Kinetic study of high-temperature removal of H_2S by novel metal oxide sorbents. Industrial & Engineering Chemistry Research 1997, vol 36, pp 846-853.

[86] Konttinen J. T.; Zevenhovem C. A. P.; Hupa M. M. Hot gas desulfurization with zinc titanate sorbents in a fluidized bed. 1. Determination of sorbent particle convertion rate model parameters. Industrial & Engineering Chemistry Research 1997, vol 36, pp 2332-2339.

[87] García E.; Cilleruelo C.; Ibarra J. V.; Pineda M.; Palacios J. M. Thermogravimetric study of regenerable sulphur sorbents for H_2S retention at high temperature. Thermochimica Acta 1997, vol 306, pp 23-30.

[88] Chen W.; Lin M.; Jiang T. L.; Chen M. Modeling and simulation of hydrogen generation from high-temperature and low-temperature water gas shift reactions. International Journal of Hydrogen Energy 2008, vol 33, pp 6644-6656.

[89] Figueiredo R. T.; Dantas Ramos A. L.; Carvalho de Andrade H. M.; G. Fierro J. L. Effect of low steam/carbon ratio on water gas shift reaction. Catalysis Today 2005, vol 107-108, pp 671-675.

[90] Hla S. S.; Park D.; Duffy G. J.; Edwards J. H.; Roberts D. G.; Ilyushechkin A.; Morpeth L. D.; Nguyen T. Kinetics of high-temperature water-gas shift reaction over two iron-based commercial catalysts using simulated coal-derived syngases. Chemical Engineering Journal 2009, vol 146, pp 148-154.

[91] Adams II T. A.; Barton P. I. A dynamic two-dimensional heterogeneous model for water gas shift reactors. International Journal of Hydrogen Energy 2009, vol 34, pp 8877-8891.

[92] Maceiras R.; Álvarez E.; Cancela M. Á. Effect of temperature on carbon dioxide absorption in monoethanolamine solutions. Chemical Engineering Journal 2008, vol 138, pp 295-300.

[93] Kuntz K.; Aroonwilas A. Mass-transfer efficiency of a spray column for CO_2 capture by MEA. Energy Procedia 2009, vol 1, pp 205-209.

[94] Issacs E. E.; Otto F. D.; Mather A. E. Solubility of mixtures of H_2S and CO_2 in a monoethanolamine solution at low partial pressures. Journal of Chemical and Engineering Data 1980, vol 25, pp 118–120.

[95] Lozza G.; Chiesa P. Cattura e sequestro della CO_2 nelle centrali a combustibili fossili. Energia 2007, vol 3, pp 84-102.

[96] Strigle R. F. Random packings and packed towers: design and applications. Gulf Publishing Company, Houston, TX, 1987.

[97] Samanta A.; Bandyopadhyay S. S. Absorption of carbon dioxide into aqueous solutions of piperazine activated 2-amino-2-methyl-1-propanol. Chemical Engineering Science 2009, vol 64, pp 1185-1194.

[98] Dang H.; Rochelle G. T. CO_2 absorption rate and solubility in mono-ethanolamine/piperazine/water. Proceedings of the 1st National Conference on Carbon Sequestration, Washington, USA, May 14-17, 2001.

[99] Sakwattanapong R.; Aroonwilas A.; Veawab A. Reaction rate of CO_2 in aqueous MEA-AMP solution: experiment and modeling. Energy Procedia 2009, vol 1, pp 217-224.

[100] Chakravarty T.; Phukan U. K.; Weiland R. H. Reaction of acid gases with mixtures of amines. Chemical Engineering Progress 1985, vol 81, pp 32-36.

[101] Liao C. H.; Li M. H. Kinetics of absorption of carbon dioxide into aqueous solutions of monoethanolamine + N-methyldiethanolamine. Chemical Engineering Science 2002, vol 57, pp 4569-4582.

[102] Bolea I.; Sampasertparnich T.; Idem R.; Tontiwachwuthikul P.; deMontigny D.; Usubharatana P. Reduction of the energy penalty of a coal-fired power plant by using a novel solvent together with an improved CO_2 capture process for post-combustion capture and compression. Proceedings of 12th Meeting of the International Post-Combustion CO_2 Capture Network, Canada, September 29-October 1, 2009.

[103] Lee J. H.; Kim J. H.; Jang K. R.; Lee I. Y.; Shim J. G. Development of amine absorbents for post-combustion capture. Proceedings of 12th Meeting of the International Post-Combustion CO_2 Capture Network, Canada, September 29-October 1, 2009.

[104] Tobiesen F. A.; Svendsen H. F. Study of a modified amine-based regeneration unit. Industrial & Engineering Chemistry Research 2006, vol 45, pp 2489–2496.

[105] Bishnoi S.; Rochelle G. T. Absorption of carbon dioxide in aqueous piperazine/methyldiethanolamine. A.I.Ch.E. Journal 2002, vol 48, pp 2788-2799.

[106] Closmann F.; Nguyen T.; Rochelle G. T. MDEA/Piperazine as a solvent for CO_2 capture. Energy Procedia 2009, vol 1, pp 1351–1357.

[107] Idem R.; Edali M.; Aboudheir A. Kinetics, modeling, and simulation of the experimental kinetics data of carbon dioxide absorption into mixed aqueous solutions of MDEA and PZ using laminar jet apparatus with a numerically solved absorption-rate/kinetic model. Energy Procedia 2009, vol 1, pp 1343–1350.

[108] Bishnoi S.; Rochelle G. T. Thermodynamics of piperazine/methyldiethanolamine/ water/carbon dioxide. Industrial and Engineering Chemistry Research 2002, vol 41, pp 604–612.

[109] Carpenter A. M. Polygeneration from coal. IEA Clean Coal Centre, Report No. CCC/139, London, United Kingdom, 2008.

[110] Consonni S.; Katofsky R. E.; Larson E. D. A gasification-based biorefinery for the pulp and paper industry. Chemical Engineering Research and Design 2009, vol 87, pp 1293-1317.

[111] Larson E. D.; Tingjin R. Synthetic fuel production by indirect coal liquefaction. Energy for Sustainable Development 2003, vol 7, pp 79-102.

[112] Cocco D., Pettinau A., Cau G. Energy and economic assessment of IGCC power plants integrated with DME synthesis processes. Proceedings of the Institution of Mechanical Engineers, Part A. Journal of Power and Energy 2006, vol 220, pp 95-102.

[113] Higman C.; van der Burgt M. Gasification. Gulf Professional Publishing/Elsevier, Burlington, Massachusetts (USA), 2003.

[114] Wang L.; Yang L.; Zhang Y.; Ding W.; Chen S.; Fang W.; Yang Y. Promoting effect of an aluminum emulsion on catalytic performance of Cu-based catalysts for methanol synthesis from syngas. Fuel Processing Technology 2010, vol 91, pp 723-728.

[115] Zhang X.; Zhong L.; Guo Q.; Fan H.; Zheng H.; Xie K. Influence of the calcination on the activity and stability of the $Cu/ZnO/Al_2O_3$ catalyst in liquid phase methanol synthesis. Fuel 2010, vol 89, pp 1348-1352.

[116] Guo X.; Li L.; Liu S.; Bao G.; Hou W. Preparation of $CuO/ZnO/Al_2O_3$ catalysts for methanol synthesis using parallel-slurry-mixing method. Journal of Fuel Chemistry and Technology 2007, vol 35, pp 329-333.

[117] Meshkini F.; Taghizadeh M.; Bahmani M. Investigating the effect of metal oxide additives on the properties of $Cu/ZnO/Al_2O_3$ catalysts in methanol synthesis from syngas using factorial experimental design. Fuel 2010, vol 89, pp 170-175.

[118] Heydorn E. C.; Stein V. E.; Tijm P. J. A.; Street B. T.; Kornosky R. M. Liquid phase methanol (LPMEOH^{TM}) project operational experience. Proceedings of the Gasification Technology Conference, San Francisco, California (USA), October 4-7, 1998.

[119] Gunda A.; Tartamella T.; Gogate M.; Lee S. Dimethyl ether synthesis from CO_2-rich syngas in the LPDME process. Proceedings of the 12th International Pittsburgh Coal Conference, Pittsburgh, Pennsylvania (USA), September 11-15, 1995.

[120] Ramos F. S.; Duarte de Farias A. M. ; Borges L. E. P.; Monteiro J. L.; Fraga M. A.; Sousa-Aguiar E. F.; Appel L. G. Role of dehydration catalyst acid properties on one-step DME synthesis over physical mixtures. Catalysis Today 2005, vol 101, pp 39-44.

[121] Sai Prasad P. S.; Wook Bae J.; Kang S.; Lee Y.; Jun K. Single-step synthesis of DME from syngas on $Cu–ZnO–Al_2O_3$/zeolite bifunctional catalysts: the superiority of ferrierite over the other zeolites. Fuel Processing Technology 2008, vol 89, pp 1281-1286.

[122] Wook Bae J.; Kang S.; Lee Y.; Jun K. Synthesis of DME from syngas on the bifunctional $Cu–ZnO–Al_2O_3$/Zr-modified ferrierite: effect of Zr content. Applied Catalysis B: Environmental 2009, vol 90, pp 426-435.

[123] Gao Z.; Huang W.; Yin L.; Xie K. Liquid-phase preparation of catalysts used in slurry reactors to synthesize dimethyl ether from syngas: effect of heat-treatment atmosphere. Fuel Processing Technology 2009, vol. 90, pp 1442-1446.

In: Gasification: Chemistry, Processes and Applications
Editor: Michael D. Baker
ISBN: 978-1-61209-681-0

Chapter 3

WATER GAS SHIFT AND MEMBRANE REACTOR STUDIES FOR THE PRODUCTION OF A HYDROGEN-RICH STREAM IN STEAM-OXYGEN GASIFICATION PROCESSES

Marta Maroño* and José María Sánchez
CIEMAT, Combustion and Gasification Division, Av. Complutense, Madrid, Spain

ABSTRACT

Gasification of different types of fuels with steam and oxygen produces a synthesis gas (syngas) which consists mainly of CO, H_2, CO_2 and H_2O. Depending on the fuel and the operating conditions, the contents of CO and H_2 in the syngas usually range from 35 % to 55 % and from 20 % to 45 % respectively. If higher hydrogen contents are desired, the Water Gas Shift (WGS) reaction allows the conversion of CO to H_2 and CO_2 in presence of excess steam and a specific catalyst. This reaction is industrially carried out in two consecutive stages, first a high-temperature step for rapid CO conversion and then a low-temperature step for achievement of lower equilibrium CO contents.

A novel approach which is being investigated to increase the production of hydrogen in gasification processes is the use of a Water Gas Shift (WGS) membrane reactor in which a single high-temperature WGS catalyst is combined with a hydrogen selective membrane. In this system, the removal of hydrogen from the reaction chamber through the membrane drives the WGS reaction to the products side and thermodynamic equilibrium can be exceeded. Catalyst and membrane must be carefully chosen and special attention must be paid to operating conditions in order to guarantee the correct performance of the system. In this work the performance of semi-commercial palladium membranes and two different high temperature WGS catalysts are described and the advantages of using a WGS membrane reactor for the production of H_2 in gasification processes versus a conventional WGS reactor are presented.

---

* marta.marono@ciemat.es

INTRODUCTION

Current energy utilization processes are still lacking from sustainability, not only regarding the use of non-renewable sources but also in terms of efficiency and environmentally benign. Hydrogen and synthesis gas (syngas) production and purification have become major areas of research for sustainable energy development worldwide. These technologies can supply clean fuels, reduce pollutant emissions including greenhouse gases and develop more efficient electrical power generation systems such as integrated gasification combine cycle (IGCC).

There are many processes, commercial or under development, for syngas and hydrogen production using gaseous, liquid or solid feedstock. Current commercial processes largely depend on fossil fuels both as source of hydrogen and as source of energy for the production process. Lately, production of hydrogen or a hydrogen-rich gas from renewable sources such as biomass has attracted a lot of attention [1]. Since biomass is renewable and has a small net CO_2 impact compared to fossil fuels, it has the potential to accelerate the realization of hydrogen as a major fuel of the future [2].

There are two main routes for biomass-based hydrogen production, namely thermo-chemical and bio-chemical, but they have major challenges to overcome [3]. Biomass has low hydrogen content, around 6 wt%, the energy content is low as well and up to now there are no completed technology demonstrations. All these facts would make it, in principle, not very interesting for biomass to produce hydrogen. Over the past several decades, however, many investigations on various hydrogen production methods have been conducted and biomass gasification is nowadays becoming a mature technology, and starts to be considered as economically viable and competitive with the conventional natural gas reforming method [4-10].

Within these research efforts outstands the CHRISGAS project [11] which prime objective was to demonstrate at the 18 MWth-scale Värnamo plant, the manufacturing of a hydrogen-rich gas from a renewable feedstock, primarily woody biomass. In the rebuilt Váxjo Värnamo Biomass Gasification Centre (VVBGC) in Sweden, the demonstration of the production of a clean hydrogen-rich synthesis gas is based on pressured steam/oxygen-blown gasification of biomass, followed by cleaning and upgrading, particularly by the use of steam reforming and water gas shift steps.

In parallel with the preparation work for demonstration testing, a comprehensive program of research and complementary product development was launched. Thus, one of the objectives of the CHRISGAS project has been to study innovative gas separation and gas upgrading systems that have not been developed sufficiently yet to be tested at a demonstration scale within the time frame of the project, but which show some attractive merits and features for further development.

In this framework CIEMAT has studied, on a bench scale, hydrogen enrichment and separation from syngas by the use of WGS reaction and membrane technology and has looked into the assessment of membrane catalytic reactors (MCR) as a valuable tool for conversion enhancement of the equilibrium limited WGS reaction by selective permeation of a reaction product, i.e. hydrogen, through the membrane. Experimental results and advantages of the proposed combined system versus conventional WGS processes are presented in this chapter.

CURRENT STATUS OF TECHNOLOGY

Within CHISGAS project scope [11], CIEMAT focused their R&D efforts on one promising novel method for the production of hydrogen from renewable sources: the combination of WGS reaction with selective separation of H_2 via inorganic membranes. This technology, based on the development of catalytic membrane reactors, arises as a sound method to generate a clean hydrogen rich stream following gasification that can be used as fuel in other processes.

In such a combined process catalyst and membrane must be carefully chosen and special attention must be paid to operating conditions in order to guarantee the correct performance of the system. A comprehensive literature review on the state of the art was performed prior to conducting experimental activities, especially regarding the following subjects:

- Catalysts for WGS reaction: Commercial and new catalysts still under study or currently being used to promote the conversion of CO under different operating conditions were reviewed.
- Membranes development for H_2 selective separation. The main types of membranes that are currently under development were reviewed, with special emphasis on the advantages and disadvantages of each type.
- Research activities on catalytic membrane reactors that combine WGS reaction with selective separation of hydrogen. Examples of proposed systems and configurations described on literature were analysed in order to identify those most promising according to the objectives of the project.

Catalysts for WGS reaction

The water gas shift (WGS) reaction is a reaction used in many industrial processes to produce a hydrogen-rich stream from CO and H_2O.

$$CO + H_2O \leftrightarrow CO_2 + H_2 \quad (\Delta H^o = -41 \text{ kJ/mol})$$

It is mildly exothermic and without a catalyst it requires operating temperatures higher than 650 °C. At temperatures above 300°C the conversion is equilibrium limited, and at lower temperatures the reaction is kinetically limited. The use of specific catalysts allows overcoming these limitations.

Some traditional examples of processes involving this reaction include coal gasification processes, H_2 production for ammonia synthesis or other industrial processes such as hydro-treating of petroleum stocks. New applications of WGS reactors have recently renewed the interest in this process. Biomass-based integrated gasification combined-cycle (IGCC) is a promising technology for power and hydrogen production. To produce a hydrogen rich gas from biomass gasification requires however the development of highly active, stable and sometimes sulphur-tolerant catalysts for the high temperature WGS reaction. The application of WGS reaction to H_2 production for mobile and small-to-medium-scale fuel cells require catalysts with high activity at a wider temperature range, no need for activation prior to use, no degradation on exposure to air or temperature cycles and availability of conventional wash-coating technologies, which may result in catalysts with reduced size and weight.

The present practice in industry is to carry out the water gas shift reaction in two steps with two types of catalysts: first a high temperature WGS reaction (300-400 °C) that is generally performed over Fe-Cr catalysts and then, a Cu-Zn catalyst, which is used in the temperature range of 200-300°C. In this two-stage approach, excess steam is usually required to achieve high CO conversion and to avoid secondary reactions.

High temperature WGS catalysts

Iron oxide is the most often used catalyst for high temperature WGS reaction. The catalytically active phase is magnetite (Fe_3O_4) that usually comes from the partial oxidation of haematite (Fe_2O_3). However, pure magnetite catalysts suffer from sintering that reduces rapidly its activity. A stabiliser, Cr_2O_3, is commonly added and the combination of Fe_3O_4 andCr_2O_3 gives commercially stabilised catalysts that can operate for several years before requiring replacement. Regarding Fe-Cr catalysts the main disadvantage is the environmental and safety problems related to chromium compounds [12].

Platinum-based WGS catalysts have recently become very attractive. Thinon et al. have screened a number of water gas shift catalysts which consist of a metal deposited on a reducible support [13]. In their study $Pt/CeO_2/Al_2O_3$ and Pt/TiO_2 were found to be the most active catalysts for the high temperature water-gas shift. Pt catalysts supported on metal oxides carriers are reported to be bifunctional, i.e. both the metal and support have a significant influence on the overall catalytic performance [14].

Moreover, Haryanto et al. found that $Pt\text{-}CeO_2/Al_2O_3$ catalysts in their unreduced form were active at ultrahigh temperature, and could be integrated downstream a biomass gasifier for synthesis gas upgrading without the need of the gas stream to be cooled, i.e. in the 500-700°C range [15].

Zirconium oxide has extensively been studied in recent years as catalyst and as support, due to its chemical and physical properties. For WGS reaction, zirconia-supported platinum catalysts emerge as an attractive catalyst, with the advantage of being non-toxic. Querino et al. highlighted the beneficial effect of incorporating zirconia as a support to platinum-ceria catalysts [16]. In their work it is claimed that Pt-ceria catalyst supported on zirconia can be a promising candidate for the WGS reaction to be carried out in just one stage at an intermediate temperature (medium temperature shift reaction). They also propose to dope the zirconium support with other metals to improve the specific surface area and the metal interaction. Among the systems studied, they found that the platinum-ceria-zirconia system resulted to be the most active for the water gas shift reaction. More recently Fonseca at al. have also compared Au- and Pt-ceria-zirconia catalysts finding that, though both are active for WGS reaction, the Au-based catalyst provides slightly better results [17]. On the other hand, other authors who have investigated $Pt\text{-}CeO_2\text{-}ZrO_2$catalysts claim that in spite of having high initial activity they are also susceptible to a high initial rate of deactivation [18].

During gasification processes, part of the sulphur contained in the fuel is released in the product gas, mainly as H_2S. Depending on the sulphur content of the biomass, H_2S concentration in the gas can vary between 50 to several hundreds ppmv. In very extreme cases (such as coal gasification) values up to 10000 ppmv are reported. Since the current industrial catalysts are extremely sensitive to sulphur the conventional approach to solve the problem with sulphur poisoning has been to remove it before the water gas shift reaction.

However, it is inevitable that trace amounts of H_2S remain in the gas stream and cause catalyst deactivation.Some of the already explored sulphur-resistant catalysts include those based on transition metal sulphide, the use of Pt/ZrO_2 materials, or the use of carbides as promising candidates such as molybdenum carbide. All these efforts show that the development of sulphur-tolerant catalysts for the water gas shift continues to be a challenge.

Low temperature WGS catalysts

For processes carried out at temperatures below 300 °C commercial WGS catalysts are typically based on a $Cu/ZnO/Al_2O_3$ or $Cu/ZnO/Cr_2O_3$ structure, where zinc oxide and/or chromium oxide are generally used as structural stabilizers and promoters. Although this low temperature stage usually follows a high temperature WGS process, the use of a Fe-Cr and Cu-ZnO combination in a two-stages process is not always suitable, especially in those applications where fast start-ups require a low catalyst volume and the use of non-pyrophoric catalysts, as for example in fuel cells [14]. Besides its pyrophoric nature, some other reported drawbacks of the Cu-ZnO catalyst include its tendency to sintering and its high susceptibility to poisoning [19].

R&D efforts deal with the development of advanced low temperature catalysts for CO conversion, for instance transition metal supported-ceria (CeO_2), including platinum, rhodium, palladium or gold [20, 21]

Membranes for H_2 separation

Membranes are physical barriers that allow the separation of a stream into two fractions on the basis of molecular or particulate size. This technology is practised on feed streams ranging from gases to colloids and started with liquid phase separation processes. The development of gas separation membranes started later, in the mid 1940s, in the field of nuclear fuels and since then huge advances have taken place, making the application of manufactured membranes a viable option.

When describing a membrane system three general terms are used: Feed side, Permeate side and the membrane itself (Fig. 1). Usually, the desired component passes through the membrane giving the permeate stream and the rest of the feed flow constitutes the retentate stream.

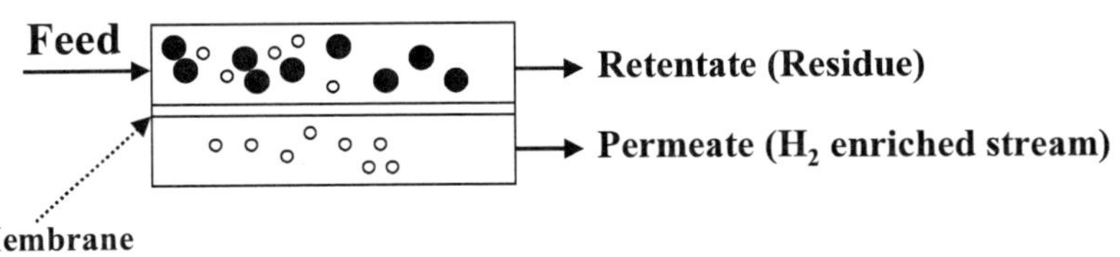

Fig. 1. Schematic representation of a membrane unit.

Gas separation by means of membranes is a concentration-driven process, which for gases is directly related to the partial pressure of gas species. In this sense, it is the differential pressure between the feed and permeate streams that influences the effective performance of a membrane. Two parameters are usually employed when characterizing the performance of a membrane: permeability and selectivity. Permeability is the rate at which any compound permeates through the membrane and selectivity is the ability of a membrane to accomplish a given separation (relative permeability of the membrane for the feed species). Both properties influence the economics and viability of a membrane for gas separation processes.

According to literature [22] membranes could be classified following several criteria: membrane material (organic, inorganic), membrane structure (porous or dense), and morphology (symmetrical or asymmetrical). Permeation mechanisms are also under discussion.Taking into account the application of the membranes to be studied in this work-separation of hydrogen at intermediate temperature- two types of membranes could be applicable:

Micro-porous ceramic membranes

Porous ceramic, particularly micro-porous materials possess high permeability and moderate to high selectivity and are chemically and thermally stable. For those reasons, they are attractive for application in syngas for hydrogen production reactions. Literature survey showed that there are several types of porous membranes which have been tested for H_2 separation or production. The most frequent components of porous ceramic membranes found in literature are alumina and silica, although titania and zirconia are also valuable options. A novel family of materials, zeolites, is currently presented as very promising in the separation field although little progress is seen in the development of small pore zeolites membranes for hydrogen separation [23] .Literature also shows limited examples of other membranes under development which are made of platinum, rhodium and iridium dispersed on porous alumina [24].

Dense phase metallic and metallic alloy membranes

They have attracted a lot of attention, because they are claimed to be commercially available. Nevertheless, for hydrogen, there have been a limited number of metallic membranes which have proved to be effective so far. These are primarily palladium-based alloys exhibiting unique permselectivity to hydrogen and good mechanical stability. R&D has evolved from thick self-supporting membranes (50-100 μm) which showed low permeance and very high cost to current membranes which consist of a thin layer (5-20μm) of palladium or palladium alloy deposited onto a porous metallic substrate. A number of alloying elements, and their content, are being studied -such as Ag, Cu, Ni, Fe, or Pt- in order to increase permeance and chemical and thermal stability.

WGS membrane reactor

According to the IUPAC definition, a membrane reactor is a device that combines the separation properties of membranes with the typical characteristics of catalytic reaction steps in only one unit. In particular, the membrane does not only play a role as a separator but also as a part of the reactor itself. In other words, a membrane reactor is an engineering device that selectively removes a product from the reaction system, giving the possibility of achieving higher conversion than a traditional process under the same operation conditions. For that reason, one of the main application opportunities of inorganic membrane reactors refers to the conversion enhancement of an equilibrium limited reaction by the selective permeation of a reaction product through the membrane. In our particular case, the equilibrium limited reaction to study is the water gas shift reaction and the reaction product to remove is hydrogen.

Publications found in the open literature show that both types of WGS commercial catalysts -low temperature (150-250 °C) or high temperature (T> 300 °C) - can be found in reported membrane reactor systems. Modifications of any of these types by doping them with different metals or new catalyst developments are attracting the attention of researchers.

Regarding membranes, basically two types of membranes are most frequently used: ceramic, Pd and/or Pd alloys supported on alumina or stainless steel. As stated before, palladium membranes are excellent candidates for hydrogen purification, especially when incorporated with a catalytic reactor combining reaction and separation in a single unit. Their theoretically infinite selectivity to hydrogen continues to draw major research efforts. However, according to literature [25], the use of Pd membranes in WGS reactors imposes several limitations on the catalyst. Among them, the most restrictive one is that the catalyst must be operated above 250°C which is typically the temperature for HT WGS catalysts to start showing activity, and whenever possible above 300°C to minimize the risk of palladium embrittlement and phase transformation which would lead to cracks in the metal film.

Another issue frequently discussed in membrane reactor technology is membrane geometry and reactor configuration. Three different geometries are reported, coils, tubes and disks. Generally, tubular membrane assemblies are preferred to flat-plate or disk configurations because they provide a much larger surface area. Concerning reactor configuration, there are two possible configurations for tubular membrane set-ups, depending which side of the tube the catalyst is placed on: an inside-outside design or an outside-inside configuration, as Fig. 2 shows. While the former requires larger inside diameters for a given wall thickness allowing self-supporting membranes, the second requires an adequate distribution of tubes in the reactor to guarantee a homogeneous flux of hydrogen.

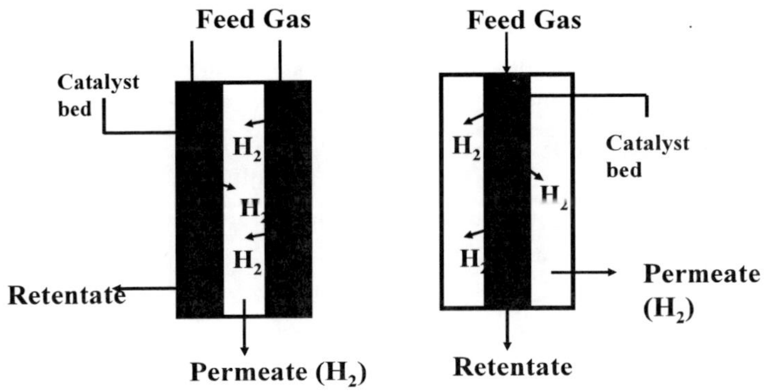

Fig. 2. Different configurations catalyst-membrane in a membrane reactor.

According to the results presented in literature, the combination of the WGS reaction with a H_2 selective membrane provides better results in terms of CO conversion and/or H_2 recovery than conventional processes. However, more research is needed to scale up and integrate these systems into industrial processes such as an IGCC gasification process or to both stationary or transportation fuel cell applications.

Water Gas Shift Studies

CIEMAT has studied on a laboratory scale the activity of commercial WGS catalysts (Fe-Cr- and Pt-based respectively). Experiments were conducted in a fixed bed reactor using simulated gases, under the operating conditions and gas composition expected at Växjo Värnamo Biomass Gasification Centre (VVBGC) for the WGS step, i.e., under conditions typical of oxygen pressurized gasification. Catalytic activity under those conditions was determined and suitability of the catalysts to be incorporated in a palladium membrane reactor was discussed. The effect of operating parameters on CO conversion and hydrogen production was evaluated. Special attention was paid to the activity of the catalyst versus temperature due to the sensitivity of Pd membranes to temperature, below 300°C. The effect of steam to CO ratio, feed gas composition and space velocity on the activity and selectivity of the catalysts for the WGS reaction were also studied in order to optimize CO conversion and H_2 production. In this chapter a summary of most relevant results are presented. Detailed results of those studies can be found elsewhere[26], [27].

Experimental

Catalysts

The Fe-Cr-based catalyst studied is a commercial high temperature WGS catalyst, industrially used in standard H_2 and ammonia production plants. It basically consists of a mixture of iron and chromium oxides. As confirmed during testing, the catalyst used in this work was supplied already in the active phase, magnetite (Fe_3O_4), so it did not require a pre-reduction step. For testing, tablets of 6x6 mm were used as shown in Fig. 3 (left). Depending on the operating conditions, 5 g or 1.3 g of catalyst were placed in the reactor.

The platinum-based catalyst was prepared and provided by CATATOR AB (Sweden). The final catalyst consists of a support of CeO_2/ZrO_2 with 4.2 % Pt as the active phase. The catalyst was supplied impregnated on 240mm x 295mm rectangular wire mesh substrates following a procedure patented by CATATOR. For testing, the catalyst was cut in circular pieces of 8 mm diameter and strung together to be fitted in the tubular reactor used in the experiments, as shown in Fig. 3 (centre). Final configuration after assembling and ready to be placed in the reactor is shown in Fig. 3 (right).

The volume of catalyst used for the experimental trials was 1.3 cm^3, and 6.21 cm^3, depending on the experiment to perform.

Fig. 3. Samples of catalyst used for WGS studies (left) Fe-Cr pellets; (center) Pt-based sheets; (right) Pt-based catalyst as mounted in the reactor.

Experimental Unit

Catalyst testing was carried out in a Microactivity Pro Unit, whose flow sheet diagram can be seen in Fig 4. It is an automatic and computerised laboratory rig for the study of catalytic reactions. The maximum operating gas flow rate is 4.5 Nl/min. The unit can work at up to 700°C and 30 bar.

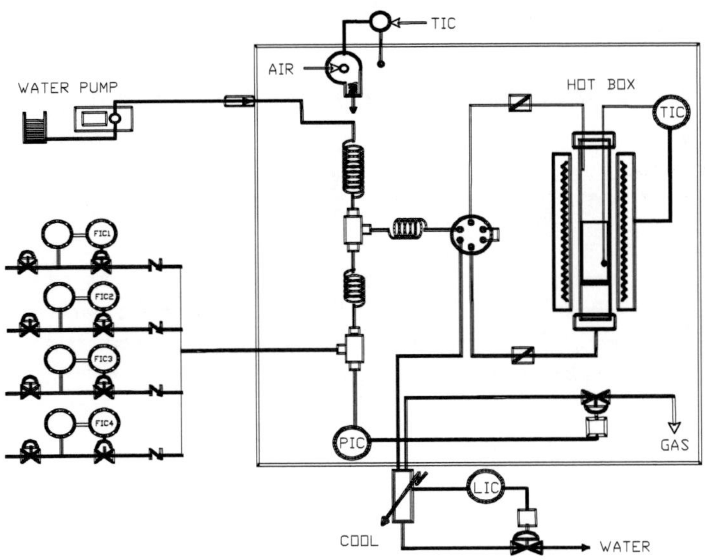

Fig. 4. Laboratory test rig for WGS studies flow sheet.

The gas mixture was produced synthetically using mass flow controllers (Hi-Tech). Deionized water was metered by a piston pump (Gilson 307) and vaporized before entering the reactor. Dry gas and water were preheated separately, in two independent loops. To this aim the entire set-up was housed in a forced air circulation oven maintained at 190°C. Dry gas and steam were then mixed before entering the reactor. A stainless steel tubular reactor manufactured by Autoclave Engineers was used for catalytic testing. It has an internal diameter of 9.2 mm, and it is 300mm long. The reactor is placed in a one single zone SS304 oven, which is able to heat the reactor up to 700°C. Gas temperature in the reactor is measured by a 1.5 mm thermocouple, directly in the catalyst bed. The reactor can be by-passed and isolated- for instance to analyse the inlet gas composition- by means of a six-way valve that connects reactor inlet and outlet. For the WGS tests presented in this work, the reactor was operated in fixed-bed, down-flow mode. In a typical run the reactor was pressurized to the operating pressure, and heated to the desired temperature under flow of nitrogen. When the desired temperature was attained, the nitrogen flow was stopped, and the simulated gasification gas was started.

Inlet and outlet gas composition were analysed using a 5890 Series II Hewlett-Packard gas chromatograph equipped with a thermal conductivity detector (TCD). Two packed 530 µm columns – a 6 Ft Porapak Q, 80/100 19001A-00 and one 6 Ft Molecular Sieve 5Å 60/80 19001A-MA2- connected in series were used to provide good gas components separation. A specific software (HP Chemstation A.6.03) was used for controlling the gas chromatograph and to analyse GC data.

Experimental tests

WGS tests were performed using three different gas compositions as shown in Table 1, a binary mixture of CO in N_2 (B) and two gas mixtures whose composition is close to the gas composition expected at the WGS unit inlet in VVBGC (M1 and M2). These latter two gas compositions were estimated based on the assumptions that the gasifier at VVBGC will operate at 900 °C (+/-50°C) and with 0.4 kg oxygen per kg biomass, and 0.2 kg steam/kg syngas (M1) or 0.6 kg steam/kg syngas (M2) will be used in the steam reformer.

Table 1. Gas composition for the experimental programme

COMPONENT	M1	M2	B
H_2	34	32	60
CO	38	23	
CO_2	13	16	
CH_4	0,5	0,1	
H_2O	15	29	
N_2	0	0	40

Results and Discussion

Catalytic activity of the WGS catalysts studied was expressed as the percentage of CO converted (% mol) X_{CO} calculated according to equation (1),

$$x_{CO}(\%) = \frac{(F_{CO})_{in} - (F_{CO})_{out}}{(F_{CO})_{in}} \times 100 \qquad (1)$$

where $(F_{CO})_{in}$, and $(F_{CO})_{out}$ are the molar flow rate of carbon monoxide at the reactor inlet and outlet, respectively. This way of expressing the catalytic activity of a catalyst is widely used.

Fe-Cr-based catalyst

High temperature WGS catalysts used industrially are mostly based on iron and chromium oxides which have been reported to be very selective for the water gas shift reaction at temperatures above 300°C maintaining stability and resistance to sintering.

Although WGS is a well studied process it cannot be assumed that the catalysts used in commercial processes will be equally suitable when used in other technologies such as gasification or membrane reactors. Very few references can be found in literature regarding the performance of high temperature WGS catalysts when used to upgrade syngas obtained from the oxygen pressurized gasification of biomass. Thus, *ad hoc* studies need to be carried out.

Hence CIEMAT conducted an experimental study under realistic conditions typical of oxygen pressurized gasification of biomass. The influence of temperature, excess steam, feed gas composition and space velocity on the activity and selectivity of the catalyst for the WGS reaction was investigated.

The effect of operating parameters on CO conversion and hydrogen production was evaluated. Special attention was paid to the activity of the catalyst versus temperature due to the limitation on operating temperature demanded by Pd membranes. The effect of steam to CO ratio, feed gas composition and space velocity on the activity and selectivity of the catalyst for the WGS reaction were studied in order to optimize H_2 production.

Catalytic activity was determined at temperatures from 250°C to 500°C and space velocities ranged from 2000 h^{-1} to 25000 h^{-1}. The majority of the experiments were carried out using the mixture M1 –see Table 1- which simulates the composition of an oxygen biomass gasification exit gas and particularly that expected at the WGS unit inlet on VVBGC. Some additional tests were done using a gas mixture with different concentration of CO in order to asses the influence of feed composition on the activity of the catalyst. All tests were run at 10 bar, which is the operating pressure at VVBGC.

Fig. 5 shows the influence of feed gas composition on CO conversion (left) and hydrogen production (right) respectively. As temperature increases also does conversion of CO and production of hydrogen. The temperature at which the catalyst started to show activity towards the WGS reaction was found to be around 300°C which would be compatible with the temperature required by palladium membranes if the catalyst is going to be incorporated in a membrane reactor.

Despite the high CO content in the feed gas (44%-60% dry basis) the catalyst has shown very good performance at intermediate temperatures 350°C-450°C, providing hydrogen content increases in the exit gas in the range of 10%-17% (dry basis) and CO concentration at the reactor outlet lower than 3% v/v.

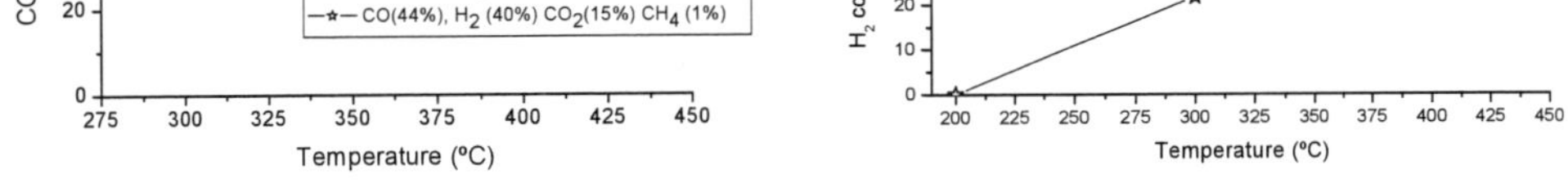

Fig. 5. Influence of feed gas composition on (left) Carbon monoxide conversion, and (right) H2 production, at space velocity of 2,885 h-1 and steam to CO ratio of 6.7.

Fig. 6 shows the influence of gas space velocity on the activity of the catalyst. Space velocity represents the contact time between gas and catalyst bed. Given the same amount of catalyst, higher space velocities result in a reduction of the contact time in the reactor. At 2885 h^{-1} and at a H_2O/CO ratio of 2, the initial content of CO in the feeding gas mixture M1 (44% v/v) was reduced to less than 10 % v/v at 380°C. When space velocity is increased, a shift of the conversion curves to higher temperatures results in lower CO conversions at the same temperature as depicted in Fig. 6 (left). In terms of CO conversion, a maximum value of

80 % v/v was reached at 2,885 h^{-1} at 380°C while at 10,000 h^{-1} a temperature higher than 450°C was required to reach the same CO conversion as Fig 6 (right) shows. Production of hydrogen followed the same trend with space velocity. Maximum H_2 content measured in the exit gas ranged between 50.1 % v/v – SV= 25000 h^{-1} and T=500°C-, and 57.6 % v/v dry basis –SV = 10000 h^{-1} and T=440°C-, which results in an increase of the H_2 content in the gas of 10-17 % v/v.

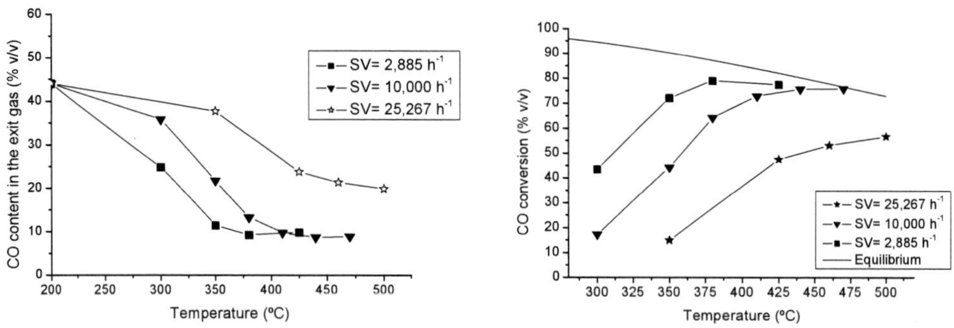

Fig. 6. Influence of space velocity (SV) on (left) evolution of CO content in the exit gas, and (right) CO conversion at a steam to CO ratio of 2, P= 10 bar, Feed gas composition 40% H2, 44% CO, 15% CO2, 1%CH4.

Concerning the effect of steam, experiments with steam to CO ratios ranging from 1 to 7 were carried out. According to the stoichiometry of the water gas shift reaction, at least a steam to CO ratio of 1 is required. Special interest was paid to results obtained at low H_2O/CO ratios because those conditions are the expected values to be used in a membrane reactor.

Regardless space velocity, increasing steam to CO ratio resulted in lower CO content in the exit gas, that is, higher CO conversion. CO content as low as 2.03 % v/v was achieved when using a steam to CO ratio of 6.7. The use of lower steam to CO ratio required higher operating temperatures to reach similar CO content in the exit gas.

Moreover, steam to CO ratio was found to play a key role to avoid secondary reactions, as shown in Fig. 7, where hydrogen and methane content evolution in the exit gas versus temperature are plotted. According to water gas shift reaction and in absence of secondary reactions, as CO is reacting with H_2O, hydrogen content in the exit gas should increase while methane content should decrease, due to the increase in overall mole number of reaction products (dry basis). Fig. 7 shows that this is what happens at steam to CO ratio equal or higher than 2. At steam to CO ratio of 1, however, methane content in the exit gas increases with temperature, whereas production of hydrogen decreases. This happens at temperatures above 350°C for the space velocity tested shown in Fig. 7, but it also happened for other values of gas space velocity not shown here. This gas composition evolution points to the possibility that at low steam to CO ratio secondary reaction are taking place, most likely the methanation reaction ($CO + 3H_2 \leftrightarrow CH_4 + H_2O$), which would be consuming CO and hydrogen to produce methane and water at temperatures below 500°C in deficiency of steam.

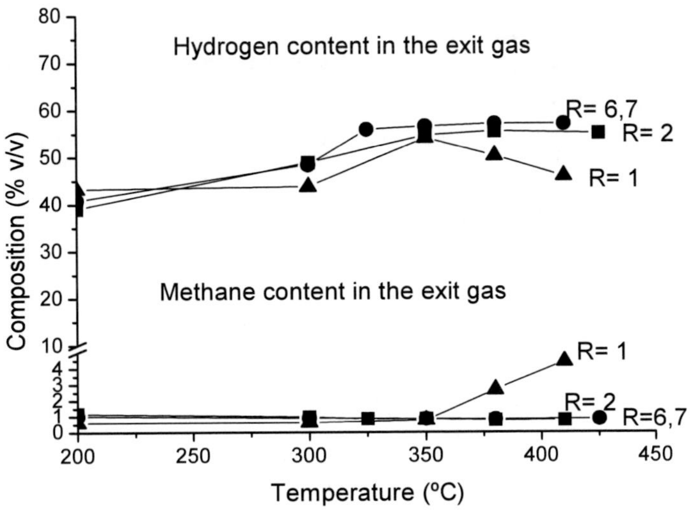

Fig. 7. Influence of steam to CO ratio (R) in the evolution of hydrogen and methane content (% v/v) in the exit gas at space velocity of 2,885 h-1, P= 10 bar and feed gas composition: 40 % H2, 44% CO, 15% CO2, 1% CH4.

Pt-based catalyst

The catalytic activity of a $Pt/CeO_2/ZrO_2$ catalyst has been investigated under realistic conditions for intermediate temperature WGS reaction. The study has included the evaluation of the effect of gas space velocity, steam to carbon monoxide ratio and feed gas composition on the performance of the catalyst. The final objective has been to determine experimental conditions which provide good catalytic performance in terms of CO conversion and selectivity to H_2 and CO_2, and to see if those conditions are compatible with the operating conditions imposed by Pd membranes in order to integrate the catalyst in a catalytic WGS membrane reactor. Detailed information on the performance of the catalyst is available in the open literature[27].

Catalytic performance has been evaluated from 200°C to 500°C. Gas composition was chosen to mimic the expected gas composition at the reformer outlet in the 18 MWth biomass-fuelled, steam-oxygen pressurised IGCC demonstration plant located in Vaxjo Värnamo Biomass Gasification Centre (VVBGC). Pressure was kept at 10 bar, which is the process pressure in the demonstration plant.

In literature it is found that Pt-based catalysts have been mainly used for the second WGS stage at low temperature. In our work, despite the high CO content in the feed gas, ranging from 32 to 60%v/v dry basis, the catalyst has shown very good performance at intermediate temperature, 300-450°C, showing potential for the first WGS stage. CO concentration at the reactor outlet reached values below 3%, what is comparable with conventional high-temperature, first-stage WGS catalysts. Equilibrium conversion was almost reached in all the tests carried out.

Regarding the influence of space velocity, higher SV values lead to lower CO conversion and a higher temperature is required to achieve a given CO conversion as shown in Fig. 8. On the other hand, lower space velocities result in higher CO conversion at lower temperatures, though the yield to hydrogen and carbon dioxide is slightly worse.

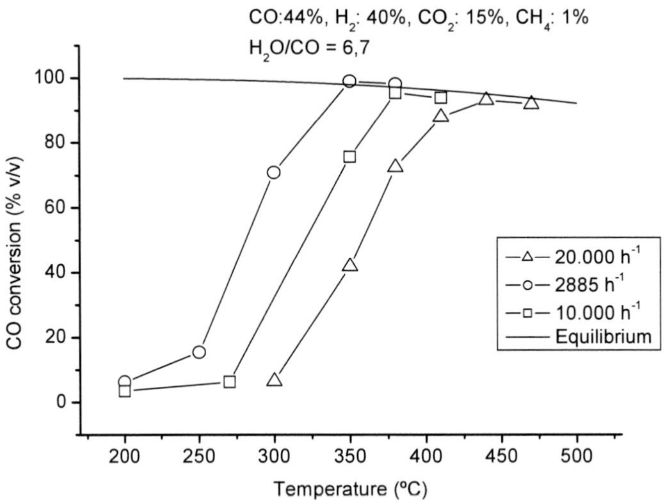

Fig. 8. Effect of space velocity (SV) on CO conversion (Feed Gas: 40% H2, 15% CO2, 44% CO, 1% CH4, P=10 bar, H2O/CO=6.7).

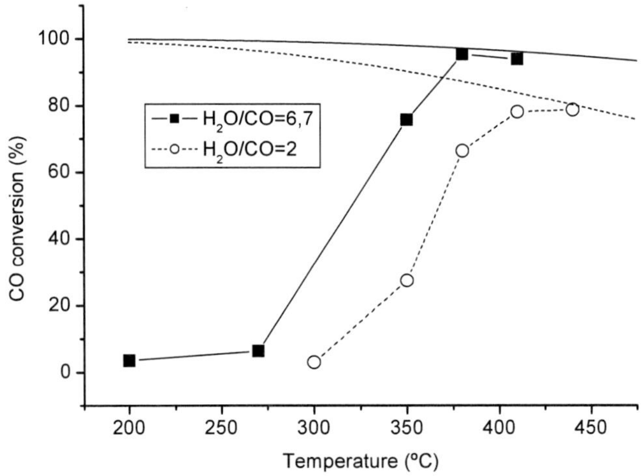

Fig. 9. Effect of H2O/CO ratio on CO conversion (Feed Gas dry basis: 40% H2, 15% CO2, 44% CO, 1% CH4, P=10 bar, SV=10.000 h-1).

With regard to steam to carbon monoxide ratio, this parameter can be considered critical on the performance of the catalyst. The activity of the $Pt/CeO_2/ZrO_2$ catalyst studied in this work depends strongly on the amount of steam available during the reaction, and it has been found that the excess of steam drives the reaction to the production of hydrogen. In order for the reaction to proceed significantly an excess of steam is needed. A higher H_2O/CO ratio gives higher carbon monoxide conversion, as can be seen looking at Fig. 9, and with better selectivity. Besides this enhanced performance, the rise in the H_2O/CO ratio shifts the conversion curves to lower temperatures, what in turn makes possible to achieve, at a given temperature, higher CO conversion to H_2 and CO_2

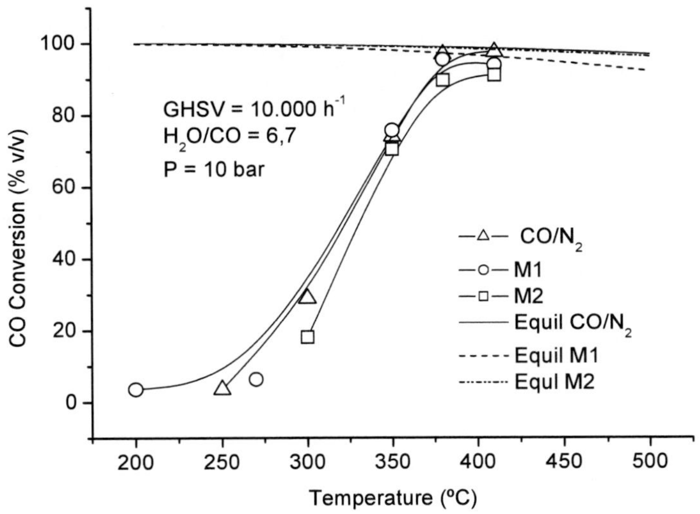

Fig. 10. Effect of gas composition on CO conversion.

As for the gas composition, no significant differences have been found in the catalyst performance when using different feed gas compositions as shown in Fig. 10. A higher content of CO in the feed leads to a slightly higher CO conversion The WGS activity of the catalyst is not significantly affected by the presence of reaction products -H_2 and CO_2- in the feed, although their presence shifts the CO conversion curves to slightly higher temperatures.

Comparison of catalysts

Both catalysts studied show high CO conversion (>90%) at space velocities around 3000 h^{-1} and steam to carbon monoxide ratios of 6.7. Under such conditions, the use of the Fe-Cr based catalyst is recommended when working temperature is in the 380°C-400°C range. If working temperature is lower, close to 350 °C, the Pt-based catalyst provides better results (see Fig. 11).

In summary, results show that different types of commercial and prepared high temperature WGS catalysts (Fe-Cr- and Pt-based) are able to convert more than 90% of the CO contained in the synthesis gas into CO_2 and H_2 under specific operating conditions (temperature, space velocity and steam to CO ratio), and therefore can be regarded as suitable to be used at VVBGC. The ferrochrome commercial catalyst studied provided good conversion of CO (less than 2-3 % v/v residual) in the temperature range of 360°C-400°C and at H_2O/CO ratios higher than 6. High conversion of CO was also obtained when using the Pt-based catalyst, in the temperature range of 350°C-380 °C and high H_2O/CO ratios. Those experiments showed very good selectivity to H_2 and CO_2.

No significant influence of gas feed composition has been found for any of the catalysts so both can be considered adequate for Växjo Värnamo Biomass Gasification Centre.

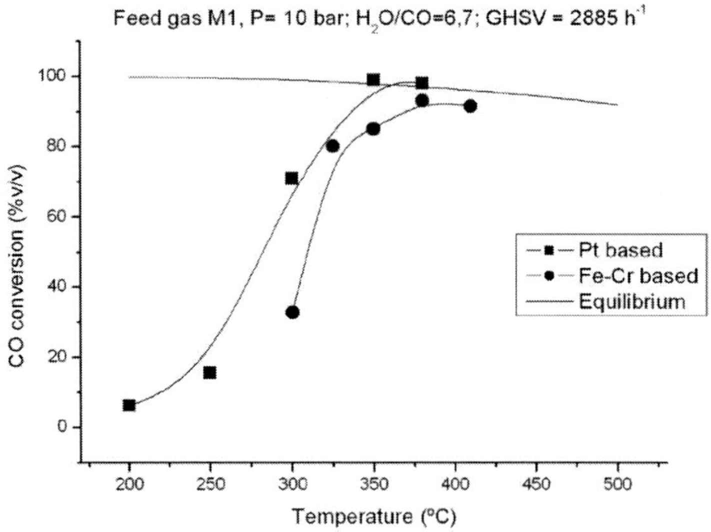

Fig. 11. Comparison of catalysts performance.

Looking at the effect of the parameters evaluated, it is seen that high values of space velocity lead to lower CO conversion and a higher temperature is required to achieve a given CO conversion. On the other hand, lower space velocities result in higher CO conversion at lower temperatures, though the yield to hydrogen and carbon dioxide is slightly worse.

Regarding the effect of steam to CO ratio, the activity of both catalysts depends strongly on the amount of steam available during the reaction. In the case of the Fe-Cr-based catalyst, in sub- or stoichiometric concentration of steam related to CO content in the gas, noticeable amounts of methane are produced. In order for the reaction to proceed significantly an excess of steam is needed. A higher H_2O/CO ratio gives higher carbon monoxide conversion, and with better selectivity to the production of hydrogen. Besides this enhanced performance, the rise in the H_2O/CO ratio shifts the conversion curves to lower temperatures, what in turn makes possible to achieve, at a given temperature, higher CO conversion to H_2 and CO_2.

Finally from the point of view of integrating the catalysts on a membrane reactor, based on the results obtained in this work and on the discussion above, from the operating conditions viewpoint it seems feasible to integrate either of them because the temperature range in which they show good catalytic performance, 300-480°C, is clearly compatible with the operating temperature required by Pd-membranes.

Studies on Separation of Hydrogen by Membranes

Taking into consideration that the final objective of our work was the production of a hydrogen-rich synthesis gas, the next step of our work dealt with the study of novel hydrogen separation systems. This approach would be able to provide a pure hydrogen stream from the gas mixture that leaves the WGS reactor. Among the available technological alternatives, a membrane reactor was chosen as very promising to produce firstly a hydrogen-rich gas and secondly to prove the advantages of combining the WGS reaction with a membrane selective to H_2 in one process unit.

In order to assess the suitability of such a system, the separation stage itself was studied first. In order to carry out hydrogen separation studies, a bench scale membrane reactor was designed and erected at CIEMAT.

The performance of palladium membranes from two sources was evaluated. The first batch of Pd-membranes was prepared for CIEMAT by a partner in another project, funded by the Ministry of Education and Science of Spain. Permeation and selectivity studies were conducted under "simplified" operating conditions. Experiments included permeation of pure gases and gas mixtures (N_2 and N_2/H_2), hydrogen partial pressures ranging from 0.1 to 1 bar, moderate trans-membrane pressures -up to 2 bar- to guarantee membrane integrity, tests at room temperature using N_2, and permeation tests for mixtures N_2/H_2 at temperatures in the range of 320°C to 400°C. A summary of the results is presented in this chapter and more details have been published elsewhere [28].

A second batch of palladium membranes was obtained from an R&D Centre in US. In this case, according to private information made available by the supplier, the membrane can be regarded as pre-commercial since it was prepared based on membrane technology, which is at an advanced technical stage, protected by patents, and ready for industrial introduction. Permeation tests were carried out for binary mixtures, H_2/N_2 and H_2/CO_2. Hydrogen content in the mixture ranged from 30% v/v to 60% v/v. The effect of gas flow rate, temperature, gas composition, and pressure on hydrogen flux was determined. Detailed results have been presented elsewhere [29], [30].

Experimental

Membranes

Two different Pd membranes have been tested. The first batch of membranes consists of a porous stainless steel tubular support (O.D. =12.7 mm; L=80 mm), -Fig. 12 (left)- provided by Mott Metallurgical, which has a pore size of 0.2 μm and a porosity of 20%. The porosity of the support was reduced by brushing. An intermediate silicalite layer was deposited onto the support, which was then coated with a Pd film by electroless plating. Thickness of the selective palladium layer after plating was estimated to be between 14 and 24 μm. Detailed information about membrane preparation procedure can be found elsewhere [31].

The membrane module consists of four tubular membranes assembled on a metal distributor that forces the feed gas to flow along the membrane. The membranes are inserted in the flanges and graphite/alumina gaskets are used in each membrane-holder joint. Four spiral rods tighten the metallic structure to prevent undesirable gas leaks. The whole structure – see Fig. 12 (right) is located in a metal cylindrical housing of 90 mm OD x 150 mm length able to withstand pressures up to 20 bar and temperatures up to 500 °C.

The second membrane studied is a pre-commercial prototype supplied by a research institute in USA. The dense palladium membrane is supported on a porous tube (O.D.=2,54 cm; L= 15 cm) and it was supplied welded to a solid-wall 316SS tube –Fig. 13-. The final membrane-tube assembly is 70 cm long and it is housed in the tubular reactor.

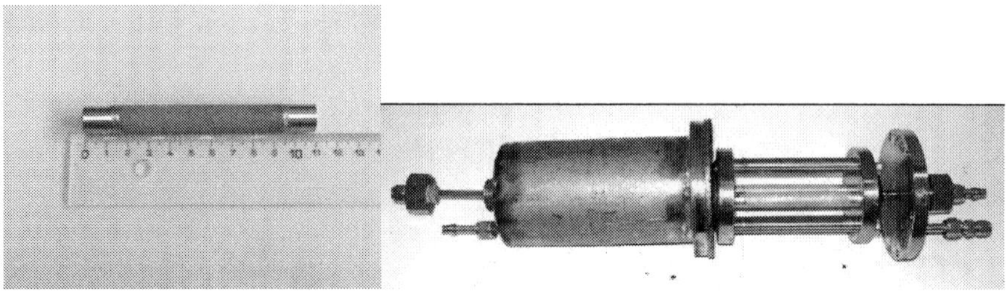

Fig. 12. (left) Prepared Membrane (right) Membrane reactor.

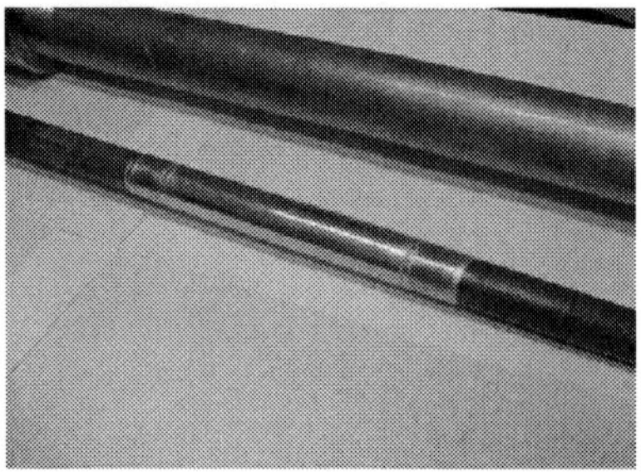

Fig. 13. Pre-commercial membrane and tubular reactor.

Test rig

Experiments have been conducted on a bench-scale level, in a facility located at CIEMAT, shown schematically in Fig. 14, where up to 2 Nm^3/h can be treated, at a temperature of 500ºC and up to 12 bar.

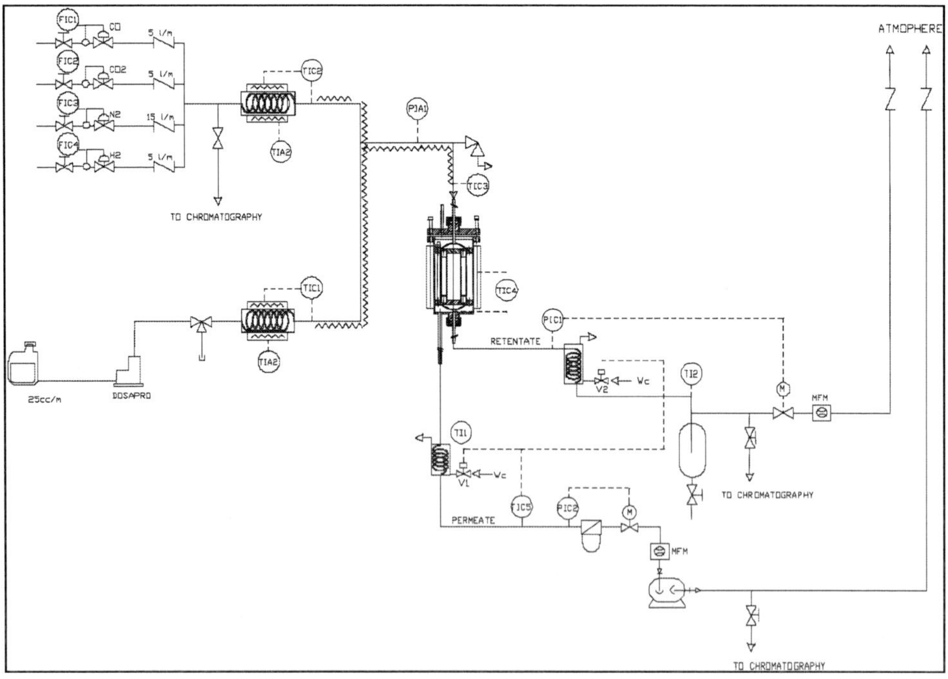

Fig. 14: Scheme of the gas separation test rig.

The facility consists of a gas feeding system based on mass flow controllers (MFCs) to feed H_2, N_2, CO, and CO_2, a vaporiser for steam feeding, a gas preheating system, the membrane reactor, and the permeate and retentate conditioning system including gas cooling, the pressure control system consisting of valves (PCVs) and flow-rate measurement by means of mass flow meters (MFMs). Inlet gas, permeate gas and retentate gas compositions are measured by gas chromatography. Depending on the membrane to study it was necessary to change the membrane reactor oven. Fig. 15 (left) shows the test rig when the pre-commercial membrane was being studied, whereas Fig. 15 (right) shows in detail the membrane reactor oven for the prepared membranes case

Fig. 15. Membrane Reactor bench-scale plant (left) oven for pre-commercial membrane installed (right) oven for prepared membranes installed.

Experimental programme

Testing programme on the prepared membranesconsisted of the study of permeation of nitrogen and hydrogen under the following conditions:Permeation with pure N_2:Temperature: 25°C, 320°C y 380°C; Feed gas flow: 500 ml/min and 750 ml/min; Trans-membrane pressure: 0.2-1.3 bar; permeation with pure $H_2$was studied without using sweep gas, using a stream of pure hydrogen, under the following conditions: Temperature: 320°C; Feed gas flow: 500 ml/min and 750 ml/min; Trans-membrane pressure: 0.1-0.7 bar.

In the case of the pre-commercial membrane prototype, permeation tests have been carried out for binary mixtures, H_2/N_2 and H_2/CO_2. Hydrogen content in the mixture ranged from 30% v/v to 60% v/v. The effect of gas flow rate, temperature, gas composition, and pressure on hydrogen flux has been determined. Gas flow rate (G) was set at, 0.75, 1.83 and 3.75 Nl/min. Two permeation temperatures -380°C and 430°C- have been studied. Retentate pressure was varied from 1 to 3.5 bar (rel). Permeate side pressure was kept atmospheric and sweep gas was not used to enhance permeation.

Results and Discussion

Performance of a membrane is usually determined in terms of permeability and selectivity.Permeate flux of a species through a gas separation membrane can be calculated using the following general equation:

$$J = \frac{Q}{l}\left(P_{xf}^{n} - P_{xp}^{n}\right) \tag{2}$$

Where:

J= flux (mol/m^{2}s)
Q = permeability (mol m/m^2 s Pa^n)
l = membrane thickness (m)
P $_{x,,f}$ = Partial pressure of species x at feed stream (Pa)
P $_{x,p}$ = Partial pressure of species x at permeate stream (Pa)
n = pressure exponent (usual values between 0.5 and 2)

In the case of porous membranes, where molecules are transported by diffusion through the pores, all gas components can permeate to some degree and the driving force is the difference of pressure between both sides of the membrane. On the other hand, for dense membranes-as Pd-based membranes- which do not have any pores, the prevailing transport mechanism is the so called solution-diffusion mechanism and only some components are able to permeate. Of all the gas components present in syngas only hydrogen is able to permeate through dense Pd membranes. If the diffusion of hydrogen through the membrane is considered the rate-limiting step, hydrogen flow will follow the Sievert´s law –equation 3- being directly proportional to the difference of the square root of the hydrogen partial pressure on both sides of the membrane:

$$J_{H_2} = \frac{Pe}{l}\left(P_{H_2f}^{0,5} - P_{H_2p}^{0,5}\right) \tag{3}$$

Where:Pe_{H2} = permeability (m^3 m/m^2 s Pa$^{0.5}$)l = membrane thickness (m)P $_{H2,f}$ = Partial pressure of H_2 at feed stream (Pa)P $_{H2,p}$ = Partial pressure of H_2 at permeate stream (Pa)

Then, *permeability* through the membrane can be obtained from equation (4) as follows:

$$Pe_{H_2} = \frac{J_{H_2}}{l\left(P_{H_2f}^{0,5} - P_{H_2p}^{0,5}\right)} \tag{4}$$

Besides permeability, another parameter must be considered when studying the performance of a membrane for gas separation. This parameter is called selectivity. In general, porous membranes provide high permeability but selectivity depends on the relative size ratio between molecules and pores. In this type of membranes separation of components is only possible when pore size is smaller than the mean free path of the molecules of gas. The theoretical separation factor is calculated as the square root of the ratio between molecular weights (eq. 4). In these conditions the transport mechanism follows the Knudsen diffusion.

$$\alpha = \sqrt{\frac{M_A}{M_B}} \quad (5)$$

Conversely, the solution-diffusion mechanism provides an infinite selectivity to H_2 and this characteristic makes palladium membranes very interesting for their potential application in equilibrium limited reactions.

Prepared Membranes

Permeation using pure nitrogen was determined at different temperatures, 25°C, 320°C y 380°C for two feed gas flow-rates, 0.5 Nl/min and 0.75 Nl/min, changing trans-membrane pressure between 0.2 and 1.3 bar in a stepwise mode. Permeation of H_2 was determined at 320°C, feed gas flow rate was set at 0.5 Nl/min and 0.75 Nl/min and trans-membrane pressure was changed in the range from 0.1 to 0.7 bar. It was found that nitrogen and hydrogen were able to permeate through the membrane.

At room temperature values of permeation of N_2 between 0.02 mol/m^{2}s and 0.12 mol/m^{2}s were obtained when feed gas flow-rate FN_2 was set at 0.75 Nl/min. For permeation of nitrogen through porous stainless steel tubes at room temperature, values within the same range have been reported by other authors [32]. A linear dependence of nitrogen flux through the membrane with the difference of pressure between both sides of the membrane was observed, which suggests that Knudsen diffusion can be the prevailing transport mechanism.

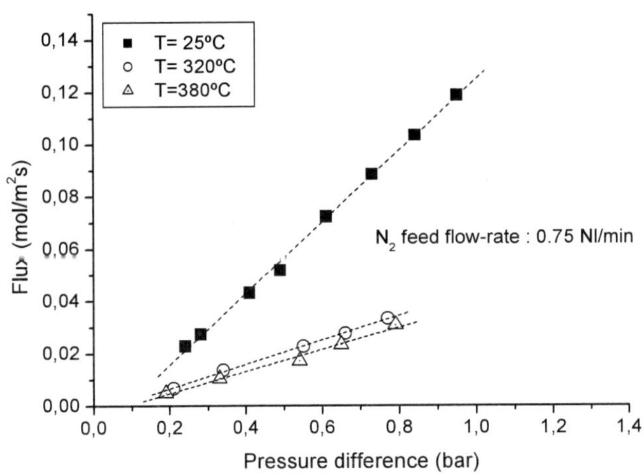

Fig. 16. Effect of temperature on N2 permeation for prepared membranes. (FN2=0.75 Nl/min).

The effect of temperature on nitrogen permeation for the same value of feeding gas flow-rate (FN_2=0.75 Nl/min) is shown in Fig. 16. If nitrogen transport through the membrane is governed by Knudsen diffusion, then a decrease in N_2 flux with increasing temperatures will be observed as reported by other authors [33]. Results shown in Fig. 16 confirm that permeation of N_2 decreases with temperature reaching values of N_2 flux between 0.0067 and

0.033 mol/m^{2}s at 320°C and 0.005 and 0.03 mol/m^{2}s at 380°C with pressure ranging from 0.2 to 1.3 bar.

Hydrogen and nitrogen permeation flux and the influence of feed gas flow-rate on permeation flux are shown in Fig. 17. A comparison between hydrogen and nitrogen fluxes shows that at 320°C and F= 0.75 Nl/min, while values between 0.025 mol/m^{2}s and 0.08 mol/m^{2}s are obtained for H_2 in the pressure range of 0.2 to 0.7 bar, flux of N_2 remained between 0.006 mol/m^{2}s and 0.025 mol/m^{2}s. These values allowed us to calculate the theoretical selectivity (estimated as the ratio between H_2 and N_2 fluxes), giving values between 2.85 and 3.75 which is approximately the expected value for Knudsen diffusion. As happened for permeation of nitrogen, a linear relationship between hydrogen flux and driving force -differential pressure- was found which suggests that H_2 is basically transported through the membrane pores. This explains the low values of selectivity. Nevertheless, looking at the effect of feed gas flow-rate on permeation pattern, no clear difference in performance is observed in Fig. 17 between pure N_2 and pure H_2 permeation. For both gases, permeation flux fits with pressure difference. For each gas, despite using two different feed gas flow rates, 750 and 500 ml/min, values of permeation flux fall on the same line. Thus, permeation flux seems to be a function of pressure difference -driving force for transport across the membrane-- and it does not seem to depend on feed gas flow-rate. Nitrogen is able to permeate through the pores due to the lack of a continuous, homogeneous palladium layer. The smaller size of the molecules of H_2 seems to favour, however, their transport through the narrowest pores of the membrane, and therefore its flux is higher than the nitrogen one. Besides, the presence of the layer of palladium might be providing some additional transport for hydrogen by means of other mechanisms, such as superficial diffusion and solution-diffusion which can contribute to increase H_2 flux.

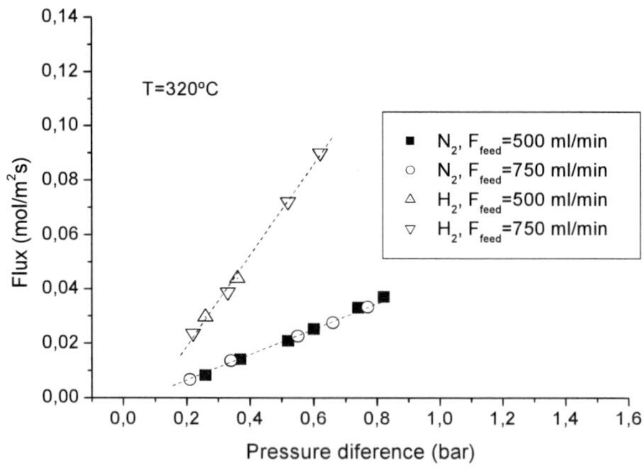

Fig. 17. Effect of Feed Gas Flow-rate on Hydrogen and nitrogen flux for prepared membranes (T=320°C).

After testing, membranes were characterised by Optical Microscopy, SEM (Scan Electron Microscopy) and by EDX (Energy Diffusion Spectroscopy). Optical Microscopy images- Fig. 18- of the cross section of the membrane showed the presence of a continuous layer of palladium deposited onto the support on some parts of the membrane -Fig 18a-, while others showed discontinuities, pinholes or even small nodules -Fig 18b-. The intermediate silicalite layer was not detected looking at the cross-section membrane´s rim image. SEM analysis showed that palladium was covering the porous support but allowing the presence of some discontinuities and holes. EDX analysis showed palladium as the only species present on the deposited layer, even when analysing the interior of some pinholes. Low selectivity shown by the membranes was therefore ascribed to defects found on the palladium layer.

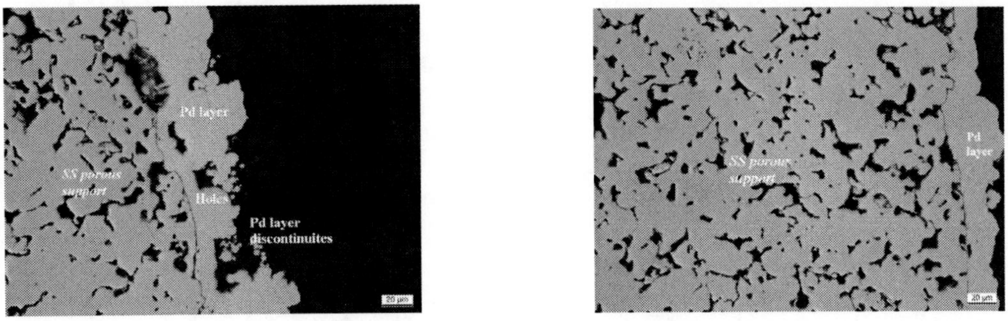

Fig. 18. Optical Microscopy images of prepared Pd-membranes after testing (a) continuous layer of Pd (b) discontinuities of Pd-layer.

Summarising, the experiments carried out with pure N_2 show that the Pd layer on the prepared membranes is not defect-free but there are pinholes or discontinuities which allows nitrogen to be transported through the membrane. The prevailing separation mechanism was found to be Knudsen diffusion, instead of the solution-diffusion mechanism which should have been provided by a continuous, dense Pd layer. This means that the technique used to prepare the membranes was not efficient enough to guarantee the continuity of the layer of palladium deposited, favouring the presence of pinholes and pores which reduced dramatically the selectivity of the membrane to H_2. The membranes showed high permeation flux but very low selectivity, close to Knudsen separation factor. Therefore, these membranes did not show potential for carrying out WGS studies in a membrane reactor since the main requirement for their application is that hydrogen selectivity must be as high as possible and ideally infinite so that only hydrogen is removed from the reaction chamber.

Pre-commercial Membrane

In order to determine defects or pinholes present on the palladium layer, permeation of nitrogen was measured in first place. To this aim pure nitrogen was fed at two different gas flow-rates, 0.75 and 1 N l/min. Reactor temperature was set at different values, room temperature, 320°C, 380°C, and 430°C. Pressure was varied between 1 and 3.5 bar. Permeate flow-rate was measured by means of a volumetric gas meter, with an accuracy of 1 ml/min. In contrast with the prepared membranes, in none of the tests nitrogen was detected in the permeate stream.

After checking that the membrane was leak-free and nitrogen was not permeating, hydrogen permeation and selectivity studies were conducted using two-component mixtures (H_2/N_2; H_2/CO_2). During testing, evolution of hydrogen flux at different values of retentate pressure, (P ret), was typically determined as Fig. 19shows.

Hydrogen flux slightly increases with time, reaching a maximum, then decreases and finally reaches a stationary value. This effect is independent of temperature and hydrogen content in the gas mixture and was more noticeable when low gas flow-rate, F= 0.75 Nl/min, was used. When gas flow rate was increased, stationary state was achieved faster. Maximum flux of hydrogen that permeated through the membrane is presented in Table 2 for different operating conditions. As can be seen, hydrogen flux is clearly affected by process operation conditions.

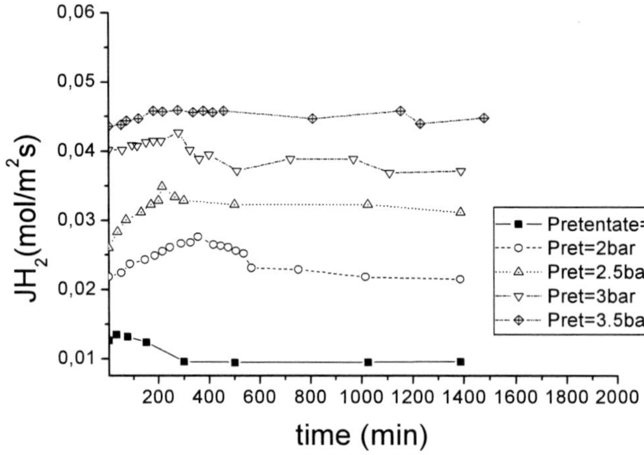

Fig. 19. Typical evolution of hydrogen flux over time observed for the pre-commercial membrane when using binary mixtures FH2=0,45Nl/min; FN2=0,23 Nl/min; T= 430°C.

Table 2. Measured H_2 permeate flux under different operating conditions using the pre-commercial membrane

Mixture	$[H_2]$ (v/v %)	F_{feed} (Nl/min)	Driving force ($kPa^{0.5}$) $PH_2feed^{0.5}$-$PH_2perm^{0.5}$	Max P feed (bar)	Maximum Pret (bar)	Maximum H_2 flux (mol/m^2s)
H_2/N_2	30	0.75	2.1	3.79	3.5	0.0055
H_2/N_2	30	1.88	2.7	4.12	3.5	0.0155
H_2/N_2	30	3.75	2.1	5.72	3.5	0.0461
H_2/N_2	30	0.75	2.1	3.77	3.5	0.0053
H_2/N_2	60	0.75	6.7	3.65	3.5	0.0212
H_2/N_2	60	1.88	6.1	3.97	3.5	0.057
H_2/N_2	60	0.75	6.5	3.58	3.5	0.020
H_2/N_2	70	1.88	3.5	1.67	1	0.058
H_2/CO_2	30	1.88	2.5	4.59	3.5	0.0051
H_2/CO_2	30	1.88	3.05	4.37	3.5	0.0098
H_2/CO_2	30	3.75	4.35	5.86	3.5	0.039
H_2/CO_2	30	1.88	5.26	6.00	3.5	0.034
H_2/CO_2	60	1.88	6.35	4.15	3.5	0.040
H_2/CO_2	60	1.88	5.6	4.66	3.5	0.0324
$H_2/CO_2/CO/H_2O$	16	10.3	3.8	10.3	10	0.0043
$H_2/CO_2/CO/H_2O$	16	10.3	2.6	9.41	8	0.0054

In Fig. 20experimentally measured H_2 flux has been plotted against the difference of the square root of hydrogen partial pressure on both sides of the membrane, which corresponds to transport through a thick palladium layer following the mechanism of solution-diffusion and when diffusion of atomic hydrogen through the dense metal layer is the rate-limiting step.

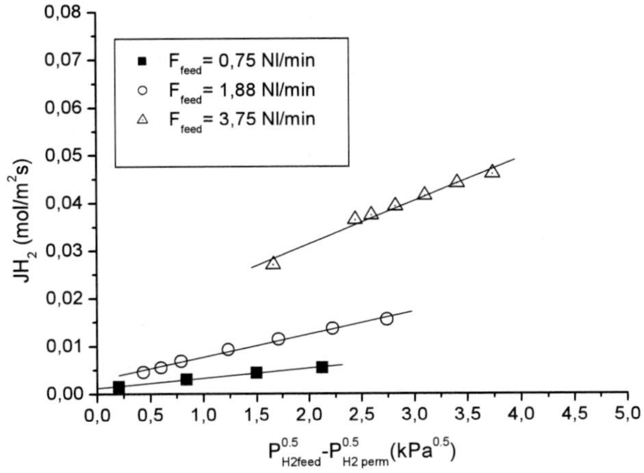

Fig. 20. Hydrogen flux for the pre-commercial membrane as a function of feed gas flow-rate ([H2] =30% v/v, [N2] =70% v/v, T=380°C.

As can be observed, for the operation range studied, a good correlation between the square root of the hydrogen partial pressure difference and hydrogen flux is found what suggests that solution-diffusion might be the main mechanism governing hydrogen transport for the membrane under study. However, linear portions clearly extrapolate to a non-zero intercept; above all for the highest feed flow rate studied, what indicates that in the smaller driving force range other mechanisms may be playing a role on hydrogen permeation. Experimental points for higher values and more importantly for smaller values of driving force should be obtained to confirm to what extent solution-diffusion and other mechanisms are governing hydrogen transport.

Regarding the effect of feed gas flow-rate on hydrogen permeation it was found that, for H_2/N_2 binary mixtures and for the experimental range evaluated, an increase of gas flow-rate has a positive effect on hydrogen flux as Fig. 20shows. A similar behaviour has been observed by other authors. For instance Peters et al. [34] reported an increase of hydrogen flux in 50%H_2/50%N_2 with total feed flow rate at 400°C and 26 bar. Ma et al [35] have reported a similar effect for a 20%CO_2/80%H_2 mixture at 400°C, and two pressures 4.8 and 14.3 bar. According to those studies, the effect of feed flow rate on flux is significant, but levels off at high rates, when gas phase diffusion and hydrogen depletion along the membrane are overcome. Such a flow rate value was not reached in our case, what means that feed flow rate could be increased further until reaching the "*level-off*" value.

When assessing the feasibility of a membrane for hydrogen separation, another important parameter is hydrogen recovery, which gives the ratio between hydrogen permeating through the membrane to hydrogen fed to the membrane reactor. Fig. 21 shows hydrogen recovery for the CRI membrane under different experimental conditions. Hydrogen recovery ranged from 20% for a hydrogen-lean (30%) mixture in CO_2 (F=1.88 Nl/min, T= 380°C) to 80% for a hydrogen-rich (70%) mixture in N_2 for the same total feed gas flow-rate and temperature.

When feeding mixtures of hydrogen with other gases into the membrane reactor, permeated hydrogen flux obtained -and therefore hydrogen recovery –is influenced by several factors such as gas mixture, H_2 content in the feed, and temperature as shown in Fig. 21 and also in the results summarised in Table 2. Higher hydrogen content in the feed gas improves H_2 recovery due to a higher driving force.

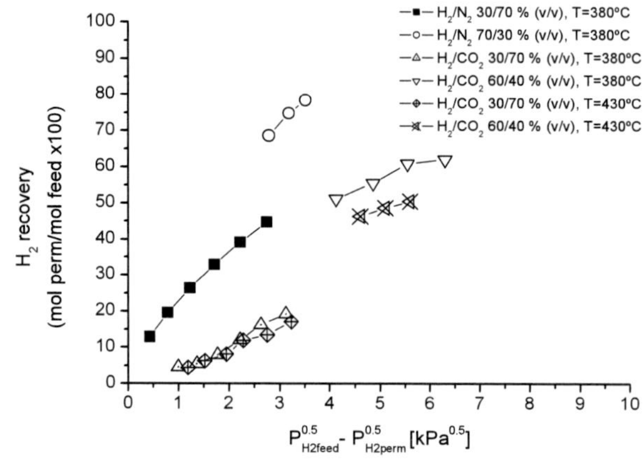

Fig. 21. Hydrogen recovery from binary mixtures for different experimental conditions.

Looking at the effect of gas components, CO_2 presence in the mixture reduces hydrogen recovery. This effect is more important at 430°C than at 380°C. In literature it is widely acknowledged that hydrogen flux is reduced due to the presence of other gas components. A number of reasons are given, such as dilution of hydrogen by other gas components, hydrogen depletion along the membrane length, concentration polarization adjacent to the membrane surface, and effects due to surface adsorption [34], [36-42]. Although out of the scope of this work, it is therefore deemed necessary for the membrane here studied, to conduct and report in future works a thorough analysis on the effect on hydrogen permeation of the different gas components involved in particular in WGS reaction and in general in synthesis gas (CO, CO_2, H_2O, N_2 and light hydrocarbons). It would be worth it as well to explore the occurrence of reactions that could be catalyzed by the Pd membrane itself, such as methane formation, which although not discussed here, was found to be forming to a small extent – less than 1% v/v- when feeding CO_2/H_2 mixtures

It is important to highlight that during testing, regardless operating conditions neither N_2 nor CO_2 have been detected in the permeate stream. This result was expected because, according to transport mechanism, dense palladium membranes feature exclusive selectivity to hydrogen provided that the membrane has no pinholes. This means that the membrane studied is defect-free. In this sense, results are very encouraging since they show that it is feasible to separate H_2 from CO_2 using Pd-based membranes like the one under study, yielding a pure H_2 permeate stream and a CO_2-rich retentate stream at high pressure. Hydrogen flux is low (approximately 15%) compared to hydrogen separation technical targets set by DOE for the period 2009-2016, but this is basically due to the fact that permeation results have been obtained for a quite small difference of hydrogen partial pressure between feed and permeate gas (1.45 bar, maximum), in comparison with 100 psi ΔP (6.89 bar) which is the reference value set by DOE [43]. This means that there is still room for improvement. Hydrogen flux and hydrogen recovery can be enhanced increasing feed pressure.

WATER GAS SHIFT MEMBRANE REACTOR STUDIES

According to the IUPAC definition, a membrane reactor is a device that combines the separation properties of membranes with the typical characteristics of catalytic reaction steps in only one unit. In particular, the membrane does not only play a role as a separator but also as a part of the reactor itself. In other words, a membrane reactor is an engineering device that selectively removes a product from the reaction system, giving the possibility of achieving higher conversion than a traditional process under the same operation conditions. For that reason, one of the main application opportunities of inorganic membrane reactors refers to the conversion enhancement of an equilibrium limited reaction by the selective permeation of a reaction product through the membrane. In our particular case, the equilibrium limited reaction to study is the water gas shift reaction and the reaction product to remove is hydrogen. Hydrogen removal from the reactor might shift the equilibrium of the WGS reaction to the right side, enhancing products formation, H_2 and CO_2, what would mean working at lower temperatures or lowering the steam to CO ratio. This last statement is somehow controversial, however, since a quasi-stoichiometric ratio of steam to carbon monoxide may enhance secondary reactions to occur, for instance Boudouard reaction or methane formation.

On this research field, Lu et al. recently published an in depth critical review and perspective on inorganic membranes for hydrogen production and purification [44]. According to this and other scientific review works, [45], [46], palladium and Pd-alloys membranes are excellent candidates for hydrogen purification, especially when incorporated with a catalytic reactor combining reaction and separation in a single unit. Palladium-based membrane´s theoretically infinite selectivity to hydrogen continues to draw major research efforts. There are a good number of research groups investigating the field of catalytic WGS membrane reactors from the research, technological and economical viewpoints [47-52].

However, the use of Pd membranes in WGS reactors imposes several limitations on the catalyst. Among them, and besides high selectivity towards hydrogen separation, the most restrictive one is that the catalyst must be operated above 250°C when CO is present, and whenever possible above 300°C to minimize the risk of palladium embrittlement and phase transformation which would lead to cracks in the metal film. In that sense only high temperature WGS catalysts can be regarded as suitable to be combined with palladium membranes.

Following the above constraints, the prototype described in this chapter consisted of the combination of a hydrogen-selective palladium membrane and a high temperature WGS catalyst. The comparison with WGS in traditional fixed bed reactors (TR) was also undertaken.

Experimental

Membrane and catalyst

The membrane reactor concept proposed in our study combines the use of a membrane selective to hydrogen permeation and a highly active catalyst for water gas shifting. Based on the results obtained in previous stages the pre-commercial Pd membrane was selected to provide an efficient way for hydrogen removal. As said before, the membrane is tubular having a diameter of 1" and it is 6" long. It is housed in a 3-zone electric furnace with temperature control.

The commercial iron-chromium catalyst, suitable for high temperature WGS under the operating conditions expected for synthesis gas after the steam reforming unit in VVBGC was selected for WGS studies in the membrane reactor. As described before, the catalyst density is 1.1 g/cc and it was supplied in pellets of 6x6 mm.

Test rig

Experiments were conducted on a bench-scale level, in the same facility in which hydrogen separation studies were performed. For WGS studies in the membrane reactor, samples of the iron chromium catalyst from the same batch used for WGS studies in a conventional fixed bed reactor were placed on the shell side of the reactor (reaction side) to make up a catalyst bed of 4.9 cm length (39 cm^3 volume), matching with the membrane zone. This configuration was the only one possible because one end of the membrane tube was a blind-end cap and the other one was provided welded to the reactor head. Consequently it was not feasible to place the catalyst inside the membrane, which would have been the preferred choice to prevent damage of the membrane, e.g. due to local hot points on the vicinity of the membrane related to catalyst activity. Inert alumina spherical pellets were also placed on the shell side of the reactor to support the catalyst. The reactor configuration chosen is depicted schematically in Fig. 22.

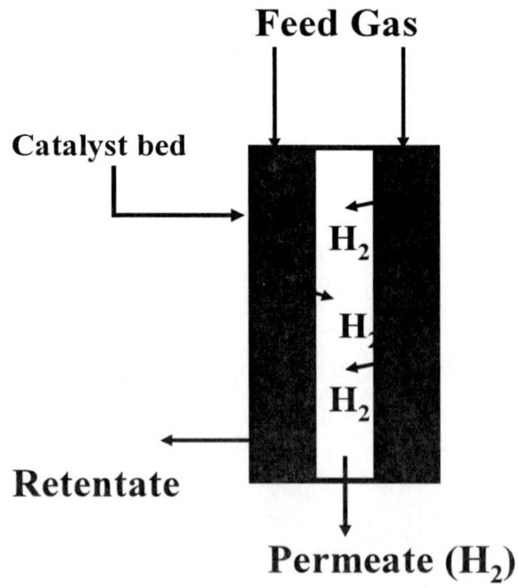

Fig. 22. Reactor configuration for Water Gas Shift Membrane Reactor (WGSMR) experiments.

Experimental programme

Based on the results obtained in our water gas shift studies in fixed bed, for the WGS membrane reactor experiments, space velocity, temperature and steam to CO ratio were set purposely at values which were known beforehand that would not fully convert CO into CO_2 and H_2. Thus, it would be possible to see if there is any effect on CO conversion enhancement due to removing hydrogen from the reaction environment.

Gas composition used for the tests was CO, 44%v/v; H_2 40% v/v and CO_2 16%v/v, which is, except for CH_4, similar to the dry gas composition expected at VVBGC after the steam reformer.

WGS experiments in the membrane reactor were conducted at two different temperatures (320°C and 380°C). For each temperature, the reaction was carried out at several pressures on the retentate side, (1.5; 5; 8 and 10 bar), relative to atmospheric pressure. Permeate pressure was kept at atmospheric pressure and sweep gas was not used. H_2O to CO ratio was set at 3, and gas hourly space was kept at 16771 h^{-1}.

Results and Discussion

From the WGSMR tests two types of results are obtained: H_2 permeation for complex mixtures, under WGS reaction conditions, and CO conversion under membrane reactor conditions, which can be compared to the activity of the catalyst in absence of permeation, that is, catalytic activity in the so-called traditional WGS reactor.

Permeation

In addition to CO conversion, from the WGS studies in the membrane reactor the following results are obtained:

- Flux of hydrogen as a function of driving force for complex system ($H_2/CO/CO_2/H_2O$) under WGS conditions
- Effect of temperature on hydrogen flux for complex system ($H_2/CO/CO_2/H_2O$)

As obtained for binary mixtures, only hydrogen flows through the membrane even for a complex gas mixture under water gas shift conditions (CO, 19%v/v; H_2 17% v/v, CO_2 7%v/v, H_2O, 57 %v/v). Maximum hydrogen permeated flux achieved for the synthesis gas mixture ranged between 0.00429 molH$_2$/m^{2}s at 320°C and 10 bar feed pressure and 0.00541 molH$_2$/m^{2}s at 380°C and 8 bar feed pressure, which are on average one order of magnitude smaller than the values obtained for H_2/N_2 and H_2/CO_2 mixtures as can be seen comparing hydrogen permeation flux shown on Table 4. This may be an indication that the presence of CO and H_2O influences negatively membrane efficiency due either to dilution by the presence of other gas components or to some inhibition effect. Moreover, operating conditions for which hydrogen permeation under WGS conditions has been determined differed significantly from the studies presented for binary mixtures in terms of total feed gas flow-rate and hydrogen content, which can therefore result in quite different values for hydrogen permeation flux.

Flux of hydrogen as a function of retentate pressure for a complex system ($H_2/CO/CO_2/H_2O$) under WGS conditions is presented in Fig. 23. For the test shown in the figure reaction temperature was 320°C.

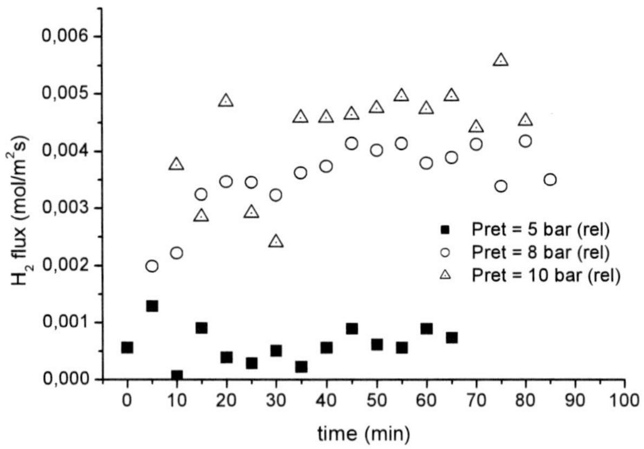

Fig. 23. Flux of hydrogen permeated through the pre-commercial membrane under WGS conditions (T=320°C, Pperm= 0 bar. (rel); F feed gas (dry)= 4,71 Nl/min; Feed Gas composition (dry basis)=44% v/v CO, 40% v/v H2, 16%v/v CO2; R (H2O/CO)= 3 mol/mol; GHSV= 16771 h-1).

Looking at Fig. 23 it can be noted that, despite fluctuations, hydrogen permeated flux is higher on increasing pressure at the shell side, i.e. retentate pressure, due to an enhancement of pressure driving force. No permeation was observed for the lowest retentate pressure studied, 1.5 bar. Estimation of driving force -the difference of the square root of hydrogen partial pressure on both sides of the membrane- shows that for that retentate pressure, and under those reactor conditions, the resulting driving force is not enough for hydrogen to permeate. A similar behaviour was found at 380°C. As was the case for binary mixtures, an increase on pressure on the retentate side resulted in more hydrogen flowing through the membrane to the permeate side.

In Fig. 24the effect of temperature on H_2 flux is plotted. Permeation fluxes at two temperatures, 320°C and 380°C are shown in the graph.

As can be observed, permeation increases with temperature which is often the case reported in literature for permeation of pure hydrogen. Hydrogen transport through a dense palladium film is an activated process and concerning the influence of temperature on hydrogen permeation an Arrhenius-type law is commonly used to describe the relationship between hydrogen permeation rate and temperature. Since only two temperatures were studied in this work such a relationship cannot be estimated. Concerning the effect of driving force, the term $(P_{feedH_2}^{0.5} - P_{permH_2}^{0.5})$ represents that the diffusion of atomic hydrogen through the palladium layer is the rate-controlling step, which is usually the case for thick palladium membranes. In our case, a good linear relationship between driving force and flux of hydrogen is found for 320°C. However at 380°C, though the number of experimental data is quite limited, relationship between driving force and flux of hydrogen is somehow poorer, which suggests that other elementary steps of the complex mechanism phenomenon of hydrogen permeation through dense palladium might be playing a role under those experimental conditions. It should be highlighted that because of the experimental operating conditions at the membrane reactor (wet feed gas composition and pressure) driving force is very small and therefore deviations from the openly agreed driving force describing hydrogen permeation can be expected.

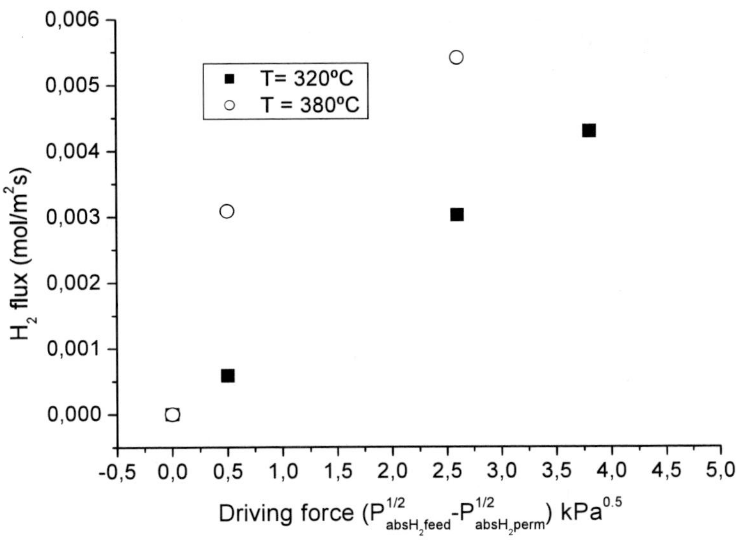

Fig. 24. Effect of temperature on Hydrogen flux under WGS conditions (Pperm= 0 bar. (rel); F feed gas (dry)= 4,71 Nl/min; Feed Gas composition (dry basis)=44% v/v CO, 40% v/v H2, 16%v/v CO2; R (H2O/CO)= 3 mol/mol; GHSV= 16771 h-1).

CO conversion

The beneficial effect of removing hydrogen from the reaction side on the conversion of CO can be seen in Fig. 25. It is known that WGS reaction in principle is not affected by the reaction pressure in a traditional packed bed reactor, because the overall number of moles does not change. On the other hand, as the driving force for H_2 permeation increases with the higher pressure on the reaction side- retentate side- in-situ removal of hydrogen enhances CO conversion in the case of the membrane reactor as Fig.25 shows. As long as hydrogen does not start to permeate, i.e. when there is not enough driving force for hydrogen transport, it can be assumed that the membrane reactor will work as a fixed bed reactor. As soon as hydrogen begins to permeate, there is an enhancement of carbon monoxide conversion. Thus at 320°C, Fig. 25, CO conversion from 4.11 % when hydrogen is not removed, i.e. under fixed bed, traditional reactor conditions to 15.75%, when pressure at the retentate side is 10 bar rel. (feed pressure 11.3 bara), in the membrane reactor concept.

Removal of hydrogen through the membrane also enhances CO conversion at the second temperature studied as Fig. 26 shows. It can be seen that as hydrogen permeate flux increases as a result of a higher pressure and hence higher driving force, conversion of carbon monoxide also increases. At 380°C there is an enhancement of CO conversion from 34.25% (traditional reactor, no permeation) to 60.5% when retentate pressure was set at 8 bar rel.(feed pressure 9.39 bara).

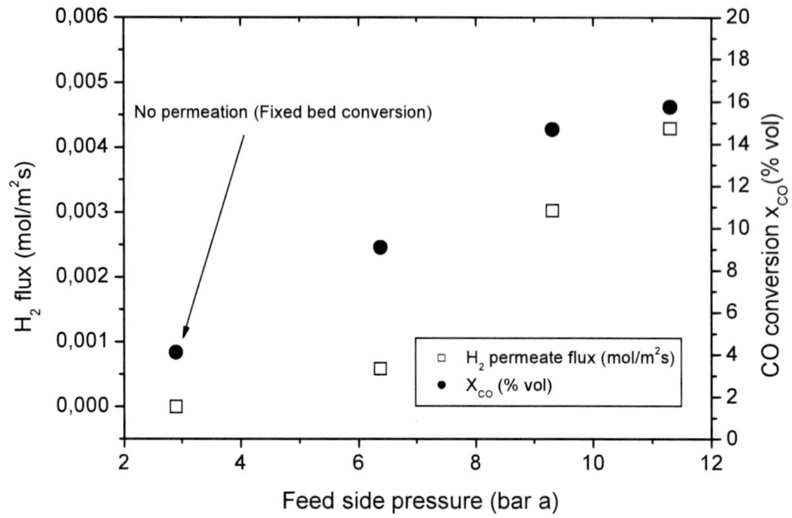

Fig. 25. CO conversion and hydrogen permeate flux in the membrane reactor at 320°C versus retentate pressure (Pperm= 0 bar. (rel); F feed gas (dry)= 4,71 Nl/min; Feed Gas composition (dry basis)=44% v/v CO, 40% v/v H2, 16%v/v CO2; R (H2O/CO)= 3 mol/mol; GHSV= 16771 h-1).

A similar beneficial effect on water gas shift conversion enhancement is reported by other authors working on this field. Despite the numerous experimental data available, comparison of results is difficult because many of them were obtained under simplified conditions (CO/H_2O only) [47-49], with a different type of catalyst, low temperature CuO-ZnO catalyst, [48], [54], with different membrane-type or support [55-56], or although with similar membranes and catalysts under very different experimental conditions [35], [57-58].

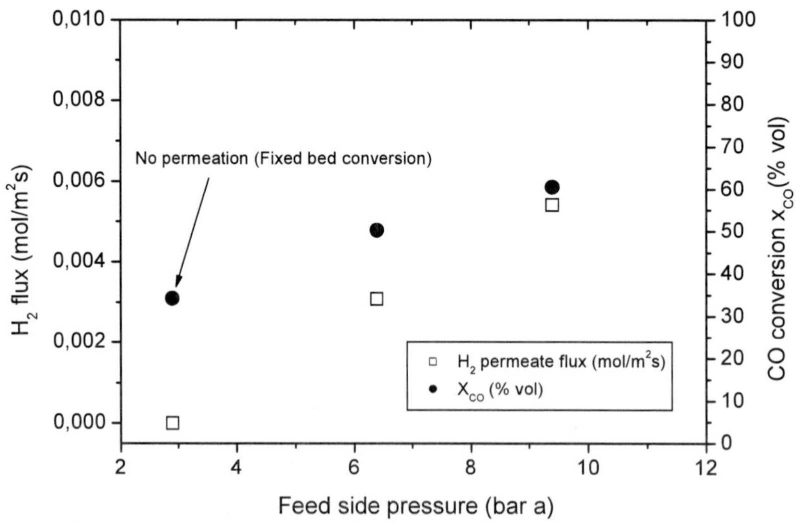

Fig. 26. CO conversion and hydrogen permeate flux in the membrane reactor at 380°C versus retentate pressure (Pperm= 0 bar. (rel); F feed gas (dry)= 4,71 Nl/min; Feed Gas composition (dry basis)=44% v/v CO, 40% v/v H2, 16%v/v CO2; R (H2O/CO)= 3 mol/mol; GHSV= 16771 h-1).

CONCLUSION

In order to produce a H_2-rich gas, within the CHRISGAS project CIEMAT has accomplished the following experimental studies:

- Water Gas Shift Studies in Traditional Packed Bed Reactor
- Separation of Hydrogen using Membranes
- Water Gas Shift Studies in Membrane Reactor

Water Gas Shift studies dealt with selection of catalysts and optimisation of operating conditions to maximise CO conversion and hydrogen production. CIEMAT studied on a laboratory scale the activity of a commercial WGS catalyst (Fe-Cr-based) and a novel Pt-based catalyst provided by CATATOR AB Sweden, in a traditional fixed bed reactor using simulated gases, under the operating conditions and gas composition expected at VäxjoVärnamo Biomass Gasification Centre (VVBGC). Results prove that both types of high temperature WGS catalysts (Fe-Cr- and Pt-based) are able to convert more than 90% of the CO content of the synthesis gas into CO2 and H2 under specific operating conditions (temperature, space velocity and steam to CO ratio), and therefore can be regarded as suitable to be used in VVBGC. The ferrochrome commercial catalyst studied provided good conversion of CO (less than 2-3 % v/v residual) in the temperature range of 360°C-400°C and at H2O/CO ratios higher than 6. High conversion of CO was also obtained when using a Pt-based catalyst, in the temperature range of 350°C-380 °C and high H2O/CO ratios. Those experiments showed very good selectivity to H2 and CO2.

Studies on separation of hydrogen by means of membranes dealt with evaluation of permeation and selectivity to hydrogen of prepared and pre-commercial membranes. CIEMAT conducted those studies on a bench scale level, in a dedicated facility built within the time frame of the CHRIGAS project. The performance of palladium membranes obtained from two sources was evaluated. The first batch of Pd-membranes was prepared for CIEMAT by a partner in another project, funded by the Ministry of Education and Science of Spain. The second membrane studied is a pre-commercial prototype supplied by a research institute in USA. Prepared membranes showed very poor selectivity, which was basically attributed to discontinuities of the layer of palladium deposited on the surface of the porous stainless steel tube. It was concluded, therefore, that those membranes were not suitable for further studies in a WGS membrane reactor.

On the other hand, permeation studies conducted with the pre-commercial membrane show that by means of a membrane reactor it is possible to completely separate hydrogen from the other gas components and produce pure hydrogen as a permeate stream. The membrane studied shows infinite selectivity to hydrogen since neither N_2 nor CO_2 nor CO permeated through it. Hydrogen is able to permeate, as long as there is enough driving force, and is removed from the reactor as a pure permeate stream. In our study many parameters have been found to play a role on hydrogen permeation, as for instance, gas composition, temperature, feed gas flow-rate, pressure, etc. Nevertheless, hydrogen flux has been found to depend mainly on the difference of hydrogen partial pressure on both sides of the membrane. From an industrial point of view this fact poses some restrictions on the use of membranes for hydrogen separation. Hydrogen partial pressure on the feed side should always be higher than hydrogen partial pressure on the permeate side so that forward permeation takes place. In order to achieve that, a hydrogen-rich gas stream or syngas at high pressure is required as membrane reactor feed gas stream. For hydrogen-diluted or low pressure feed gases the use of sweep gas is deemed necessary.

Finally, CIEMAT has studied the advantages of using a water gas shift membrane reactor over a traditional fixed bed reactor, in terms of potential for producing a high purity hydrogen stream and for enhancing CO conversion at set operational conditions. The experimental device included a Pd-based membrane and a commercial high temperature Fe-Cr-based, WGS catalyst, which was packed in the annulus between the membrane and the reactor outer shell. Results prove feasible to produce a pure hydrogen permeate stream, even in the case of complex reaction system ($H_2/CO/CO_2/H_2O$) under WGS conditions mixtures. Regarding comparison of CO conversion for WGS in a traditional reactor (TR) and in a membrane reactor (MR), results show that in the MR concept, removal of hydrogen from the reaction side has a positive effect on the conversion of WGS reaction at a given temperature. On increasing pressure on the reaction side, permeation is enhanced what pushes the reaction towards further product formation, due to removal of one of the reaction products, i.e., hydrogen, and hence increasing carbon monoxide conversion.

REFERENCES

[1] Babu, S.P. Biomass Gasification for hydrogen production-process description and research needs. Revised draft; IEA, Gas Technology Institute, IL, U.S.A. 2002

[2] Kalinci, Y.; Hepbasli, A.; Dincer, I. Biomass-based hydrogen production: A review and analysis Int. J. Hydrogen Energy 2009, 34, 8799-8817.

[3] Meng, N.; Dennis, Y.C.; Leung, M. K.H.; Leung, K. S. Fuel Proc. Tech 2006, 87, 461.

[4] Saxena, R.C.; Seal, D.; Kumar, S.; Goyal, H.B. Thermo-chemical routes for hydrogen rich gas from biomass: a review. Ren.Sust.Ene. Rev.2008, 12, 1909–27.

[5] Zhang, R.;Cummer, K.; Suby, A.; Brown, R.C. Biomass-derived hydrogen from an air-blown gasifier Fuel Proc. Tech. 2005, 86, 861-874.

[6] Mahishi, M.R.; Goswami, D.Y. Thermodynamic optimization of biomass gasifier for hydrogen production. Int. J. Hydrogen. Energy 2007, 32, 3831-3840.

[7] Florin, N.H.; Harris, A.T. Hydrogen production from biomass coupled with carbon dioxide capture: the implications of thermodynamic equilibrium Int. J. Hydrogen Energy 2007, 32, 4119-4134.

[8] Aznar, M.P.; Caballero, M.A.; Corella, J.; Molina, G. and Toledo, J.M. Hydrogen production by biomass gasification with steam-O2 mixtures followed by a catalytic steam reformer and a CO-shift system Ene. & Fuels 2006, 20, 1305-1309

[9] Lu, P.; Yuan, Z.; Ma, L.; Wu, C.; Chen, Y.; Zhu, J. Hydrogen-rich gas production from biomass air and oxygen/steam gasification in a downdraft gasifier Renew. Ene. 2007, 32, 2173-2185.

[10] Hemmes, K.;De Groot, A.;Den Uil, H.BIO-H2. Application potential of biomass related hydrogen production technologies to the Dutch energy infrastructure of 2020-2050. ECN-C-03-028, April 2003

[11] Bengtsson, S.; Ståhl, K.; Waldheim, L.; Gårdmark, L. The CHRISGAS Project – Clean Hydrogen-Rich Gas through Gasification and Hot Gas Upgrading. Bioenergy in Wood Industry 2005 Conference, Jyväskylä, Finland,12-15 September, 2005.

[12] Carneiro de Araujo, G.; Rangel, M.C. An environmental friendly dopant for the high-temperature shift catalysts, Catal Today 2000, 62,pp 201.

[13] Thinon, O.; Diehl, F.; Avenier, P.; Schuurman, Y. Screening of bifunctional water-gas shift catalysts, Catal. Today 2008, 137, pp. 29

[14] Panagiotopoulou, P.; Kondarides D.I. A comparative study of the water-gas shift activity of Pt catalysts supported on single (MOx) and composite (MOx/Al2O3, MOx/TiO2)metal oxide carriers, Cat. Today 2007, 127, pp. 319.

[15] Haryanto, A.; Fernando, S.; Adhikari, S. Ultrahigh temperature water gas shift catalysts to increase hydrogen yield from biomass gasification, Catal. Today 2007, 129, pp. 269.

[16] Querino, P.T.; Bispo, J.R.C.; Rangel, M.d.C The effect of cerium on the properties of Pt/ZrO2catalysts in the WGSR, Cat. Today 2005, 107-108, pp. 920.

[17] Fonseca, A.A.; Fisher, J.M.; Ozkaya, D.; Shannon, M.D.; Thompsett, D. Ceria-zirconia supported Au as highly active low temperature Water-gas shift catalysts, Topics in Catal. 2007, 44, pp. 223.

[18] Häggblad, R., Hulteberg, C., Brandin J., Stabilization and Regeneration of CeO2 and CeO2/ZrO2 based Pt catalyst for the water gas shift reaction, Com2005 /First International Symposium on Fuel Cell and Hydrogen Technology, Volume Fuel Cell and Hydrogen Technologies, p 641, 2005

[19] Twigg, M.V. Catalysts Handbook; 2nd edition, Wolfe Press, London, 1996.

[20] Choung, S.Y.; Ferrandon, N ;.Krause, T. Pt-Re bimetallic supported CeO2-ZrO2 mixed oxides as water gas shift catalsts, Cat. Today 2005, 99, pp. 257-262.

[21] Fiorot S.; Galleti, C.; Specchia, S.; Saracco, G.; Specchia, V. Development of water gas shift supported catalysts for fuel processor units IntJChem.Re.Eng.2007 vol. 5 art. A113.

[22] Keizer, K.; Uhlhorn, J.R.; Burggraaf, T.J. Gas Separation using inorganic membranes. In Membrane Separations Technology. Principles and Applications; Noble, R.D; Stern, S.A.; Ed. Elsevier, Membrane Science and Technology Series 2, Elsevier B.V., 1995, pp.553-588.

[23] Caro, J.; Noak, M.; Kolsch, P. Zeolite membranes. From the laboratory scale to Technical Applications. Adsorption 2007, 11, pp. 215.

[24] Kajiwara, M.; Uemiya, S.; Kojima, T.; Kikuchi, E. Rhodium and Iridium dispersed porous alumina membranes and their hydrogen permeation properties. Cat. Today.2000, 56, issues 1-3, pp. 83-87.

[25] Ma, Y. H Palladium Membranes for Hydrogen Separation. In Membranes for Hydrogen Conversion, Peinemann, K.V; Pereira, S.; Ed. Wiley-VCH, Membrane Technology, 2008, pp. 255-324. 2008, Vol 2, pp. 245-261.

[26] Maroño, M.; Sánchez-Hervás, J.M.; Ruiz, E. Hydrogen-rich gas production from oxygen pressurized gasification of biomass using a Fe–Cr Water Gas Shift catalyst Journal of Hydrogen Energy 2010, Vol. 35, n 1, pp. 37-45.

[27] Maroño, M.; Sánchez-Hervás, J.M.; Ruiz, E.; Cabanillas, A. Study of the suitability of a Pt based catalyst for the upgrading of a biomass gasification syngas stream via the WGS reaction Catalysis Letters 2008, 126, pp 369-426.

[28] Sanchez J.M.; Maroño, M. and Barreiro M.M. Permeability and selectivity to H_2 of Pd membranes supported on porous stainless steel tubes, Euromembrane 2009, Montpellier, France 6 -10 September 2009

[29] Sanchez J.M.; Maroño, M.; Barreiro M.M. Separation of CO_2 and H_2 using palladium membranes (for integration in power generation with CO_2 capture) Euromembrane 2009, Montpellier, France 6 -10 September 2009
[30] Sanchez J.M.; Maroño, M.; Barreiro M.M. Preliminary Studies of Water Gas Shift catalysts and a Pd-based membrane for H_2 and CO_2 production & separation in a Catalytic Membrane Reactor, 9th International Conference on Catalysis in Membrane Reactors, ICCM9, 28 June- 2 July, 2009, Lyon, France.
[31] Calles, J.A.; Sanz, R.; Alique, D. Study of preparation of Pd-based membranes by electroless plating over PSS supports, Euromembrane 2009, Book of Abstracts, p. 216, Sept 2009
[32] Mardilovich, M.P.; She, Y.; Hua Ma, Y;Rei, M-H. Defect-free Palladium Membranes on Porous Stainless-Steel Support. AIChE Journal 1998, vol. 44, N° 2, pp. 310-322.
[33] Uemiya, S.; Kajiwara, M.; Kojima, T. Composite Membranes of Group VIII Metal Supported on Porous Alumina. Ceramics Processing 1997, vol. 43, 11A, pp. 2715-2723.
[34] Peters, T.A.; M. Stange, M.; Klette, H.; Bredesen, R. High pressure performance of thin Pd–23%Ag/stainless steel composite membranes in water gas shift gas mixtures; influence of dilution, mass transfer and surface effects on the hydrogen flux, J. Memb. Sci. 2008, 316, pp. 119–127.
[35] Ma, Y.H.; Ayrtuk, M.E.; Augustine, A.S.; Kazantzis, n. K. Palladium Membrane Reactor for Water Gas Shift Reaction. In 9th International Conference on Catalysis in Membrane Reactors, Lyon, 2009, Book of Abstracts pp. 41-42.
[36] Li, A.; Liang, W.; Hughes, R. The effect of carbon monoxide and steam on the hydrogen permeability of a Pd/stainless steel membrane. J. Memb. Sci. 2000, 165, pp. 135-141.
[37] Hou, K.; Hughes, R. The effect of external mass transfer, competitive adsorption and coking on hydrogen permeation through thin Pd/Ag membranes. J. Memb. Sci. 2002, 206, pp. 119-130.
[38] Gielens, F.C.; Knibbeler, R.J.J.; Duysinx, P.F.J.; Tong, H.D.; Vorstman, M.A.G.; Keurentjes, J.T.F. Influence of steam and carbon dioxide on the hydrogen flux through thin Pd/Ag and Pd membranes. J. Memb. Sci. 2006, 279, pp. 176-185.
[39] Guazzone F.; Engwall E.E.; Ma Y.H. Effects of surface activity, defects and mass transfer on hydrogen permeance and n-value in composite palladium-porous stainless steel membranes. Cat. Today 2006, 118, Issues 1-2, pp. 24-31.
[40] Galluci, F.; Chiaravalloti, F.; Tosti, S.; Drioli, E.; Basile, A. The effect of mixture gas on hydrogen permeation through a palladium membrane: Experimental Study and theoretical approach. Int. J. Hydrogen. Energy 2007, 32, pp. 1837-1845.
[41] Caravella A.; Barbieri G.; Drioli E. Concentration polarization analysis in self-supported Pd-based membranes.Sep. and Purif. Tech. 2009, 66, pp. 613–624.
[42] Catalano J.; Baschetti M.G.; Sarti G.C. Influence of the gas phase resistance on hydrogen flux through thin palladium–silver membranes. J. Memb. Sci. 2009, 339, pp. 57–67.
[43] Hydrogen from Coal Programme: Research, Development and Demonstration Plan for the period 2009 through 2016 External Draft, DOE Department of Energy sep 2009, p. 34.
[44] Lu, G.Q.; Diniz da Costa, J.C.; Duke, M.; Giessler, S.; Socolow, R. Inorganic Membranes for hydrogen production and purification: A critical review. J. Coll.&Interf. Sci. 2007, 314, pp. 589-603.

[45] Basile, A.; Galluci, F.; Tosti, S. Synthesis, Characterization and Applications of Palladium Membranes. In Inorganic Membranes Synthesis, Characterization and Applications; Mallada, R.; Menendez, M.; Ed. Elsevier, Membrane Science and Technology Series 13, Elsevier B.V., 2008, pp. 255-324.
[46] Brunetti, A.; Barbieri, G.; Drioli, E.; Lee, K.H.; Sea, B.; Lee, D.W. WGS Reaction in a membrane reactor using a porous stainless steel supported silica membrane. Chem. Eng. and Proc. 2007, 46, pp. 119-126.
[47] Criscuoli, A.; Basile, A.; Drioli, E. An analysis of the performance of membrane reactors for the water–gas shift reaction using gas feed mixtures. Cat. Today 2000, 56, pp. 53–64.
[48] Basile, A.; Chiappetta, G.; Tosti, S. and Violante, V. Experimental and simulation of both Pd and Pd/Ag for a water gas shift membrane reactor. Sep. & Purif. Tech. 2001, 25, pp. 549-571.
[49] Peters, T.A.; M. Stange, M.; Klette, H.; and Bredesen, R. High pressure performance of thin Pd–23%Ag/stainless steel composite membranes in water gas shift gas mixtures; influence of dilution, mass transfer and surface effects on the hydrogen flux. J. Memb. Sci. 2008, 316, pp. 119–127.
[50] Boutikos, P.; Nikolakis, V.A simulation study of the effect of operating and design parameters on the performance of a water gas shift membrane reactor. J. Memb. Sci. 2010, 350, pp. 378–386.
[51] Amelio, M.; Morrone, P.; Gallucci, F.; Basile, A. Integrated gasification gas combined cycle plant with membrane reactors: Technological and economical analysis. Ene. Conv. &Manag.2007, 48, pp. 2680–2693.
[52] Iyoha O.; Enick R.; Killmeyer R.; Howard, B.; Morreale, B.; Ciocco, M. Wall-catalyzed water gas shift reaction in multi-tubular Pd and 80wt%Pd-20wt%Cu membrane reactors at 1173 K. J. Membr. Sci. 2007, 298, pp. 14-23.
[53] Arstad, B.; Venvik, H.; Klette H.; Walmsley, J.C.; Tucho, W.M.; Holmestad, R.; Holmen, A.; Bredesen, R. Studies of self-supported 1,6 um Pd/23wt%Ag membranes during and after hydrogen production in a catalytic membrane reactor. Cat. Today 2006, 118, pp. 63-72.
[54] Uemiya, S.; Sato, N.; Ando, H.; Kikuchi, K. The water gas shift reaction assisted by a palladium membrane reactor.Ind. Eng. Chem. Res. 1991, 30 (3), pp. 585.
[55] Tosti, S.; Violante, V.; Basile, A.; Chiappetta, G.; Castelli, S. de Francesco, M.; Scaglione, S.; Sarto, F. Catalytic membrane reactors from tritiated water in the ITER furl cycle. Fusion Eng. Des. 2000, 49-50, pp. 953.
[56] Ayturk, M.E.; Kazantzis, N.K.; Ma, Y.H. Modelling and Performance Assessment of Pd- and Pd/Au-based Catalytic Membrane Reactors for Hydrogen Production. Energy & Environ. Sci. 2009, 2, pp. 430-438.
[57] Pinacci, P.; Broglia, M.; Valli, C. Evaluation of the water gas shift reaction in a palladium membrane reactor. In 9th International Conference on Catalysis in Membrane Reactors, Lyon, 2009, Book of Abstracts, p. 77-78
[58] Barbieri, G.; Brunetti, A.; Scura, F.; Drioli, E. CO inhibition effect on Pd-based membrane reactors performance for water gas shift reaction. In 9th International Conference on Catalysis in Membrane Reactors, Lyon, 2009, Book of Abstracts p. 63-64.

In: Gasification: Chemistry, Processes and Applications
Editor: Michael D. Baker
ISBN: 978-1-61209-681-0

Chapter 4

MEMBRANE AND MEMBRANE REACTOR TECHNOLOGIES IN THE TREATMENT OF SYNGAS STREAMS PRODUCED FROM GASIFICATION PROCESSES

Angelo Basile[1], Alessandro Blasi[2], Giuseppe Fiorenza[2], Adolfo Iulianelli[1], Tiziana Longo[1] and Vincenza Calabrò[3]

[1] Institute on Membrane Technology, ITM – CNR, Via P. Bucci, Cubo 17/C c/o University of Calabria, Rende (CS), Italy

[2]ENEA: Italian National agency for new technologies, Energy and sustainable economic development, Department of Energy Technologies, Efficiency and Renewable Sources, C.R. Trisaia S.S. Km, Rotondella (MT), Italy

[3] University of Calabria, Department of Engineering Modeling, University of Calabria, Via P. Bucci, Cubo, Rende (CS), Italy

ABSTRACT

Gasification of different types of fuels with steam and oxygen produces a synthesis gas (syngas) which consists mainly of CO, H_2, CO_2 and H_2O. Depending on the fuel and the operating conditions, the contents of CO and H_2 in the syngas usually range from 35 % to 55 % and from 20 % to 45 % respectively. If higher hydrogen contents are desired, the Water Gas Shift (WGS) reaction allows the conversion of CO to H_2 and CO_2 in presence of excess steam and a specific catalyst. This reaction is industrially carried out in two consecutive stages, first a high-temperature step for rapid CO conversion and then a low-temperature step for achievement of lower equilibrium CO contents.

A novel approach which is being investigated to increase the production of hydrogen in gasification processes is the use of a Water Gas Shift (WGS) membrane reactor in which a single high-temperature WGS catalyst is combined with a hydrogen selective membrane. In this system, the removal of hydrogen from the reaction chamber through the membrane drives the WGS reaction to the products side and thermodynamic equilibrium can be exceeded. Catalyst and membrane must be carefully chosen and special attention must be paid to operating conditions in order to guarantee the correct performance of the system. In this work the performance of semi-commercial palladium membranes and two different high temperature WGS catalysts are described and the advantages of using a WGS membrane reactor for the production of H_2 in gasification processes versus a conventional WGS reactor are presented.

Keywords: Gasification, inorganic membrane, polymeric membrane, hydrogen separation, carbon dioxide separation.

1. INTRODUCTION

Biomass is already widely exploited as a source of renewable energy, being mostly used for low efficiency applications for heating. A fundamental target for both scientific and industrial community in the field of bioenergy is to shift the traditional use of biomass to high-efficiency applications for the generation of combined heat and power (CHP) or the production of valuable by-products (biofuels, chemicals, etc.).

Gasification is a thermo-chemical process allowing the conversion of biomass (or any other carbonaceous solid or liquid fuel) into a gaseous fuel, which main fuel components are carbon monoxide, hydrogen, methane and other hydrocarbons, normally in lower percentages. This conversion occurs by reacting at a high temperature, usually ranging from 700 to 1200 °C, with an opportune gasification agent, such like air, steam, oxygen or a mixture of them.

Thermal energy required by the process is obtained by the partial oxidation of biomass itself (autothermal gasification) or supplied by an external source (indirect or allothermal gasification). In any case, the energy consumption related to this conversion process on average corresponds to the 20-30% of the biomass chemical energy, being therefore higher than the typical energy fraction which is not converted during the combustion process. On the other hand, the gaseous energy carrier offers several advantages with respect to the original solid biomass, such like:

- easy transportation and distribution;
- high combustion efficiency;
- reduction in the emissions from the conversion process;
- utilization in internal combustion engines for small scale power generation;
- high conversion into power efficiency for small to medium capacity applications;
- feed of advanced conversion devices such like micro gas turbines or fuel cells;
- utilization in synthesis or upgrading processes, for example aimed at the production of liquid or gaseous biofuels.

Although, technical viability of biomass gasification (BMG) and the related environmental benefits are widely acknowledged, the extent of its commercial utility has mostly been limited so far to heat production (generally for district heating), co-firing with coal, oil or natural gas for power or CHP production. In recent past, the development of advanced processes and broader application of BMG have been impeded by competition with low-cost fossil fuels and an inadequate market pull. Other negative impacting factors are the lack of an infrastructure for sustainable supply of quality controlled biomass and the need of adequate incentive driven policies and partnerships with industry so as to develop, demonstrate and deploy bioenergy technologies, in particular advanced high conversion efficiency processes. The last two to three years have witnessed unprecedented volatility in oil and gas prices and raised concerns about security of energy supply. Further, the need to attain several environmental targets is fast approaching, notably related to climate change and the reduction of primary energy usage. Against this background many of the national renewable energy plans are constantly reviewed and revised to implement measures encouraging the commercialization of biomass energy technologies. In this framework, BMG may play a central role in producing biofuels, substitute natural gas (SNG), value-added chemicals, heat, and power.

2. Biomass Gasification Principles

2.1 Processes during gasification

As schematized in Figure 1, BMG takes place via three subsequent processes according to the increase in the temperature due to the heat input:

1) drying in the range of 100-150 °C;
2) devolatilization from 250 to 550 °C;
3) gasification over 700 °C if a proper gasification agent is present (maximum practical value is usually up to 1200 °C).

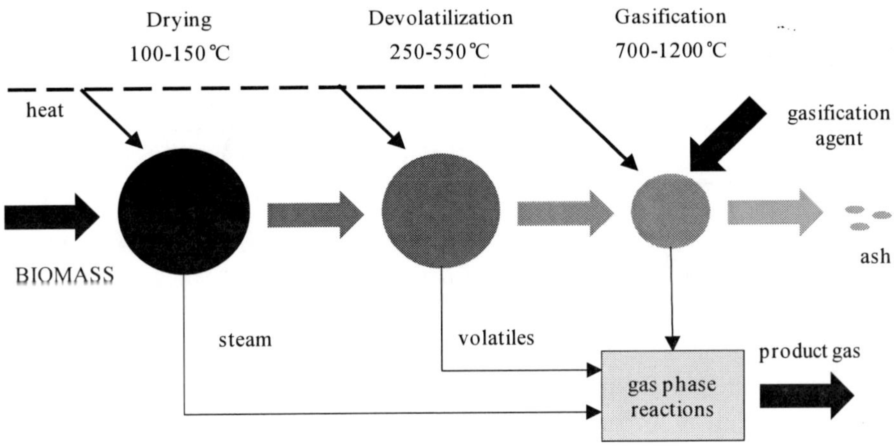

Figure 1. Diagram of processes during biomass gasification.

The steam, volatiles and gases released during drying, devolatilization and carbon gasification, respectively, participate in the gas phase reactions originating the final product gas. The fraction of the solid fuel which is not gasified is discharged as the ash. In case of not previously processed biomass, ash is usually in the order of 2-3 % on dry basis [Reed, 1981].

2.2 Gasification reactions

The chemical reactions during gasification concerns initially the carbon gasification due to the presence of the gasification agent; subsequently, a number of gas phase reactions occurs. The main reactions together with the associated enthalpy variations are listed in Table 1.

As it can be observed, endothermic reactions are predominant: this explains why an adequate heat input is necessary for the correct development of the process. Moreover, it is to be noted that chemical equilibrium is never reached in industrial processes where the target is to achieve the best cost-effectiveness. Generally, an incomplete conversion (80-90%) of the carbon fraction in the biomass is accepted. In effect, either a greater heat input or a longer residence time of the biomass in the reactor are required in order to increase the gasified carbon fraction [Reed, 1981].

Table 1. Main reactions during gasification and associated enthalpy variations

Carbon gasification reactions		
$C + \frac{1}{2} O_2 \rightarrow CO$	(-110.52 kJ/mole)	Partial oxidation
$C + CO_2 \rightarrow 2CO$	(+172.6 kJ/mole)	Boudouard reaction
$C + H_2O \rightarrow CO + H_2$	(+131.4 kJ/mole)	Carbon reforming
$C + 2H_2 \rightarrow CH_4$	(-74.9 kJ/mole)	Methanation
Gas phase reactions		
$CH_4 + H_2O \rightarrow CO + 3H_2$	(+201.9 kJ/mole)	Methane reforming
$CO + H_2O \rightarrow CO_2 + H_2$	(-41.2 kJ/mole)	Water Gas Shift

3. Gasification Process Layout

In Figure 2 the general scheme of BMG process is summarized. At high temperature biomass reacts with the gasification agent as illustrated in the previous sections producing a raw fuel gas. When a clean biomass is used, main substances in the product gas are particles, tars, acid compounds (chlorine and sulfur), nitrogen components. These substances may be detrimental for the gas utilization device, therefore a preliminary gas cleaning is required in order that the product gas is adequate for its final application.

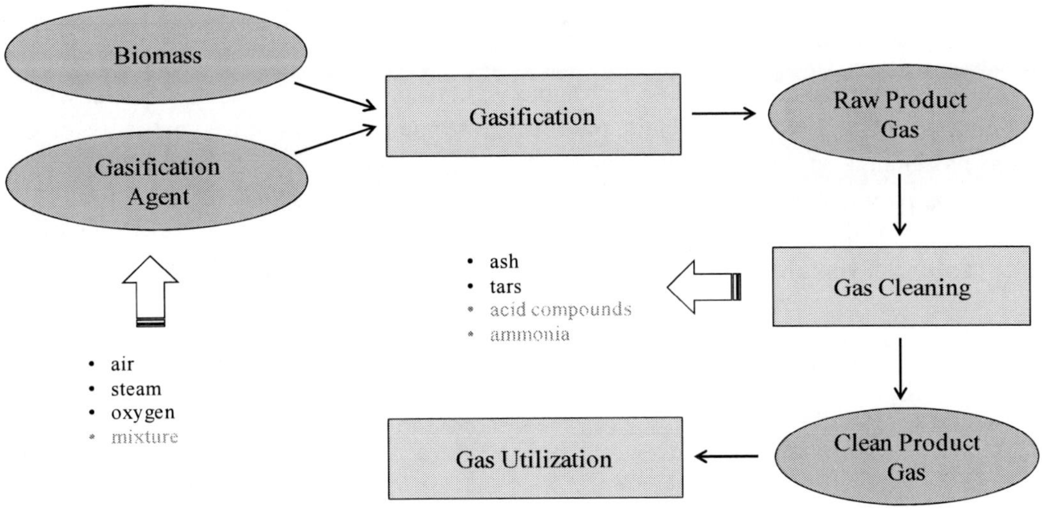

Figure 2. Biomass gasification processes general layout.

4. Gasification Process Variables

Gasification efficiency and product gas composition are influenced by a large number of variables, which make extremely difficult the theoretical analysis of the process and its operational control. The main parameters to be considered, when investigating BMG, are:

a) biomass composition and physical properties (size, moisture content);
b) gasification agent (type, input temperature and velocity of the stream);
c) temperature (more correctly temperature profile) in the gasifier;
d) operational pressure;
e) equivalence ratio (for air or oxygen gasification);
f) steam to biomass ratio (for steam gasification);
g) gasification ratio (for gasification with air or oxygen mixed with steam);
h) residence time of the biomass into the gasifier;
i) possible use of catalysts and the related composition and size;
j) type and characteristics of the gasification reactor (i.e. gasification technology).

4.1 Gasification agent

The gasification agent is one of the most relevant variable in the biomass gasification process, strongly affecting product gas quality. Furthermore, the use of special technologies or the modification of conventional ones is required in order to use gasifying agents different from air. In effect, the air-blown gasifier is the most simple technology, but the product gas is diluted by a high percentage of nitrogen, thus proving to be a low calorific value fuel gas. This can be increased by the use of oxygen: the absence of nitrogen allows the production of an average calorific value fuel gas, but the additional costs related to the production and supplying of oxygen must be included. Obviously an intermediate solution can be adopted, by using an oxygen-enriched air.

Steam can be used as well in order to obtain an average calorific value fuel gas, which is in addition richer in hydrogen than the product gas from an oxygen-blown gasifier. Moreover, steam production is less onerous than oxygen production. On the other hand steam gasification requires more complex technological solutions (indirect gasification).

Table 2.Typical product gas composition according to the gasification agent

	air	oxygen	steam
CO	20	26	22
CO_2	16	30	22
H_2	15	34	42
CH_4 (C_2H_6/C_3H_8)	2	9	12
N_2	47	1	2
LHV (MJ/Nm^3)	5	11	13

In Table 2, the typical composition of product gas and its low heating value (LHV) is reported according to the use of the different gasification agents.

4.2 Gasification temperature

Gasification temperature is the most relevant process variable, since it strongly affects the kinetics of chemical reactions and, as a result, the conversion rates and yields.

In general, product gas quality improves as the gasification temperature increases according to the following main issues:

- fraction of both H_2 and CO grows to the detriment of CH_4 and other hydrocarbons;
- specific energy (enthalpy) of the product gas increases;
- tar concentration is reduced.

In Figure 3 the effect of process temperature (T_g) on the product gas composition as well as enthalpy (h_{gas}) is shown [Mathieu et al., 2002].

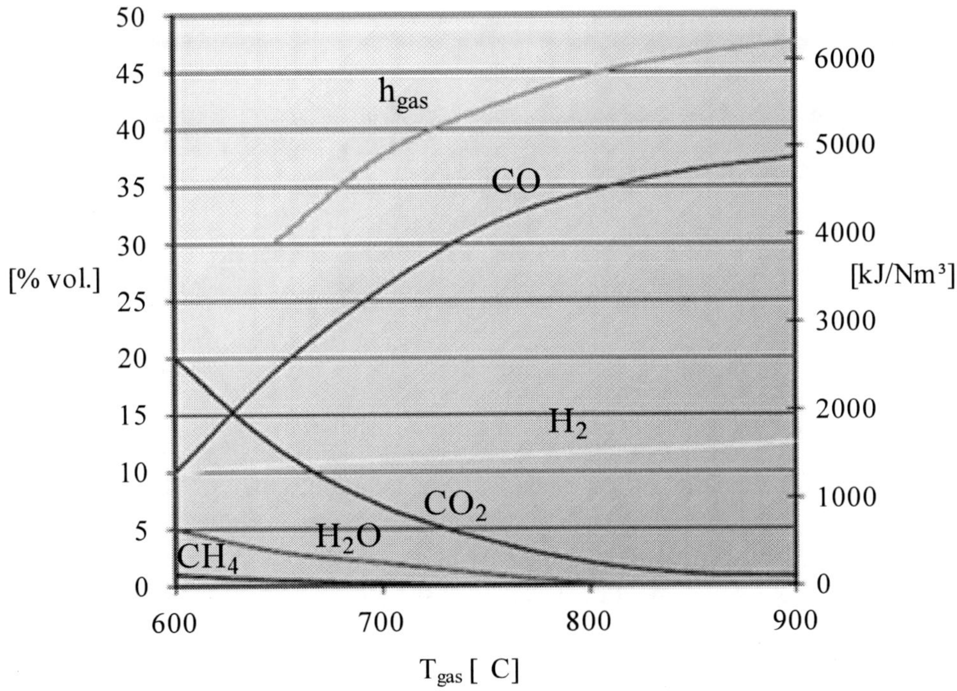

Figure 3. Effect of gasification temperature on product gas quality.

On the other hand, a higher gasification temperature implies:

- challenging plant management and need of using more valuable materials;
- rise in the process energy requirements together with a decrease in the global conversion efficiency;
- high sensible heat of the product gas recoverable on the condition of a further increase in plant complexity (insertion of heat exchangers).

4.3 Gasifier operational pressure

Pressurized biomass gasification is attractive in case of applications where a high pressure product gas is required, such like gas turbines, pressurized fuel cells, synthesis processes.

When the operational pressure grows, the specific capacity of the gasifier increases also and, as it can be observed in Figure 4, process efficiency slightly improves up to a certain pressure value (12-13 bar in the case of Figure 4), since the effect associated to the increase in the CH_4 fraction prevails on the consequences of the reduction in both CO and H_2 percentages. After that, if pressure continues to be increased, efficiency begins to lessen [Mathieu et al., 2002].

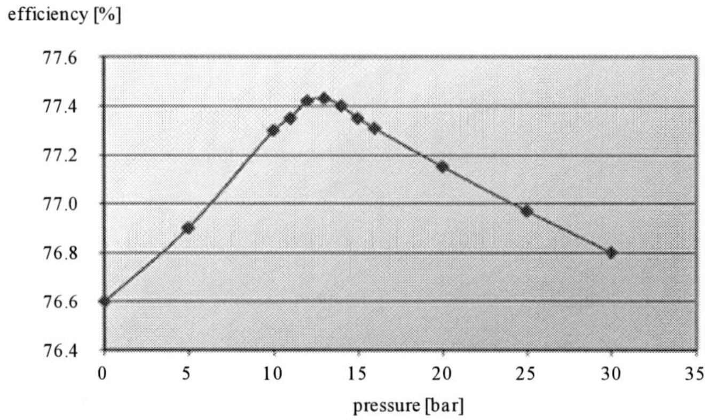

Figure 4. Effect of gasifier operational pressure on process efficiency.

On the other hand, gasifier operation under pressure implies:

- higher investment and running costs;
- increase in the amount of entrained dust;
- more critical conditions for system and workers safety.

For these main reasons, a very limited number of pressurized biomass gasifiers is currently in operation worldwide.

4.4 Equivalence ratio

Equivalence ratio (ER) is a key process variable in air (or oxygen) gasification. ER is defined as the ratio of the actual oxygen flow rate to the stoichiometric oxygen flow rate, which is required for the complete combustion of the biomass. For a given gasification temperature, the value of ER must be high enough in order to allow an autothermal process, where no additional heat provided by an external source is needed after the initiation stage.

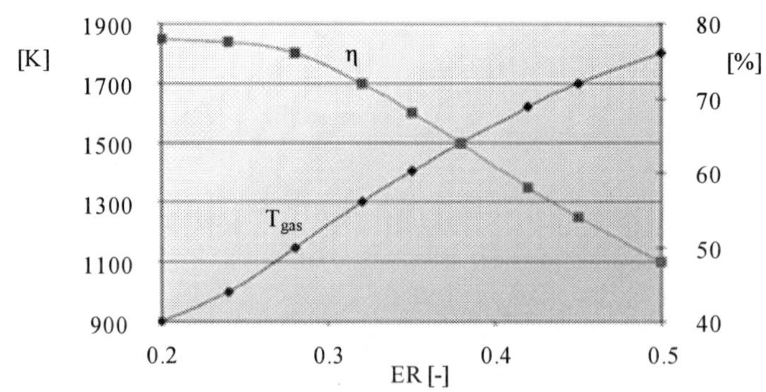

Figure 5. Effect of ER on process efficiency and temperature.

On the other hand, as it is shown in Figure 5 [Mathieu et al., 2002], an increase in the equivalence ratio implies:

- reduction in the process efficiency;
- growth of both N_2 and CO_2 fractions with a consequent diminution of product gas LHV.

4.5 Steam to biomass ratio

Steam to biomass ratio (SBR) is a key process variable in steam gasification. SBR is defined as the ratio of the water flow rate to the dry biomass feed rate. The water flow rate includes both the biomass moisture content and the additional steam supplied to the reactor as the gasification agent.

Due to the partial conversion of water into product gas fuel components, when SBR increases both quantity and quality of the product gas improve. In particular, hydrogen fraction grows, while tar concentration is reduced.

On the other hand, an increase in the steam to biomass ratio implies:

- rise of the heat requirements of the process with a resulting reduction in its efficiency;
- diminution of the product gas LHV;
- decrease in the specific water conversion to fuel gaseous components, as it is shown in Figure 6[Hofbauer et al., 2000].

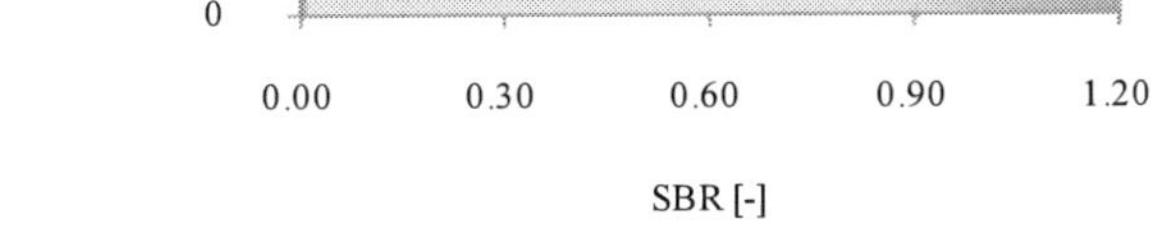

Figure 6. Effect of SBR on water conversion ratio (WC).

5. Gasification Technologies

The most common classification of gasification technologies is based upon the type of reactor, as it is summarized in Figure 7 [Hofbauer et al., 2007]. In the next section a short description of the most common technologies is provided.

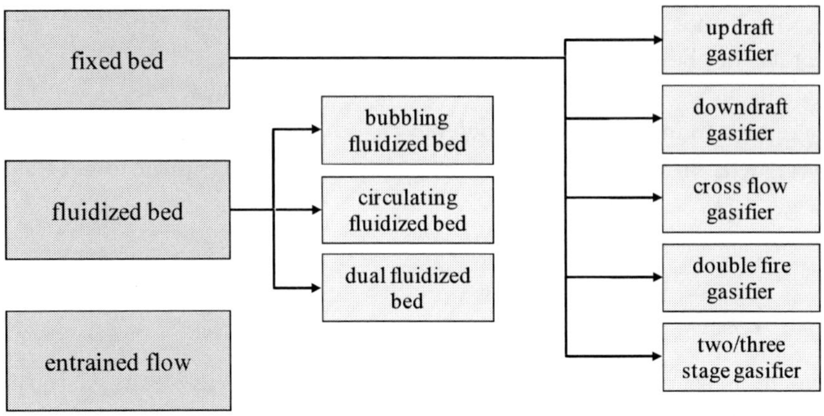

Figure 7. Classification of gasification technologies.

5.1 Updraft gasifier

In the fixed bed updraft gasifier (Figure 8) product gas leaves the reactor from the top, thus moving in opposite direction with respect to the feeding biomass. It is a simple and established technology with very low cost for the equipments. Gasification reactor is highly flexible to different feedstocks and can be rather easily operated under pressure. The scaling up to a fuel capacity that can exceed 10 MW thermal is possible. A fuel of nearly uniform size but in a fairly wide range (5 to 100 mm) is required. It also offers a good tolerance to ash (up to 25%) and moisture (up to 60%) content in the feedstock. Product gas is characterized by a high concentration of tar (around 100 g/Nm³) and a limited presence of dust, while its outlet temperature is rather low (100-400 °C). This technology appears to be particularly suitable for thermal uses.

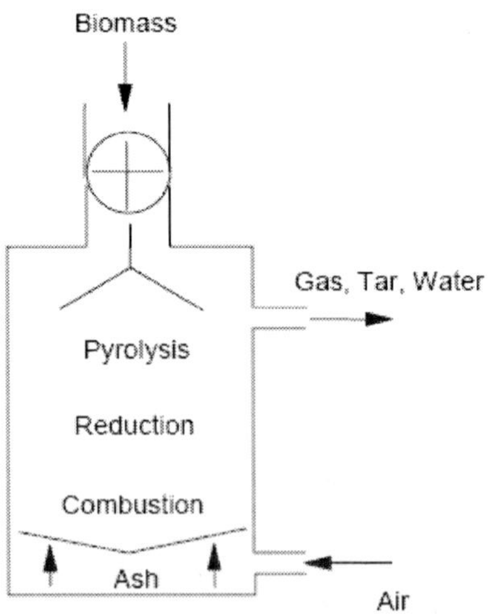

Figure 8. Updraft gasifier scheme [Hofbauer et al., 2007].

5.2 Downdraft gasifier

In the fixed bed downdraft gasifier (Figure 9) product gas leaves the reactor from the bottom, even if it usually flows again upstream in a cavity wall around the reactor in order to use its sensible heat to dry the feeding biomass. Hence, in this system biomass and product gas move in the same direction. This technology is more complex and, as a result, more expensive with respect to an updraft gasifier. Furthermore, it is not yet fully mature for commercial applications. The reactor is very sensitive to load fluctuations and requires a fuel of uniform size, but in a rather wide range (from 20 to 100 mm). It is also less tolerant to ash (up to around 5%) and moisture (up to 25%) content in comparison with updraft technology. In return product gas has a very low tar concentration (even less than 1 g/Nm^3) although a significant particle content, which is, however, much simpler to remove. Outlet temperature can be very high (up to 1000 °C), if sensible heat is not recuperated inside the reactor to dry the incoming biomass, otherwise it becomes similar to the updraft gasifier (around 400 °C). Owing to the improved product gas quality, this technology is considered to be suitable for the production of electricity. On the other hand, it is characterized by a limited scalability, with a maximum achievable capacity that usually does not exceed 1 MW electrical.

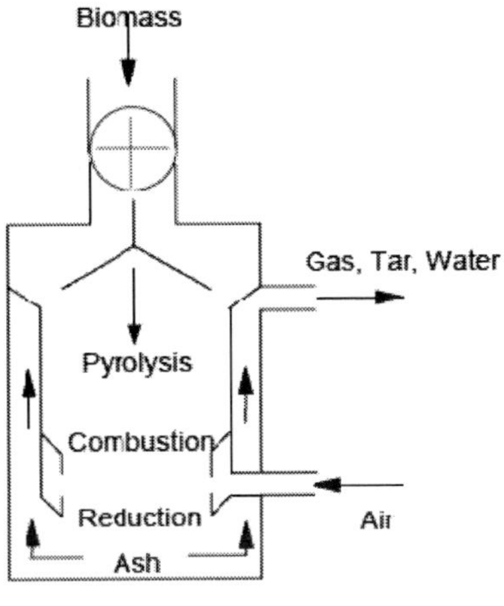

Figure 9. Downdraft gasifier scheme [Hofbauer et al., 2007].

5.3 Two/three stage gasifier

In a two (or three) stage gasifier (also called pyro-gasifier) devolatilization (i.e. pyrolysis) and gasification occur in two different reactors (Figure 10). This technology is more complex in comparison with either updraft or downdraft gasifier and, consequently, is more costly. Furthermore, it still needs a strong development. On the other hand, the separation of pyrolysis and gasification steps leads to several advantages. In effect, very high chemical efficiencies (over 90%) can be achieved in the gasification process with a product gas substantially free of tar (less than 10 mg/Nm^3) and even with a low dust concentration. In addition, the reactor offers a good flexibility to the type and the conditions of the feeding biomass; in particular, it has a good tolerance with respect to moisture content (up to 45%).

The high quality of the product gas makes this technology to have a great potential for CHP production applications.

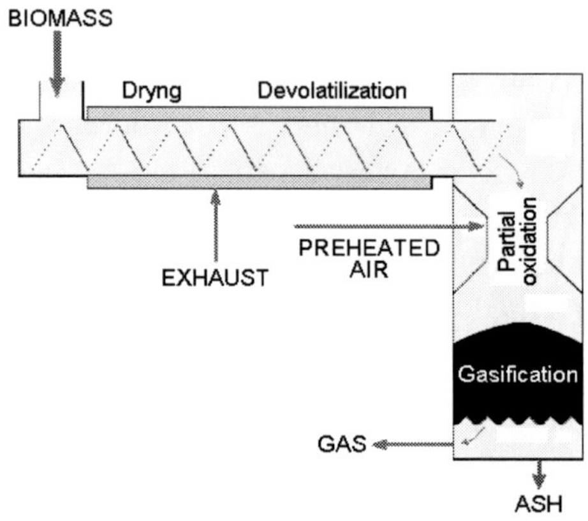

Figure 10. Two-stage gasifier scheme [Gøbel et al., 2004].

5.4 Bubbling fluidized bed gasifier

In a fluidized bed gasifier the intimate contact between the solid fuel and the bed particles, which are fluidized at high temperature, facilitates the heat transfer, thus accelerating the reactions with a resulting increase in the reactor specific capacity. In particular, in a bubbling fluidized bed gasifier (Figure 11) the fluidization velocity (1-2 m/s) is slightly higher than the minimum required value with the aim of keeping the bed in suspension. The more complex system design and management in comparison to fixed bed gasifiers limit the application of this technology to medium to large capacities (2 to 50 MW thermal) in order to achieve a real cost-effectiveness.

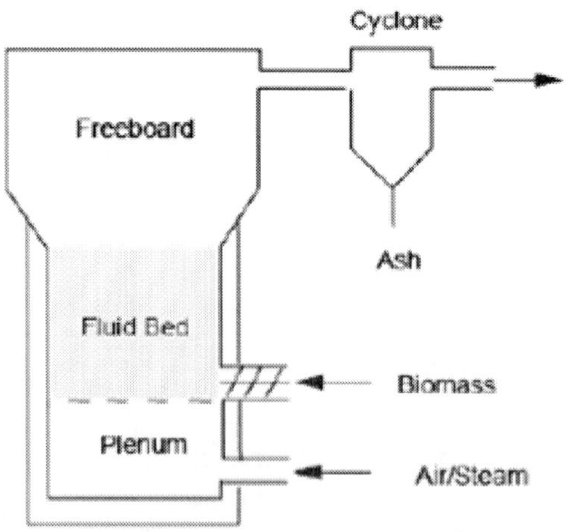

Figure 11. Bubbling fluidized bed gasifier scheme [Hofbauer et al., 2007].

These systems are currently in operation for co-firing and, even though a lesser degree, for CHP production. The material of the bed is typically silica sand, but it is also possible to add a catalyst in order to reduce tar formation or, in general, to improve product gas composition, according to its final use. This technology offers a high tolerance to the type and size of the solid fuel (from 1 to 100 mm) as well as its ash and moisture content. On the other hand, either the use of too small size biomass or the presence of low melting ash may be challenging. As far as product gas quality is concerned, a medium concentration of both tar (in the range from 5 to 20 g/Nm³) and dust is usually found.

5.5 Circulating fluidized bed gasifier

In the circulating fluidized bed gasifier (Figure 12) higher values for the fluidization velocity (4-6 m/s) are adopted in comparison with a bubbling fluidized bed in order to establish the recirculation of the bed particles through an apposite cyclone. The increased turbulence enhances the heat and mass transfer determining a higher specific capacity even with respect to a bubbling fluidized bed gasifier. Therefore this technology appears especially suitable for large capacity installations (from 10 to 200 MW thermal) and it is currently used especially in large scale co-firing systems for power production. Furthermore, the circulation of biomass and, after the devolatilization step, of char increases its residence time into the reactor leading to a higher conversion rate. Due to its operational principle, circulating fluidized bed gasifier requires a smaller sized solid fuel (less than 50 mm) with respect to a bubbling fluidized bed. On the other hand, it offers a higher operational flexibility and ease of control.

5.6 Dual fluidized bed gasifier

In a dual fluidized bed gasifier (Figure 13) the reactor is divided into two different chambers intended for gasification and combustion, respectively. The heat required by the gasification process is generated inside the combustor by the combustion of both the residual char from the gasification chamber and an additional biomass amount. The transfer of both char to the combustion zone and heat to the gasification zone is obtained via the circulation of the hot bed material, which also acts as a barrier between the gaseous mixtures present in the two chambers. Therefore this technology allows to perform an indirect gasification with the use of steam as both gasification agent and fluidization medium inside the gasifier. The product gas is virtually free of nitrogen and has a high fraction of hydrogen (up to over 40% vol.), while tar concentration is lower (around 10 g/Nm³) with respect to average values relevant to fluidized bed gasifiers. Despite the losses associated to the indirect gasification mechanism, chemical efficiency is rather high thanks to the full exploitation of the no-gasified carbon, which occurs inside the combustion chamber. Dual fluidized bed gasifiers are already available on the market for CHP production on the medium scale (from 2 to 5 MW electrical). Furthermore, due to the special product gas quality, this technology offers interesting perspectives for advanced applications, such like the production of bio-fuels (Fischer-Tropsch diesel, SNG) or the feeding of fuel cells (both solid oxide and molten carbonate). Obviously the technology is more complex to design and manage than a conventional circulating fluidized bed gasifier due to the presence of two gas streams in output (product and flue gas), which both need to be cleaned, even if a simpler cleaning is necessary for flue gas. Furthermore the optimization of the bed material circulation rate is critical.

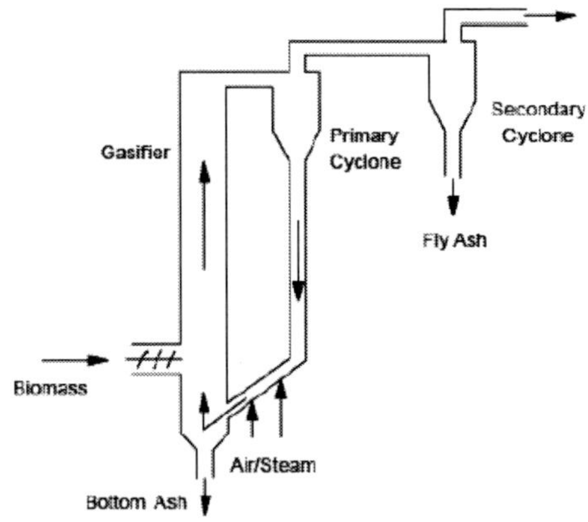

Figure 12. Circulating fluidized bed gasifier scheme [Hofbauer et al., 2007].

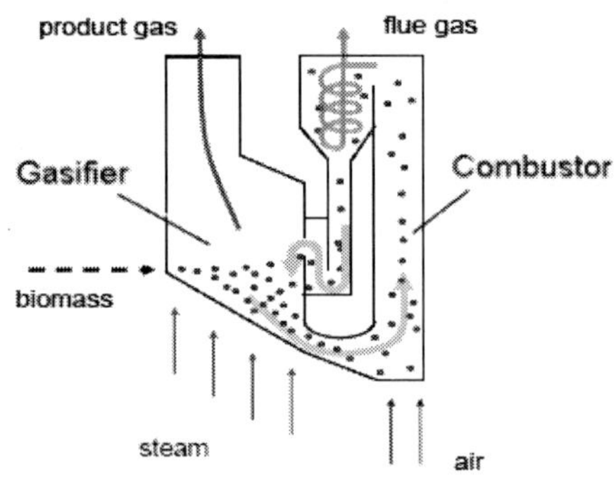

Figure 13. Dual fluidized bed gasifier scheme [Hofbauer et al., 2007].

5.7 Entrained flow gasifier

In the entrained flow gasifier (Figure 14), a high process temperature is usually achieved (up to over 1200 °C) due to the use of oxygen (mostly mixed with steam) as the gasification agent. The system requires a very small sized solid fuel (less than 1 mm). Therefore biomass needs to be pulverized before its feeding into the reactor, unless a previous devolatilization step is provided. The system is normally pressurized, being this operational mode easier to accomplish for this technology. Due to the high gasification temperature and the absence of nitrogen, an improved quality gas is obtained with a very low concentration of tar and a reduced fraction of methane and other light hydrocarbons. Moreover a high conversion efficiency around 80% can be reached. This technology is proven and reliable, but its application is practical only for large capacity installations, because of the high investment and running costs. It is currently used on the commercial scale for the gasification of coal and, at an experimental degree, for biomass and wastes, mostly aimed at the production of liquid biofuels.

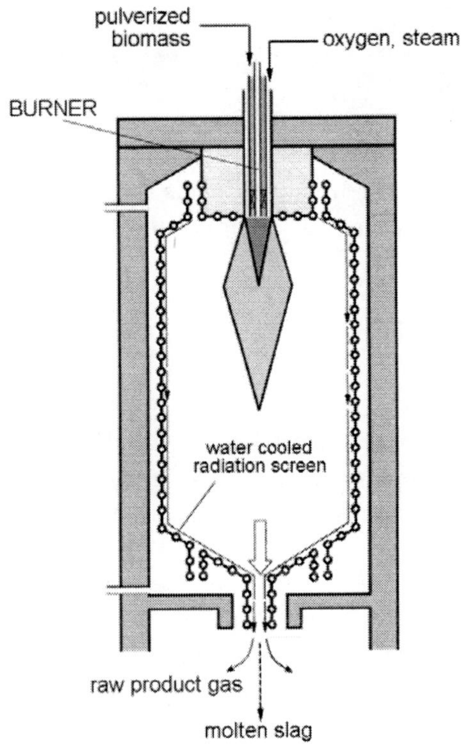

Figure 14. Pressurized entrained flow gasifier scheme [Henrich et al., 2004].

6. Product Gas Cleaning

Gas cleaning is the key step in DMG in order to improve the product gas quality so that it becomes suitable for being exploited for power generation, especially if advanced conversion devices, such like gas turbines or fuel cells, are used. Actually these systems are extremely sensitive to pollutants concentration. On the other hand gas cleaning becomes even more critical if product gas is used as the feed stream for synthesis processes aimed at the production of liquid or gaseous biofuels (FT diesel, methanol, SNG).

Amongst the several contaminants which can be present in the raw product gas, tars represent the most challenging substance to be removed. The term tar designates a set of organic compounds such as toluene, phenol, pyridine, polycyclic aromatic hydrocarbons (naphthalene, anthracene, etc.) that condense from relatively high temperatures (over 300 °C) [Devi et al., 2003]. As a result, tar condensation can create serious problems to filters, other cleaning equipments, and gas utilization devices. Therefore, tar removal is a fundamental step, since many components in a gasification plant as well as the conversion machines present a very low tolerance to these substances, especially as far as heavy tars are concerned.

On the other hand, tars have a high energy content, which is important to recover in order to avoid depressing the global efficiency of the gasification process.

Several approaches for tar reduction have been developed and tested so far. Ongoing research mostly deals with the development and demonstration of methods which allow an effective tar removal with the best trade off between system cost and process efficiency. Depending on the position where the tars are removed, all the available methods can be classified into two types:

a) primary methods: tars are removed inside the gasifier (in-situ or in bed removal);
b) secondary methods: tars are removed downstream of the gasifier(in series removal).

Secondary methods can involve a chemical or physical raw product gas treatment, including:

- tar cracking downstream the gasifier either thermally or catalytically,
- mechanical separation systems such as cyclones, baffle filters, ceramic filters fabric filters, rotating particle separators, electrostatic filters and scrubbers.

Thermal tar cracking is a proven method for the removal of tars in large-scale plants aimed at the thermal treatment of waste, where the product gas reaches temperatures of 1400-1600 °C with a residence time in the order of seconds. Under these conditions, the synthesis gas from the cracker will only contain low amounts of methane, while all other thermodynamically unstable hydrocarbons are cracked. On the other hand, catalytic tar cleaning is attractive as no additional energy input may be required. The advantages and disadvantages of different tar cracking catalysts are summarized in Table 3 [Abu El-Rub et al., 2002].

Although downstream gas cleaning methods are established and very effective in tar removal, as well as the reduction in the concentration of other substances, such like ammonia, in some cases they are not economically viable [Giordano, 1998]. On the other hand, primary methods show a growing interest, since they can allow the direct production of a clean gas optimizing the efficiency of the conversion process and reducing the requirements of gas treatment downstream.

Several approaches can be followed in order to achieve in bed tar removal, such like:

- opportune selection of the gasification technology, such as a two/three stage gasifier;
- optimal configuration of the reactor, aimed, for example, at the maximization of the gas residence time inside it;
- optimization of process variables, for example by using a gasification temperature over 900 °C or oxygen or steam as the gasifying agent;
- utilization of catalysts, such like nickel which, in case of fluidized bed gasification, can be directly added to bed material.

Table 3. Summary of tar cracking catalysts advantages and disadvantages [Abu El-Rub et al., 2002]

Catalyst	Advantages	Disadvantages
Calcined rocks	• inexpensive and abundant • attain high tar conversion around 95% with dolomite • often used as guard beds for expensive catalysts • most popular for tar reduction	• fragile materials and quickly eroded from fluidized beds
Olivine	• inexpensive • high attrition resistance	• lower catalytic activity than dolomite
Clay minerals	• inexpensive and abundant • less disposal problems	• lower catalytic activity than dolomite • most natural clays do not support the high temperatures (over 800°C) needed for tar reduction (lose pore structure)
Iron ores	• inexpensive and abundant	• rapidly deactivated in absence of hydrogen • lower catalytic activity than dolomite
Char	• inexpensive • natural production inside the gasifier • high tar conversion comparable to dolomite	• consumption because of gasification reactions • biomass char properties not fixed and depends on biomass type and process conditions
Fluid Catalytic Cracking	• relatively cheap but not cheaper than the above • more knowledge is known about it from the experience with FCC unit	• quick deactivation by coke • lower catalytic activity than dolomite
Alkali metals	• natural production in the gasifier • reduce ash handling problem when used as a catalyst	• particle agglomeration at high temperatures • lower catalytic activity than dolomite
Activated alumina	• high tar conversion comparable to dolomite	• quick deactivation by coke
Transition metals	• able to attain complete tar reduction at around 900°C • increase the yield of CO and H_2 • Ni-based 8 to 10 times more active than dolomite	• rapid deactivation because of sulfur and high tar content in the feed. • relatively expensive

7. Syngas Applications

Syngas is an important industrial production with many potential uses. For instance, syngas can be processed for generating manufacture chemicals, fertilizers, liquid fuels, substituting natural gas (SNG) or hydrogen, as reported in Table 4, where the use of each component, which can be isolated and/or purified from syngas, is reported.

Table 4. Uses of different compounds of syngas

Compounds	Uses
Hydrogen	Electricity generation and transportation fuels
Nitrogen	Fertilizers, pressurizing agents
Ammonia	Fertilizers
Carbon monoxide	Chemical industry feedstock and fuels
Carbon dioxide	Injected into sequestration wells
Steam	Turbine drivers for electricity generation
Minerals and solids	Slag for roadbeds
Sulfur	Elemental sulfur for chemical industry

However, the most known application of syngas is the *Fischer-Tropschsynthesis*, which converts syngas into a liquid synthetic fuel by means of a chemical reactions set. Indeed, thisprocess, produces a petroleum substitute to be used as a synthetic lubrication oil and synthetic fuel.

The syngas can be also used as a fuel in IGCC (Integrated gasification combined cycle) power generation configuration in order to produce electricity. In fact, IGCC systems combine the high efficiency and low emissions of gas turbines with the ability to run on syngas. In IGCC plants, coal, other low-cost solid or heavy liquid fuels are not burned directly, but they are processed with oxygen and water in a high pressure gasifier to form syngas. The clean syngas is, then, combusted in a gas turbine generator, which generates most of the electricity to be produced in a such plant. The waste heat is recovered and used to produce steam that drives a smaller steam turbine generator in a combined cycle configuration to produce additional electricity.

Moreover, syngas can be used to produce dimethylether (DME). It is primarily produced by converting hydrocarbons, sourced from gasification of natural gas or coal. In fact, syngas can be converted into methanol in the presence of a catalyst (usually Cu-based), with subsequent methanol dehydration in the presence of another catalyst (e.g. silica-alumina) and production of DME.

Currently, a growing application of syngas concerns the hydrogen production due to, mainly, the current context of environmental problems and development of "hydrogen economy". In fact, the stream coming out from a gasification plant is of great industrial interest because it contains around 50% of hydrogen (in dry condition),whichcan be recovered by means several processes such as a hydrogen perm-selective membrane reactor (MR), and between 40–45% of CO that, usually, can be reduced in an upgrading stage, producing a stream with higher hydrogen content [Brunetti et al., 2009].

In traditional applications, the upgrading of reformate streams is performed by using a multi-stage WGS process and the hydrogen-rich stream coming out from the last reactor is flowed to an unit for hydrogen separation, as shown in Figure 15.

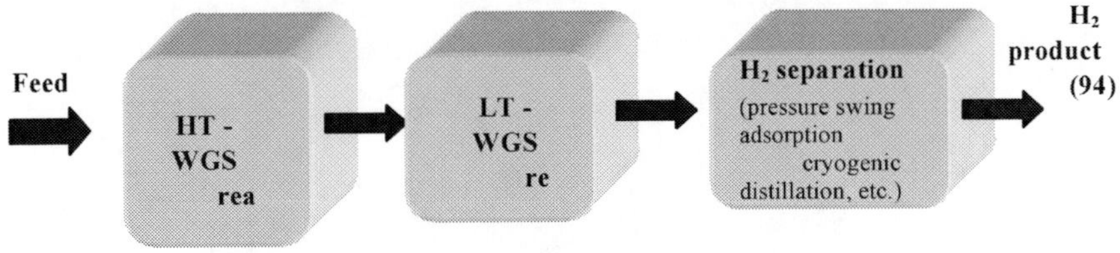

Figure 15. Scheme of a traditional process for hydrogen production via WGS reaction and next purification.

Nevertheless, this process (WGS and hydrogen separation) can be also performed in a single unit asa Pd-alloy membrane reactor (Figure 16), which allows a pure hydrogen stream to be produced owing to the full perm-selectivity of palladium-membranes to hydrogen permeation with respect to all other gases. In the meanwhile, this solution could reduce the capital and operating costs by using a only unit for both reaction and hydrogen separation processes.

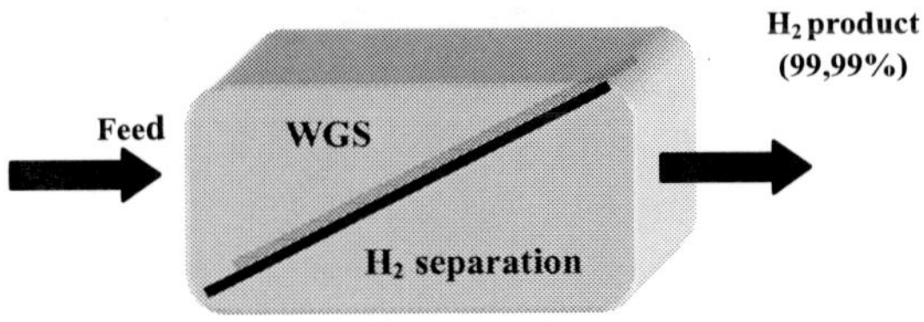

Figure 16. Scheme of pure hydrogen production by WGS reaction in Pd-alloy membrane reactor.

A more detailed description of benefits of membrane reactor for carrying out WGS reaction with respect to the conventional systems is reported in the next paragraphs.

8. HYDROGEN PRODUCTION FROM SYNGAS

Owing to the decrease of oil reserves and the increase of the environmental pollution caused by the emissions of CO_2 and other greenhouse gases, it is strongly necessary to develop new technologies and to exploit alternative renewable sources with respect to the fossil fuels. In this route, the combination of hydrogen and electricity represents an important solution to realise sustainable energy. Fuel cells represent the most efficient devices for converting hydrogen into electricity. In particular, proton exchange membrane fuel cells (PEMFCs) (Figure 17) may produce electricity directly from hydrogen and oxygen, without combustion, making the process non-polluting.

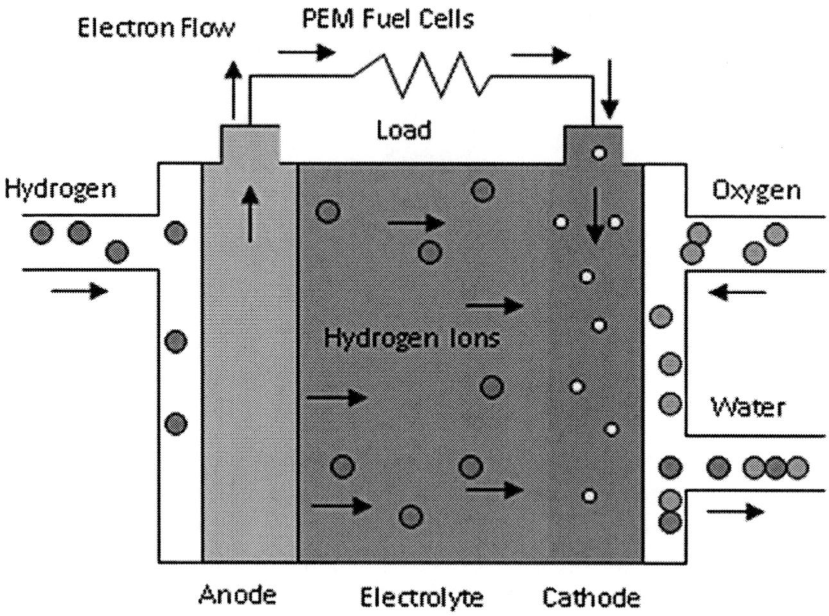

Figure 17. Diagram of a PEM fuel cell.

PEMFCs are characterized by the following benefits:

- low operative temperatures (80 – 100 °C),
- high current density,
- compactness,
- fast start-up,
- suitability for discontinuous operation.

These features make PEMFCs the most promising and attractive vehicles for a wide variety of power applications ranging from portable/micropower and transport to large-scale stationary power systems for buildings and distributed generation (Figure 18) [Costamagna et al., 2001].

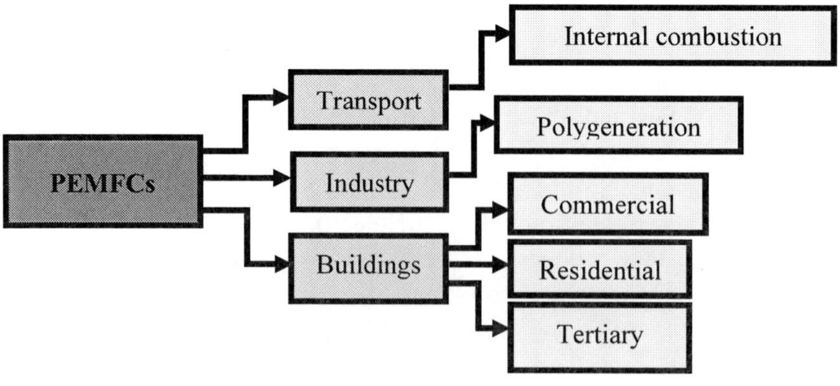

Figure 18. Summary of the PEMFC applications.

As well known, the syngas is characterized by high hydrogen content, around 50%,and when it is processed in a water gas shift (WGS) reactor, a further increase of its hydrogen content can be reached.

As above mentioned, at industrial level, as the first step the production of syngas takes place in a gasifier. Subsequently, the syngas is cooled down to remove its contaminants, especially H_2S and, then, it is reheated to be further reacted with steam in a WGS reactor to maximize its hydrogen content.

The WGS reaction is exothermic and its equilibrium conversion decreases with temperature. Therefore, typically two reactors, one operating at high temperature (HTS) and another at lower temperature(LTS), are used in order to increase CO conversion. The stream coming out from the WGS reactor, although it contains a CO concentration smaller than 0.5%, must be treated in further separation units to produce hydrogen with a CO content tolerable by PEMFCs(typically only a few ppm of CO for avoiding the poisoning of their anodic catalyst).

As an alternative, it would be more convenient to convert CO into hydrogen via WGS reaction using the inorganic membrane reactor technology. Indeed, palladium-based MRs are devices able to perform simultaneously the WGS reaction and hydrogen separation in the same unit. The choice of an inorganic membrane to be housed in a MR mainly depends on the degree of required hydrogen purity.

In the last decades, the inorganic membranes have attracted the scientific interest owing to their high hydrogen perm-selectivity. For instance, dense self-supported palladium membranes are full selective to hydrogen permeation through the membrane with respect to all the other gases.Therefore, they are, usually, assembled in a membrane reactor with the intent to recover a pure hydrogen stream. Moreover, palladium-based MRs allow to concentrate in $CO_2$the retentate stream (the stream not permeated through the membrane), which could be further separated via polymeric membrane technology. After a short introduction about membranes and membrane reactors, in the following a detailed description of the inorganic membranes used for hydrogen separation, pointing out their benefits and drawbacks, is proposed as well as a short description of polymeric membranes used for CO_2 capture.

9. Membrane and Membrane Reactor Definitions

As defined by IUPAC [Koros et al., 1996], a membrane is a structure having lateral dimensions much greater than its thickness through which mass transfer may occur under a variety of driving forces such as gradient of concentration, pressure, temperature, electric potential, etc. A schematic representation of a two-phase system separated by membrane is given in Figure 19, where the Phase 1 is usually considered as the feed, while the Phase 2 as the permeate.

In separating one or more gases from a feed mixture and in generating a permeate stream, the membrane acts as a "filter" that allows the preferential passage of certain substances.

In particular, membranes separate from the mixture only the gases able to pass through them more rapidly than others.

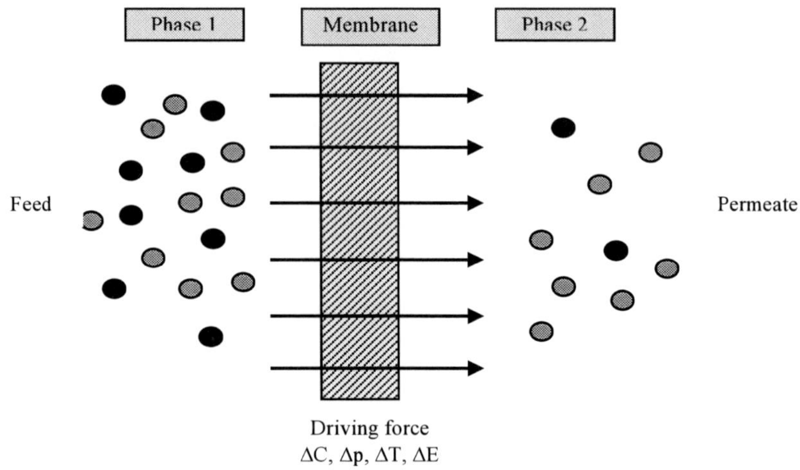

Figure 19. Schematic representation of a two-phase system separated by a membrane.

Membranes can be classified according to their nature, geometry and separation regime. The classification, based on their nature, distinguishes the membranes into *biological* and *synthetic* ones. These latter are still subdividedinto *organic* (polymeric) and *inorganic* (ceramic, metal, etc.) membranes, as shown in Figure 20.

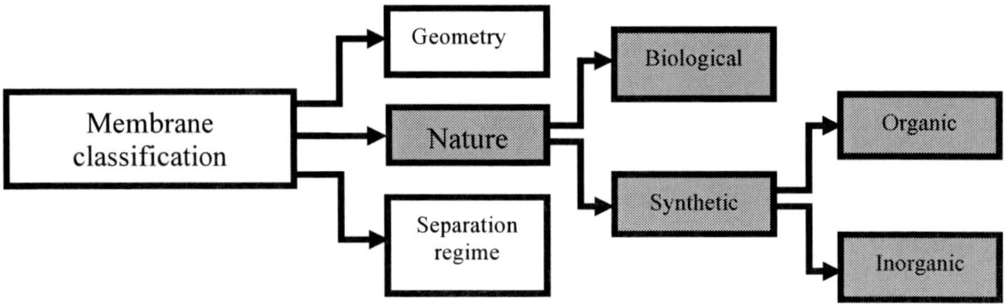

Figure 20. Scheme of a general classification of the membranes based on their nature.

The key parameters, dictating membrane performances, are:

- permeability, the flux of a specific gas through the membrane:

$$\frac{P}{l} = \frac{Q}{A\Delta p} \quad (1)$$

  where l is the effective thickness of the membrane, Q the measure of the gas permeation rate through the membrane, A the membrane surface area and Δp is the pressure difference across the membrane;
- ideal gas selectivity, the membrane preference to pass one gas species and not another:

$$\alpha_{ij} = \frac{P_i}{P_j} \tag{2}$$

For many membranes, there is a trade-off between selectivity and permeability. A highly permeable membrane tends to have low selectivity and vice versa. For example, in the polymeric membranes case, Robeson [Robeson, 2008] suggested that this trade-off may be represented by a graph, where the "log" of the selectivity is represented against the "log" of the permeability of the more permeable gas for a binary mixture, as schematically shown in Figure 21.

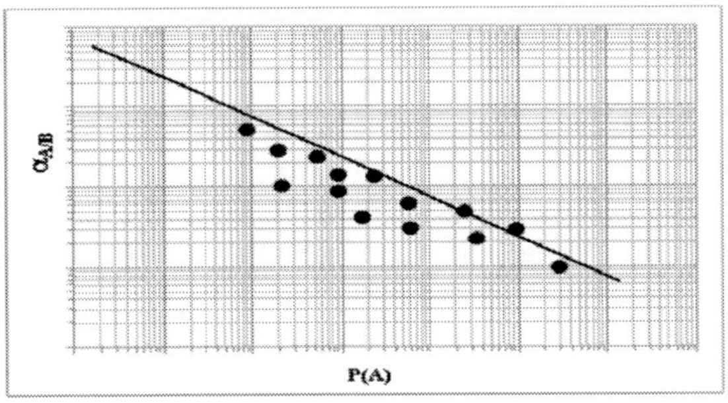

Figure 21. Upper bound correlation for general binary mixture separation.

In this figure, the upper-bound represents the limit for achieving a high selectivity combined with a high permeability. The upper-bound relationship is expressed by the following equation (3):

$$P_i = k\alpha_{ij}^{n} \tag{3}$$

where P_i is the permeability of the more permeable gas and n is the slope of the log–log limit.

However, apart from permeability and selectivity, also other membrane properties are very important such as their thermal, chemical and plasticization resistance as well as the aging effects for ensuring continual performance over long time periods.

The application of a membrane into a reactor allows an engineering device to be created for simultaneously carrying out a reaction (steam reforming, dry reforming, autothermal reforming, etc.) and a membrane-based separation in the same physical unit. Several configurations are conventionally used for the membrane housing such as plate-and-frame or tubular systems. For example, in Figure 22, a tubular MR configuration is illustrated.

Generally, a MR is an apparatus that selectively removes a product from the reaction system or introduces a reactant, reaching better performances in terms of conversion and/or yield than a fixed bed reactor (FBR) operating at the same conditions. This performance improvement is accounted for the membrane effect, which favours the shifting of conversion of an equilibrium limited reaction towards the products, by selectively removing a such product from the reaction side to the permeate side. For example, for WGS reaction, the CO conversion in a conventional system is limited by Thermodynamics. By selectively removing hydrogen from the reaction side, the thermodynamic equilibrium restrictions can be overcome and both higher conversion and hydrogen yields can be achieved. In Table 5, some of the most significant advantages of the MRs with respect to the FBRs are summarized.

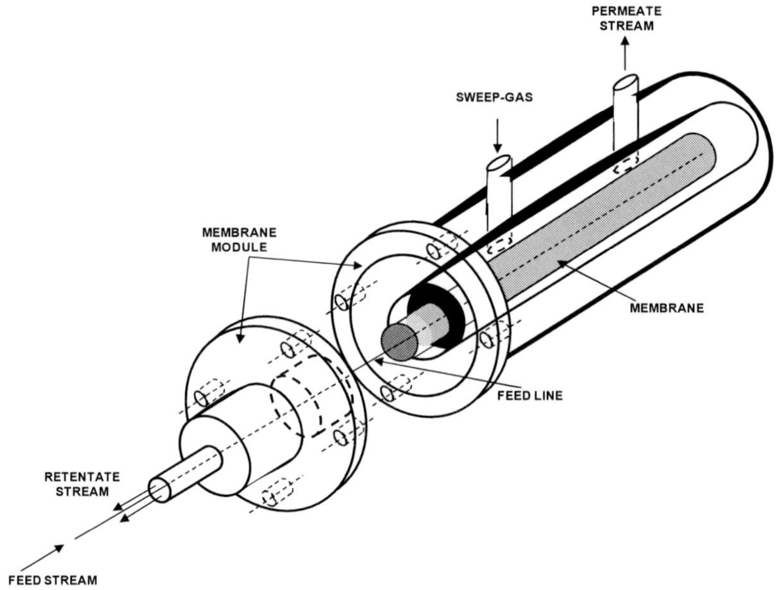

Figure 22. Conceptual scheme of a tubular membrane reactor.

Table 5. Benefits of a membrane reactor

BENEFITS OF MEMBRANE REACTOR
Conversion enhancement of the equilibrium limited reactions
Enhancement of the hydrogen yield and hydrogen selectivity (in the field of MRs used for hydrogen production)
Achievement of the same performances attained in the FBR at milder operative conditions or, instead, achievement of better performances at the same operative conditions as in the FBR
Reduced capital costs due to the combination of the reaction and separation process in only one system

10. Inorganic Membrane Reactors

In the last four decades, many efforts have been directed towards the development of inorganic membranes at scientific level because of their applicability at higher operating temperatures than polymeric membranes. As already stated above, the most used membranes in the field of high purity hydrogen production are the dense palladium-based membranes owing to their full hydrogen perm-selectivity with respect to all the other gases [Gao et al., 2004]. However, recent scientific studies also pointed out that alloys such as Zr- V- and Nb-based, Ni-Nb-Zr [Yamaura et al., 2003], V-Al alloys [Zhang et al., 2002] or V-Ni [Amano et al., 1991] represent cheaper solutions to Pd-alloys.

- As a general consideration, inorganic membranes can be subdivided on the basis of their morphology and/or structure into porous and dense. In particular, as indicated by IUPAC [Koros et al., 1996] definition, porous membranes can be classified according to their pore diameter in microporous ($d_p < 2$ nm), mesoporous (2 nm $< d_p < 50$ nm) and macroporous ($d_p > 50$ nm).

Dense metal membranes can be used in the hydrogen separation field and, generally, they are operated in the temperature range of 300 - 600 °C [Paglieri et al., 2002]. Dense palladium-based membranes can be supported or laminated.

When supported, the membrane upper temperature limits depend on the material, chemical atmosphere and support characteristics such as porosity and pore diameter [Uemiya, 1999]. Porous supports are typically made by binary oxides as SiO_2, Al_2O_3 and B_2O_3, giving excellent anchor effect and adherence [Mallada et al, 2008]. Also porous stainless steel (PSS) can be considered as a valid support due to its mechanical durability, its thermal expansion coefficient close to that of palladium and the ease of gas sealing [Mallada et al, 2008]. Unfortunately, PSS support forms an alloy with the palladium at relatively high temperatures, lowering the hydrogen permeability. Generally, the support is built up by a porous structure, where the main part has relatively large pores (commonly 2–20 μm) and one or more thin layers, with decreasing pore size.

In the case of laminated membranes, thin palladium (or palladium alloys) layer avoids the formation of oxides on the metallic surfaces resulting in a reduction of the hydrogen adsorption activation energy and, as a consequence, in an increase of the hydrogen permeation flux. Hydrogen permeation through these membranes can occur through a solution-diffusion mechanism, constituted by a complex process with several stages [Adhikari et al., 2006; Dittmeyer et al., 2001]:

1) dissociation of molecular hydrogen at the gas/metal interface;
2) adsorption of the atomic hydrogen on membrane surface;
3) dissolution of atomic hydrogen into the palladium matrix;
4) diffusion of atomic hydrogen toward the opposite side;
5) re-combination of atomic hydrogen to form hydrogen molecules at the gas/metal interface;
6) desorption of hydrogen molecules.

Therefore, the permeation of hydrogen through a metal membrane is a function of the solubility and diffusion of hydrogen in the metal in addition to the membrane thickness and temperature. At a fixed temperature, the hydrogen permeation flux through a dense palladium membrane can be expressed by means of the following equation (4):

$$J_{H2} = Pe_{H2}\,(p^{n}_{H2,ret} - p^{n}_{H2,perm})/\delta \quad (4)$$

where: J_{H2} is the hydrogen flux through the membrane; Pe_{H2} the hydrogen permeability; δ the membrane thickness; $p_{H2,ret}$ and $p_{H2,perm}$ the hydrogen partial pressures in the retentate (the reaction side) and permeate sides (the volume in which the hydrogen permeating through the membrane is collected), respectively, and n (in the range 0.5–1) is the dependence factor of the hydrogen flux on the hydrogen partial pressure. This equation even points out the inverse proportionality to the membrane thickness. The role of the membrane thickness is very important: on one hand, a thinner membrane offers a higher permeability; on the other hand, thicker membranes are necessary in order to ensure the mechanical resistance and strength.

When the pressure is relatively low, the diffusion is assumed to be the rate-limiting step and the factor n is equal to 0.5. In this case, the equation (5) becomes the Sieverts-Fick law [Dolan et al., 2006]:

$$J_{H2,Sieverts} = Pe_{H2}(p^{0.5}_{H2,ret} - p^{0.5}_{H2,perm})/\delta \quad (5)$$

On the contrary, at high pressures the hydrogen-hydrogen interactions in the palladium bulk are not negligible. In this case, n becomes equal to 1:

$$J_{H2} = Pe_{H2}(p_{H2,ret} - p_{H2,perm})/\delta \quad (6)$$

The most important problem associated with the use of pure palladium membranes is the "hydrogen embrittlement" phenomenon. When the temperature is below 300 °C and the pressure below 2.0 MPa, theβ-hydride phase may nucleate from theα-phase, resulting in severe lattice strains, so that a pure palladium membrane becomes brittle after a few cycles ofα=β transitions [Grashoff et al., 1983; Hsieh, 1989]. These transitions do not take place as a change of the lattice structure, but as a lattice dilatation. Theβ-hydride phase formation is represented as a clustering of hydrogen atoms, whose energy of attraction, being associated with the lattice, strains around the dissolved hydrogen atom [Brodowsky, 1972].

Table 6. The palladium surface contamination of some compounds

Poisoning of sulphur compounds	It is irreversible process and causes the conversion of palladium in palladium sulphide, whose lattice constant is twice than pure Pd and, thus, the structural stress leads to the formation of cracks. For example, Pd-Cu and Pd-Au increase the resistance to H_2S.
Poisoning of CO	It could cause a decrease in the hydrogen permeation flux, because the CO is adsorbed on the palladium surface blocking available dissociation sites for hydrogen. Moreover, this reduction becomes more significant at low temperature (below 150 °C) or high CO concentration.
Poisoning of H_2O	The adsorbed water molecules dissociate on the surface of the Pd film: $H_2O_{ads} \rightarrow OH_{ads} + H_{ads}$ and/or $H_2O_{ads} \rightarrow O_{ads} + 2H_{ads}$ where the H_{ads} may permeate into the bulk of the Pd film. H_2O is recombined through these reactions: $2OH_{ads} \rightarrow H_2O_{gas} + O_{ads}$ and/or $O_{ads} + 2H_{ads} \rightarrow H_2O_{gas}$ Therefore, the process of H_2O dissociation/recombinative desorption contaminates the palladium surface with adsorbed O.
Poisoning of coke	Hydrogen selectivity of a thin palladium membrane decrease after it is brought in contact with coke at elevated temperature. This phenomenon can be addressed to the fact that carbon atoms penetrate into the palladium lattice and cause the expansion of the palladium lattice.

A possible solution to avoid this phenomenon is represented by the use of a Pd-alloy containing other metals, such as silver. In fact, the role of silver is explained by its electron donating behaviour, being largely similar to the one of the hydrogen atom in palladium. Silver and hydrogen atoms would compete for the filling of electron holes [Shu et al., 1991].

As reported in Table 6, another critical problem is represented by the palladium surface contamination of some compounds such as Hg vapour, hydrogen sulfide, SO_2, thiophene, arsenic, unsaturated hydrocarbons, chlorine carbon from organic materials,. These chemical compounds can poison the membrane surface reducing the effective hydrogen fluxes by 20 to 100 % [Kluiters, 2004].

Porous membranes(ceramics, carbon, etc.) separate hydrogen from other gases such as CO_2, CO, CH_4 by allowing the smaller hydrogen molecules (diameter = 2.83 Å) to diffuse through the membrane pores and, thus, by leaving behind larger molecules [Phair et al., 2006]; namely, the membrane acts as a molecular sieve, whose efficiency is determined by the pore size and distribution relative to the molecule size. Therefore, as a result, the purity of hydrogen produced is not very high [Phair et al., 2006].These membranes can be classified in:

1) Carbon membranes
   Adsorption selective carbon membranes separate non- (or weakly) adsorbable from adsorbable gases (such as H_2S, NH_3 and CFCs) on the basis of an adsorption-surface migration mechanism. Carbon membranes can be used in non-oxidising environments with temperatures in the range of 500-900 °C. However, the performance of these membranes deteriorate severely if feed streams contain organic traces and they are brittle and, therefore, difficult to package if the membrane surfaces become larger. Furthermore, the price of carbon membranes is still high [Kluiters, 2004].

2) Ceramic membranes
   Generally, they possess a two-layer structure: the separation membrane is usually made by alumina, zirconia, titania or silica. They can operate in the range 300-800 °C, in some cases also up to 1000 °C. The main transport mechanisms for porous inorganic membranes are resumed in Table 7. In some cases, molecules can move through the membrane by means of more than one mechanism. The separation factor for these mechanisms depends strongly on pore size distribution, temperature, pressure and interactions between gases being separated and the membrane surfaces.

As a further consideration, it should be considered that inorganic membranes, although more expensive than the polymeric, are chosen in the hydrogen purification owing to their benefits such as high resistance towards solvents, high mechanical stability and elevated resistance at high operating temperatures, as resumed in Table 8.

Table 7. Description of the transport mechanisms for porous inorganic membranes

Knudsen diffusion	The Knudsen number K_n is defined as the ratio of the average free path of fluid molecules (average distance between collisions) and a representative physical length (e.g., the pore radius). The smaller the Knudsen number, the larger the pores become (relative to the average free path of fluid molecules), the lower selectivity becomes. For Knudsen numbers < 1 the dominant transport mechanism is viscous flow, which is non-selective.
Surface diffusion	Gas molecules are adsorbed on the pore walls of the membrane and migrate along the surface. Surface diffusion increases the permeability of the components adsorbing more strongly to the membrane pores. At the same time, the effective pore diameter is reduced. Consequently, transport of non adsorbing components is reduced and selectivity is increased. This positive contribution of surface diffusion only works for certain temperature ranges and pore diameters. Moreover, it can occur in parallel with Knudsen diffusion.
Capillary condensation	It occurs if a condensed phase (partially) fills the membrane pores. If the pores are completely filled with condensed phase, only the species soluble in the condensed phase can permeate through the membrane. Fluxes and selectivities are generally high for capillary condensation. The appearance of capillary condensation, however, strongly depends on gas composition, pore size, and uniformity of pore sizes.
Molecular sieving	If pore sizes become sufficiently small (3.0-5.2 Å), molecular sieving can be used to separate molecules that differ in kinetic diameter: the pore size becomes so small, that only the smaller gas molecules can permeate through the membrane.

Table 8. Benefits and drawbacks of inorganic membranes with respect to the polymeric ones

Benefits	Drawbacks
Long-term stability at high temperatures	High capital cost
Resistance to harsh environments (chemical degradation, pH, etc.)	Embrittlement phenomenon (in the case of dense Pd membranes)
Resistance to high pressure drops	Low membrane surface per module volume
Inertness to microbiological degradation	Difficulty of achieving high selectivities in large scale micro-porous membranes
Easy cleanability after fouling	Generally low permeability of the highly H_2 selective (dense) membranes at medium temperatures
Easy catalytic activation	Difficult membrane-to-module sealing at high temperature

11. POLYMERIC MEMBRANES FOR CO_2 CAPTURE

The overabundance of CO_2 in the atmosphere constitutes one of the main contributors to global climate change. In fact, it is responsible for about 64% of the greenhouse effect and global warming [Zhou et al., 2009]. This has led to scientific interest in development of new methods for reducing the amount of CO_2 emitted into the atmosphere from industries or power plants. Conventionally, there are three families of processes for CO_2 capture: sorption processes (absorption in solvents or adsorption on solid sorbents), cryogenic processes and membrane processes [e-net 2, e-net 3].

Among these processes, polymeric membrane processes could offer the potentiality of relatively higher capture temperature by eliminating the intermediate cooling and reheating steps, necessary, for example, in the amine-based absorption process. Moreover, membranes could offer several advantages such as low capital investment, compactness, ease of installation and operation, simplicity, flexibility and ability to perform separations at low energy penalties [Kohol et al., 1997; Abertz et al., 2006; Basu et al., 2004].

In CO_2 separation field, polymeric membranes may be used, for example, in the treatment of CO_2-rich streams, exiting from inorganic MRs, previously used for hydrogen production/separation by means WGS reaction, as schematized in Figure 23.

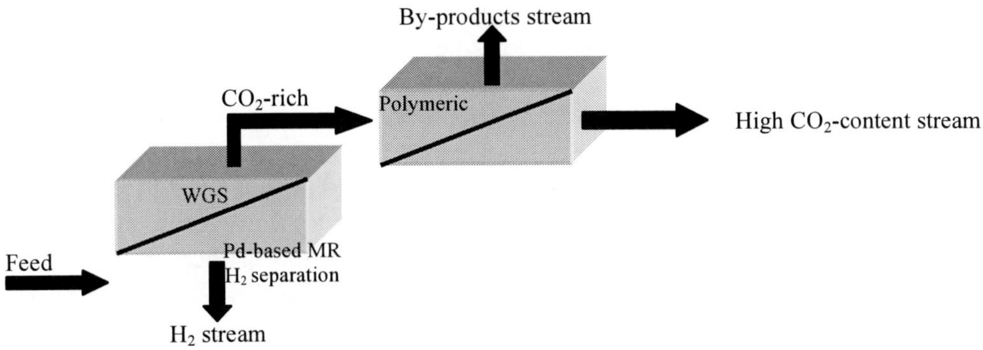

Figure 23. Scheme of treatment of a CO_2-rich stream, exiting from Pd-based MR.

Polymeric membranes can be subdivided into glassy(prepared at temperatures below the glass transition temperature) and rubbery (prepared at temperatures above the glass transition temperature) membranes. Glassy membranes possess relatively high selectivity and low flux, whereas rubbery membranes show high flux but low selectivity. Operating temperatures are limited up to around 180 °C. The most important properties of polymeric membranes are summarized in Table 9.

Table 9. Properties of CO_2 separation membranes

High CO_2 permeability	Usually, there is a trade-off between selectivity and permeability. A highly permeable membrane tends to have low selectivity and vice versa.
High CO_2/N_2 selectivity	
Thermal and chemical resistance	Long term stability at elevated temperatures is necessary to maintain the robustness of the membrane-based systems. A good example of polymers with superior intrinsic thermal stability is polybenzimidazole (PBI), which is usable at elevated temperature (> 400 °C).
Plasticization resistance	The membranes have to be resistant to plasticization phenomena, which change the polymer configuration by increasing the free volume of mixtures. The polymer matrix swells due to sorption of one specie, accelerating the permeation of the other. As a consequence, the polymer membrane loses its selectivity.
Cost effective	Ability to be cheaply manufactured into different membrane modules.

Among the polymeric membranes with reasonable CO_2 permeability and selectivity, much attention should be paid on the following membranes:

a) *Polyamides* are generally synthesized by the reaction of a diamine with a diahydride (PMDA, 6FDA and BTDA) in an aprotic solvent to form a polyamic acid. This polymer then undergoes a polycondensation reaction (either thermally or chemically) to form the desired polyimide. Polyimides combine excellent thermal and chemical stability with a wide range of CO_2 permeabilities.

b) *Polycarbonates* are commonly synthesized by reaction between a diol and phosgene under a variety of conditions. The majority of the polycarbonates tends to have a CO_2 permeability under 40 barrers and selectivities range from 15 to over 25. One notable exception to this is the polycarbonate TMHFPC, which possesses a CO_2 permeability of 111 and a CO_2/N_2 selectivity of 15 [Costello et al., 1994].

c) *Polyarylates* are synthesized by reaction of a diol, such as bisphenol A, with an acid chloride. This reaction can be performed under a variety of conditions such as interfacial or a stepwise polycondensation. The majority of polyarylates possesses a CO_2 permeability lower than 25 barrer; however, a small number shows permeabilities as high as 85 [Pixton et al., 1995], such as the sterically bulky polyarylate, TBHFBPA/tBIA, which exhibits the highest CO_2 permeability with a CO_2/N_2 selectivity equal to 19 [Kumar,2009].

d) *Polypyrrolones* are conventionally synthesized by reaction of dianhydride and a tetraamine functionalized monomer [Yang et al., 2005]. Polypyrrolones are structurally similar to polyimides, although they are considerably more rigid [Liu et al., 2006]. They are characterized by high thermal and chemical resistance. Nevertheless, owing to their high rigidity, the polypyrrolones can act as a molecular sieve.

e) *Polysulfones* are generally synthesized by condensation reaction between bisphenol and dihalogenateddiphenylsulfone. The most extensively studied polysulfone is PSF formed using bisphenol A [Barbari et al., 1995]. Other polysulfones are structurally related to this polymer. In fact, by substitution of bisphenol A with a different diol, large numbers of PSF derivatives have been synthesized, which display a wide range of CO_2 permeabilities and selectivities.

12. Status of Conventional Biomass Gasification Development and Commercialization

Basing upon several status reports concerning BMG [Kwant et al., 2004; Braccio et al., 2006; Bain, 2007] as well as personal communications, in this section an updated review of BMG state of the art is presented.

The early biomass gasifiers were developed based upon the principle of moving (or fixed) bed coal gasifiers. These systems include the eight automated Bioneer biomass gasifiers, with a thermal capacity ranging from 4 to 5 MW, that were built in Finland and Sweden during mid-1980s for heating purposes. The most established updraft fixed bed biomass gasifiers were developed by Babcock & Wilcox Vølund since 1988. A 4 MW thermal capacity plant was erected in 1996 at Harboøre in Denmark and is still in operation. This plant was originally designed for district heating, while a CHP capability was added in 2000. Basing upon the same concept three more plants (8 to 11 MW thermal) were installed in 2007 in Japan by JFE and are now in operation. Downdraft fixed bed gasification have been widely investigated also. Several plants for CHP production by XYLOWATT sa, CTU/Pyroforce, Ankur Scientific Energy Technologies Pvt. Ltd. are currently in operation worldwide, even if their capacity normally does not exceed few hundreds of kW electrical.

As far as fluidized bed technology is concerned, the successful high throughput circulating fluidized bed combustors (CFBC) designs, developed during the 1970's, were later modified to operate effectively as circulating fluidized bed gasifiers (CFBG). The early CFBGs in Scandinavia produced a fuel gas that could readily replace fuel oil in industrial burners. Given that there was little or no gas cleaning involved in these applications, because of the final destination of the product gas as a fuel for industrial burners, these plants were successfully scaled-up to 100 MW thermal capacity, processing about 600 tonnes per day of biomass. Since the early 1980s, Ahlstrom Pyropower Inc. (the present Foster Wheeler Energia OY) has succeeded in building seven CFBGs (3 to 70 MW thermal) in Finland, Sweden, Portugal, and Belgium. The recent additions in the last decade to successful Ahlstrom/Foster Wheeler Energia biomass CFBG for co-firing include the plants in Lahti, Finland, and Ruien, Belgium, having a thermal capacity of 60 and 50 MW, respectively. The Lahti plant has operated for over 40,000 hours with gasifier availability in excess of 95%. Lurgi has built three CFBG units of which the 100 MW thermal plant in Rüdersdorf, Germany, is successfully gasifying waste to produce fuel gas for firing a cement kiln. After some plant modifications, the 83 MW thermal co-firing plant by Essent/Lurgi in Geertruidenberg, the Netherlands, is now back in operation. In 1988, Götaverken (now Metso Power) has built a 25 MW thermal CFBG in Sweden, which is still in operation. During the same period, Termiska Processer AB (TPS) has developed CFBG technology. A 15 MW thermal capacity CFBG for CHP production using refuse derived fuel (RDF) pellets as the feedstock was erected at Greve in Chianti in Italy. The plant worked for around 5,000 hours in the 1997-98 two-year period converting to gas 9,000 tonnes of RDF.

In the field of power or CHP production, Sydkraft and Ahlstrom were involved in the development of a second generation pressurized CFBG process for integrated gasification combined cycle (IGCC) application at Värnamo in Sweden. The 18 MW thermal capacity plant was operated at 18 bar pressure, raw gas was cleaned without condensation employing high temperature filters, and the low calorific value fuel gas was successfully combusted in a closely integrated Typhoon gas turbine, to generate 6 MW of electricity and 9 MW of district heat. The Värnamo plant was mothballed in 2000, after more than 8,000 hours of gasifier stand alone operation and 3,600 hours of integrated operation with the gas turbine. Since October 2005, efforts are in progress to reactivate the Värnamo plant as the centerpiece for demonstration in a multinational development project for synthesis gas production.

Ahlstrom/Foster Wheeler Energia also tested the bubbling fluidized bed gasification technology. A 40 MW thermal capacity plant was constructed in 2001 at Corenso United Oy Ltd paper mill in Varkaus, Finland. The system was intended for gasifying the mixture of polyethylene plastics and aluminium coming from the separation process for the recovery of wood fibre from recycled carton board. After gasification, non-oxidized aluminium was separated from the product gas and then recovered, while gas was exploited for CHP production. This plant was successfully operated for about five years, producing energy and recovering metals from the waste feedstock. Another example of bubbling fluidized bed gasifier is the 28 MW thermal capacity plant for CHP demonstration recently started by Carbona/Andritz in Skive, Denmark. The system produce now 6 MW of power and 12 MW of district heat. It was initially operated as district heating plant and only at the end of 2008 the full power capacity was achieved.

Basing upon the dual fluidized bed gasifier concept the Technical University of Vienna has developed since early 1990's a fast internal circulating fluidized bed (FICFB) biomass gasification process. In 2002 the former pilot plants were scaled-up to a commercial gasifier for CHP production with a fuel capacity of 8 MW and 2 MW electrical in Güssing, Austria, in cooperation with REPOTEC. This plant is now in continuous operation, being its current availability over 8,000 hours per year. Since a hydrogen-rich gas is obtained via this process, a wide range of slipstream investigations are being conducted at this facility in order to evaluate the production of liquid bio-fuels (FT diesel) and SNG. Based on these studies, 15% of the raw gas stream is completely conditioned and converted to SNG in a closely coupled demonstration plant. Moreover, a second plant for CHP production (2.8 MW electrical) has started to work in 2009 in Oberwart, Austria, and a third one (3.6 MW electrical) is under construction at Villach, Austria. In both plants, constructed by Ortner, the exhaust gas from the internal combustion engines is used to feed an organic Rankine cycle (ORC) turbine allowing to achieve a global conversion efficiency from biomass to electricity over 30%.

The field of application for BMG concerning synthesis gas production has shown an upsurge of activities in recent years. Based upon the entrained flow gasification technology, CHOREN has successfully tested on the pilot scale the Carbo-V process for production of synthetic biofuel or biomass to liquids (BTL). The 1 MW thermal "alpha" plant was constructed in Freiberg, Germany, in 1998 and the process was extensively and successfully tested over an operating time of 22,500 hours. In 2001, a methanol synthesis stage and later Fischer-Tropsch synthesis were added to the Carbo-V pilot plant. Starting from the results achieved on the pilot scale, a 45 MW thermal "beta" plant fully integrated with a FT unit based on the Shell technology for BTL production on commercial scale has just been constructed in Freiberg and is currently almost ready to work. At Forschungszentrum Karlsruhe (FZK), Germany, a 5 MW thermal load plant for the conversion of biomass residues to fuel by flash pyrolysis and subsequent entrained flow gasification (bioliq® process) is under construction in cooperation with Lurgi. The pyrolysis unit is already operational, while the start up of the gasification unit is scheduled within the end 2011. Range Fuel is in the process of building a large scale plant for the thermo-chemical conversion of biomass to ethanol (or methanol) in Georgia, USA. The process include a devolatilization step and a subsequent steam reforming in order to produce the syngas. In the forest industry, Stora Enso, UPM and Norske Skog are all engaged in plans for biofuels, the two former already started pilot plant scale tests for gas cleaning in 2009.

In conclusion, BMG can be considered as an established technology for heat production or co-firing with fossil fuels, where gas cleaning requirements are not so critical. As far as power generation is concerned, BMG has a good potential in the segment of medium capacity systems (from 2 to 5 MW electrical), as dual fluidized bed gasifiers installed in Austria are demonstrating. For large capacities installations (over 10 MW electrical), biomass combustion conventional systems still are assessed as more reliable and easy to manage. Similarly, gasification must deal with a hard competition with combustion combined to ORC turbines for small scale installations (from 500 kW to 1.5 MW). This explains why, in recent years, interest in BMG is shifting to applications to the production of biofuels, even if an extensive work of RD&D is still necessary in this field.

Conclusion

The main technical barriers to commercialize biomass gasification include reliable handling and feeding of mixed feedstocks (in particular, the low-density herbaceous biomass), reduction of tar formation in gasification reactors, particulate entrainment and its removal, managing carbon, tar, alkali, chlorides and ammonia in fuel gases, ash withdrawal, waste water handling in conventional gas cleaning. These factors have limited reliable scale-up and successful demonstration of high-efficiency advanced BMG. Furthermore, it has been frequently observed that improvements made in the reliability of individual process steps do not necessarily guarantee successful scale-up and integrated operation of the overall process. Therefore, system integration remains an important issue in process scale-up, demonstration, and commercialization.

At present, limited but significant research is continuing in Europe and USA to address many of the technical hurdles listed above. These activities include fundamental research in understanding and modeling the overall biomass gasification process, investigating the formation of tar and its destruction inside the gasification reactor as well as on catalytic surfaces. Consequently, research is continuing to find safe, reliable, and environmentally acceptable tar decomposition catalysts and to explore innovative BMG technologies capable to produce condensable hydrocarbon-free fuel or synthesis gases. On the other hand, it is worth exploring the tolerable limits of raw gas contaminants and the types of condensable hydrocarbons in selected gas processing or energy conversion devices.

However, the potential of hydrogen as an energy carrier will certainly have a considerable role in the future energy system. Therefore, many companies, governments and laboratories focus their efforts trying to produce hydrogen in a more technically, environmentally and economically attractive way, based on an idea of sustainable development.

Although, currently, hydrogen upgrading in refinery applications can be achieved by the pressure swing adsorption or cryogenic separation processes, membrane technology constitutes an excellent candidate for hydrogen purification, especially when incorporated in a membrane reactor, combining the reaction/separation processes in a single unit. Moreover, today the scientific interest towards hydrogen production from syngas is increasing, especially when coming from inorganic membrane reactor technology. Future research should be devoted to develop low cost effective inorganic membranes and to prepare defect-free and homogeneous membranes able to work for a long period at high temperatures and pressures, in order to validate them as good alternative to the conventional systems.

References

Abertz V., Brinkmann T., Dijkstra M. (2006), “Developments in membrane research: from material via process design to industrial application”, Adv. Eng. Mater. 8, 328-358.

Abu El-Rub Z.Y., Bramer E.A., Brem G. (2006), “Modeling of tar reduction in biomass fuelled gasification using biomass char as a catalyst”, Science in Thermal and Chemical Biomass Conversion, Volume 1, CPL Press, Newbury, UK, 530-541.

Adhikari S., Fernando S. (2006), "Hydrogen membrane separation techniques", Ind. Eng. Chem. Res. 45, 875-881.

Amano M., Komaki M., Nishimura C. (1991), "Hydrogen permeation characteristics of palladium-plated V---Ni alloy membranes", J. Less-Common.Met. 172, 727-731.

Bain R.L. (2007), "World Biofuels Assessment; Worldwide Biomass Potential: Technology Characterizations (Milestone Report)", NREL Report No. MP-510-42467.

Barbari T.A., Datwani S.S. (1995), "Separation properties of polysulfones membranes treated with molecular bromine", J. Membrane Sci. 107, 263-266.

Basu A., Akhtar J., Rahman M., Islam M. (2004), "A review of separation of gases using membrane systems", Petrol. Sci. Techn. 22, 1343-1368.

Braccio G., Canneto G., Fiorenza G. (2006), "Biomass gasification: state of the art and the activity of ENEA", Proceedings of Technologies and Perspectives of Energy Production from Biomass Workshop, Piacenza, Italy, November 20-22.

Brodowsky H. (1972), "On the non-ideal solution behavior of hydrogen in metals", Ber.BunsengesPhysik. Chem. 76, 740-749.

Brunetti A., Barbieri G., Drioli E. (2009), "Upgrading of a syngas mixture for pure hydrogen production in a Pd–Ag membrane reactor", Chem. Eng. Sci. 64, 3448-3454.

Costamagna P., Srinivasan S. (2001), "Quantum jumps in the PEMFC science and technology from the 1960s to the year 2000 Part II. Engineering, technology, development and application aspects", J. Power Sou. 102, 253-269.

Costello L.M., Koros W.J. (1994), "Effect of structure on the temperature dependence of gas transport and sorption in a series of polycarbonates", J. Polymer Sci. Part B: Polymer Phys. 32, 701-713.

Devi L., Ptasinski K.J., Janssen F.J.J.G. (2003), "A review of the primary measures for tar elimination in biomass gasification processes", Biomass and Bioenergy, 24, 125-140.

Dittmeyer R., Höllein V., Daub K. (2001), "Membrane reactors for hydrogenation and dehydrogenation processes based on supported palladium", J. Mol. Catal. A: Chem. 173,135-184.

Dolan M.D., Dave N.C., Ilyushechkin A.Y., Morpeth L.D., McLennan K.G. (2006), "Composition and operation of hydrogen-selective amorphous alloy membranes", J. Membrane Sci. 28, 30-55.

e-net 1: "Syngas production from coal", IEA ETSAP–Technology Brief S01-May 2010, http://www.etsap.org/E-techDS/PDF/S01-Coal%20gasification-GS-gct-AD.pdf

e-net 2: International Energy Agency (IEA), "Prospects for CO_2 capture and storage", IEA Publications, Paris, France (2004).

e-net 3: "IPCC special report on carbon dioxide capture and storage, Working group III of IPCC", B. Metz, O. Davidson, H.C. de Coninck, M. Loos and L.A. Meyer (eds.), Cambridge University Press, Cambridge, UK (2005).

Gao H., Lin Y.S., Li Y., Zhang B. (2004), "Chemical stability and its improvement of palladium-based metallic membranes", Ind. Eng. Chem. Res. 43, 6920-6930.

Giordano P. (1998), "Production of electricity from wood: IISc-Dasag gasifier with IC-engine application", Swiss Federal Office of Energy, Berne.

Gøbel B., Henriksen U., Ahrenfeldt J., Jensen T.K., Hindsgaul C., Bentzen J.B., SørensenL.H.(2004), "Status - 2000 hours of operation with the Viking gasifier", Proceedings of the 2nd World Conference and Technology Exhibition on Biomass for Energy and Industry, Rome, Italy, May 10-14.

Grashoff G.J., Pilkington C.E., Corti C.W. (1983), "The purification of hydrogen – A review of the technology emphasing, the current status of palladium membrane diffusion", Plat. Met. Rev. 27, 157-168.

Henrich E., Weirich F. (2004), "Pressurized entrained flow gasifiers for biomass", Env. Eng. Sci., 21,53-64.

Hofbauer H., Rauch R. (2000), "Stoichiometric water consumption of steam gasification by the FICFB-gasification process", Proceedings of Progress in Thermochemical Biomass Conversion Conference, Innsbruck, Austria, September 17-22.

Hofbauer H., Fürnsinn S. (2007), "Gasification - Quo Vadis ?", Proceedings of the Success and Visions for Bioenergy Workshop, Salzburg, Austria, March 21-23.

Hsieh H.P. (1989), "Inorganic membrane reactors – A review", AIChESymp. Ser. 85, 53-67.

Kohol A.L., Nielsen R. (1997), "Gas purification", 5th ed. Gulf Publishing, Houston, TX.

Koros W.J., Ma Y.H., Shimidzu T. (1996), "Terminology for membranes and membrane processes", J. Membrane Sci. 120, 149-159.

Kumar P. (2009), "Inclusion chemistry in periodic mesoporous hosts: growth of quantum-confined materials and gas separation membranes", http://etd.ohiolink.edu/send-pdf.cgi/KUMAR%20PARVEEN.pdf?ucin1186772727.

Kwant K.W., Knoef H. (2004), "Status of Gasification in countries participating in the IEA and GasNet activity - August2004",IEA Bioenergy Gasification/ EU Gasification Network.

Liu J.G., Zhao X.J., Yang H.X., Li H.S., Fan L., Yang S.Y. (2006), "Novel high temperature polypyrrolones with asymmetric biphenyl moieties in the main chain: Synthesis and characterization", Macr. Chem. Phys. 207, 1272-1277.

Mallada R., Menéndez M. (2008), "Inorganic membranes: synthesis, characterization and applications", Technology & Engineering, pp. 460.

Mathieu P., Dubuisson R. (2002), "Performance analysis of biomass gasifier", Energy Conversion and Management, 43, 1291-1299.

Paglieri S.N., Way J.D. (2002), "Innovations in palladium membrane research", Sep. Purif. Reviews 31, 1-169.

Phair J.W., Badwal S.P.S. (2006), "Materials for separation membranes in hydrogen and oxygen production and future power generation", Sci. Tech. Adv. Mat. 7, 792-805.

Pixton M.R., Paul D.R. (1995), "Gas transport properties of polyarylates: substituent size and symmetry effects", Macromolecules 28, 8277-8286.

Priest S.H. (2010), Encyclopedia of Science and Technology Communication, SAGE Publications Inc., pp. 1144.

Reed T.B. (1981), "Biomass gasification principles and technology", Noyes Data Corporation, Parkridge, New Jersey, USA.

Robeson L.M. (2008), "The upper bound revisited", J. Membrane Sci. 320, 390-400.

Shu J., Grandjean B.P.A., Van Neste A., Kaliaguine S. (1991), "Catalytic palladium-based membrane reactors: A review", Can. J. Chem Eng. 69, 1036-1060.

Uemiya S. (1999), "State-of-art-the-art of supported metal membranes for gas separation", Sep. Pur. Methods 28, 51-85.

Yamaura S.I., Shimpo Y., Okouchi H., Nishida M., Kajita O., Kimura H., Inoue A. (2003), "Hydrogen permeation characteristics of melt-spun Ni-Nb-Zr amorphous alloy membranes", Mater. Trans., The Japan Institute of Metals (JIM) 44, 1885-1890.

Yang K., Liu Z., Mao M., Zhang X., Zhao C., Nishi N. (2005), "Molecularly imprinted polyethersulfone microspheres for the binding and recognition of bisphenol A", Anal. Chim.Acta 546 30-36.

Zhang Y., Ozaki T., Komaki M., Nishimura C. (2002), "Hydrogen permeation characteristics of vanadium–aluminium alloys", Scripta Mater. 47, 601-606.

Zhou Q., Chan C.W., Tontiwachiwuthikul P. (2009), "A monitoring and diagnostic expert system for carbon dioxide capture", Exp. Systems Appl. 36, 1621-1631.

In: Gasification: Chemistry, Processes and Applications ISBN: 978-1-61209-681-0
Editor: Michael D. Baker

Chapter 5

A DYNAMIC MODEL FOR PREDICTING TAR CONTENT IN GASES FROM A STRATIFIED DOWNDRAFT GASIFIER: PART I –MODEL DEVELOPMENT

Lu Aye[1,*], San Shwe Hla[1], M. A. Connor[2] and D. F. Stewart[1]

[1]International Technologies Centre (IDTC), Department of Civil and Environmental Engineering

[2]Department of Chemical and Biomolecular Engineering

The University of Melbourne, Victoria 3010, AUSTRALIA

ABSTRACT

This paper describes the development and the predictions of a dynamic model for a stratified downdraft wood gasifier. The numerical model, which incorporates a two-step pyrolysis mechanism, uses mass and energy balances, together with information about rates of chemical reactions and physical transport processes, to describe the complex physical and chemical processes taking place in the gasifier. The model is capable of predicting the gas composition, primary and secondary tar contents, and gas and solid surface temperature profiles along the axis of the gasifier. The model predicts an outlet gas tar content of between 20 and 200 mg/Nm³ for the range of input parameters used in the model simulation, assuming that prior to ignition the section of the gasifier below the ignition ports contains charcoal rather than wood. The stability of the reaction zones was also investigated and for stable operation it was found that the mass flux of supplied air needed to be in the range 0.158-0.167 kg/m^{2}s.

Keywords: Biomass; Gasification; Tar content; Modelling

* Corresponding author. e.mail: lua@unimelb.edu.au, Tel.: +613-8344-6879, fax: +613-8344-6868

1. INTRODUCTION

Biomass is a widely available resource with the potential to contribute significantly to the energy demands of both developed and developing communities [1]. Thermochemical biomass conversion technologies have attracted much attention in recent years. Of these thermal conversion processes, gasification has received particular attention.

Large-scale gasification systems are of interest because they offer a way of producing electricity more efficiently than conventional direct combustion systems [2]. Small-scale systems are also of interest, with downdraft gasifiers looking especially attractive for small-scale energy supply applications (<1.5 MWth), for which there is a big market in both developed and developing economies [3]. According to an extensive review of 50 gasifier manufacturers in Europe, USA and Canada, 75 % of the commercial gasification plants on offer were of the downdraft type [4]. Despite the high level of interest in gasification, one major barrier remains to be overcome before successful commercialization of biomass gasification technologies into the world energy market can be achieved on a large scale. This is finding a way to efficiently remove the tar present in the gasifier exit gases.

Tar can be eliminated by auxiliary systems such as scrubbers, but these are not reliable for long term operation and, in addition, cause a serious environmental problem due to the large quantities of wastewater produced. This has stimulated interest in the concept of reducing tar levels inside the gasifier and several attempts have been made to do this by modifying the traditional downdraft gasification system [5]-[9].

When developing a cleaner gasification unit, it is helpful to have access to a numerical model that can describe gasifier behaviour and provide insights into the processes taking place within the gasifier. A number of gasification models have been developed, most of which are for gasifier systems using coal as a fuel rather than biomass. The usefulness of coal-oriented models is limited as there are significant differences between the ways that coal and biomass behave during gasification; this is a result of differences in physical properties, chemical composition and types and levels of impurities. Besides, most of the coal gasification models are for the countercurrent (updraft) process since updraft fixed-bed coal gasification is also the most important commercial gasification process: around 90% of the coal gasifiers in the world operate in countercurrent mode [10].

Such models are also of no use for estimating outlet gas tar contents since no tar formation, burning and cracking processes are considered within these models. Omitting to include these processes is important in the case of the downdraft gasifier since the levels of carbon monoxide and hydrogen, the two components contributing most of the heating value in producer gas, are influenced by the extent to which tars (oxygenates and hydrocarbons) are cracked within the gasifier. Another neglected factor in most biomass gasification models is how fully the biomass leaving the pyrolysis and oxidation zones is carbonised; this is of significance since it also influences the tar content of the final gas product.

Of the limited number of numerical models of fixed bed biomass gasification available in the literature, most have been formulated based on highly simplifying assumptions. For example, some of the steady state models[11], [12] neglect the kinetic mechanisms of pyrolysis and cracking of volatile products and the first three zones of the downdraft gasifier (drying, pyrolysis and combustion) are usually lumped together. This assumption is only acceptable when it is just the levels of major gas components in the outlet gas that are of interest; it is inappropriate if information is needed on how products of pyrolysis vary within the gasifier. For an understanding of the tar formation, cracking and oxidation processes inside the downdraft gasifier, a numerical model must include kinetic expressions for these processes. In a recent stratified downdraft gasification model, kinetics of biomass pyrolysis and oxidation of gaseous species are included, enabling patterns of change of gaseous species and temperatures in the pyrolysis and combustion zones to be predicted [13]. However, the results obtained using that model seem unrealistic. For example, predicted solid surface temperatures are lower than the minimum temperature required for char gasification reactions to proceed at an acceptable rate but levels of heterogeneous reaction products (CO and H_2) are nevertheless predicted to keep increasing all the way to the grate at the base of the gasifier. In addition, the gas temperature, which would be expected to be higher than the solid surface temperature due to the heat released during gas phase oxidation of volatiles in the combustion zone, is shown as never exceeding the solid surface temperature.

In this study, a dynamic model of a throat-less downdraft gasifier has been developed that contains fewer simplifying assumptions than previously published models. The factors taken into account in this model include: temperature differences between the gas and solid surface; particle shrinkage during heterogeneous reactions; the effect of reaction temperature on the CO/CO_2 ratio for heterogeneous char combustion; and the effective thermal conductivity, which takes into account the importance of radiation at high temperatures. A two step pyrolysis process is assumed and the oxidation of volatile gases and secondary tars is accounted for. This paper describes the development of the numerical model and the general results obtained from it. These include predictions of the tar profile along the body of a stratified downdraft gasifier. The sensitivity of the model to changes in operating and model parameters is described in a second paper (Part II) [14] while a third paper (Part III) [15] describes an experimental investigation undertaken to validate the predictions of this model.

2. Model Development

In stratified downdraft gasification, solid fuel is introduced, together with air, at the top of the gasifier. It flows downwards through the gasifier, undergoing drying, pyrolysis, char oxidation, volatiles oxidation, and char gasification. For modelling purposes it is convenient to divide the gasifier up into several zones, each dominated by reactions typical of one of the above processes. However, it is important to note that in practice boundaries between adjacent zones are not clearcut since in these regions many types of reactions can be occurring simultaneously. Prior to construction of the model, some simplifying assumptions also needed to be made about particle sizes in the fuel bed. In the drying and pyrolysis processes, moisture and volatiles are released from the solid fuel and the density of the solid fuel decreases. However, even though in practice some shrinkage occurs during these processes, particularly during pyrolysis [16], the anisotropic nature of this process (in materials like wood) makes it very difficult to describe accurately. Therefore the fuel particle size was assumed to remain constant in the drying and pyrolysis zones. In the combustion and gasification zones, particle size was assumed to decrease as predicted by the shrinking core reaction model.

Drying

Drying of the solid fuel is normally formulated in reactor models by one of three different approaches:

- it is assumed that moisture from the moist biomass or charcoal (in the case of charcoal gasification) evaporates instantaneously as the fuel enters the combustion and gasification zone [17];
- the process of drying is considered to be a diffusion limited process [10], [13];
- the drying rate is assumed to be kinetically controlled [18], [19].

In this model, the latter approach was followed and the drying process was described using an Arrhenius type kinetic equation developed by Chan *et al.* [20]:

$$R_m = (1-\varepsilon)A_m \exp\left(\frac{-E_m}{RT_s}\right)\rho_m \qquad (1)$$

Pyrolysis

*(Note: yields of Acetylene (C_2H_2) and Ethane (C_2H_6) are omitted since they are present only in very small quantities.)

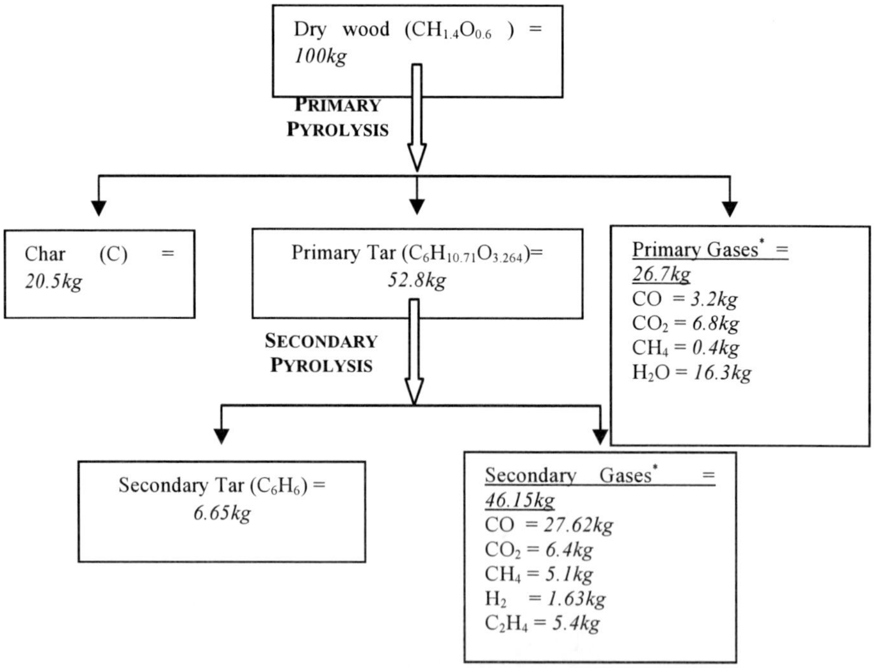

Figure 1. The pyrolysis mechanism used in the numerical model and assumed product yields (adapted from [26]).

A two-step pyrolysis process, shown in Fig. 1, was used in this model. Experimental investigations of primary wood pyrolysis kinetics have been undertaken by various researchers. Almost all of these experimental studies have employed finely divided particles or used sample particles less than 1.5 mm in size, so kinetic data that can be applied directly to the gasification of large wood particles are scarce. When large particles are decomposed by application of an external heat source, a thermal wave moves from the particle surface to its centre. This involves a complex combination of heat and mass transfer processes that is very hard to model accurately, especially in an anisotropic material such as wood. The most suitable kinetic expression found from the literature was that of Chan *et al.* [21] and this was the expression selected for use in the model:

$$R_{p1} = (1-\varepsilon)A_{p1}\exp\left(\frac{-E_{p1}}{RT_{avg}}\right)\rho_v \tag{2}$$

Before this expression can be applied an average particle temperature needs to be found. This is a function of wood surface temperature, primary pyrolysis kinetics and particle size and was determined using Eq. 3, a correlation developed by Reed and Markson [22], who determined the time required for completion of the flaming pyrolysis process and correlated this with particle size, temperature, moisture content and oxygen fraction. It would be expected that the behaviour of a particle in a gasifier fuel bed would differ from that of an isolated single particle. However, no published information was found on how the decomposition of a single particle is affected by the presence of surrounding particles and so the single particle pyrolysis rate expression described below was used without modification.

$$T_{avg} = \frac{1}{\left(\frac{1}{T_s} + \frac{R}{E_{p1}} \times \ln\left(953.138d_p + 46861.984d_p^2\right)\right)} \tag{3}$$

Secondary pyrolysis is the process of cracking of primary tar into lower molecular weight gases and secondary tar. Recently a limited number of experimental studies of vapour phase cracking of wood pyrolysis tar has been undertaken [23]-[29]. These investigations employed a variety of wood types and experimental processes. Liden *et al.* [27], Diebold [28] and Kosstrin [29] used a fluidised-bed technique where both the solid and tar decomposition reactions took place simultaneously in the same reactor. Morf *et al.* [23], Rath & Staudinger [25] and Boroson *et al.* [26] studied tar decomposition processes separately by using a set-up comprising two distinct reactors, the first one devoted to solid pyrolysis, the second to tar cracking under a flow of carrier gas with no packing present. Fagbemi *et al.* [24] investigated the effect of biomass type by determining decomposition rates for tar derived from three different biomass materials (wood, coconut shell and straw).

Arrhenius constants determined by Boroson *et al.* [26] show very good agreement with those in a recent report [23]. Moreover, the secondary pyrolysis kinetic data obtained when using two separate experimental reactors seem more appropriate than those obtained when using one reactor. Therefore the Arrhenius kinetic data provided by Boroson *et al.* [26] were used in this model for the secondary pyrolysis (primary tar cracking) process:

$$R_{p2} = (\varepsilon) M_{tar1} A_{p2} \exp\left(\frac{-E_{p2}}{RT_C}\right) C_{tar1} \tag{4}$$

Table 1. List of drying, devolatilization, heterogeneous and homogeneous reaction rate constants used in the mathematical model

Reaction	Kinetic rate constants	Ref:
Drying	**A_m = 5.13 × 10^6 (1/s)** E_m = 87,900 (J/mol)	**[20]**
Primary pyrolysis	**A_{p1} = 3.20 × 10^5 (1/s)** E_{p1} = 113,000 (J/mol)	**[21]**
Secondary pyrolysis	**A_{p2} = 9.55 × 10^4 (1/s)** E_{p2} = 93,300 (J/mol)	**[26]**
$C + (\gamma)O_2 \xrightarrow{k1} (2-2\gamma)CO + (2\gamma-1)CO_2$ **(5)**	**A_C = 4.3 (-)** **E_C = 28,185 (J/mol)**	**[36]**
	A_1 = 0.685 (m/K s) **E_1 = 74,830 (J/mol)**	**[36]**
$C + CO_2 \xrightarrow{k2} 2CO$ **(6)**	**A_2 = 589 (m/K s)** **E_2 = 222,825 (J/mol)**	**[10]**
$C + H_2O \xrightarrow{k3} CO + H_2$ **(7)**	**A_3 = 5.7144 (m/K s)** **E_3 = 129,700 (J/mol)**	**[37]**
$C + 2H_2 \xrightarrow{k4} CH_4$ **(8)**	**A_4 = 3.42 ×10^{-3} (m/K s)** **E_4 = 129,700 (J/mol)**	**[10]**
$CO + H_2O \xleftrightarrow{k5} CO_2 + H_2$ **(12)**	**A_{5F} = 0.03 (m³/kmol s)** **E_{5F} = 60,279 (J/mol)**	**[38]**
$C_6H_6 + 3O_2 \xrightarrow{k6} 3H_2 + 6CO$ **(13)**	A6 = 2.07 × 10$^{4.9}$ (m$^{1.5}$/kPa$^{0.3}$ kmol$^{0.5}$ K s) **E_6 = 80,230 (J/mol)**	**[40]**
$CO + 0.5O_2 \xrightarrow{k7} CO_2$ **(14)**	**A_7 = 1.3 × 10^{11} (m^3/kmol s)** **E_7 = 125,580 (J/mol)**	**[41]**
$CH_4 + 1.5O_2 \xrightarrow{k8} CO + 2H_2O$ **(15)**	**A_8 = 1.0 × 10$^{11.7}$ (m^3/kmol$^{0.5}$ s)** **E_8 = 202,600 (J/mol)**	**[42]**
$C_2H_4 + O_2 \xrightarrow{k9} 2H_2 + 2CO$ **(16)**	**A_9 = A_6** **E_9 = E_6**	**[40]**
$H_2 + 0.5O_2 \xrightarrow{k10} H_2O$ **(17)**	**A_{10} =3.53×10$^{8.4}$(m$^{3.6}$/kmol$^{1.2}$ s)** **E_{10} = 30,514 (J/mol)**	**[43]**

Tar cracking can occur as a result of heterogeneous vapour-char reactions as well as by homogeneous gas-phase reactions. Neglecting the contribution of heterogeneous cracking reactions can lead to tar transformation rates being underestimated. There appear to be no data available on heterogeneous tar cracking mechanisms so these had to be accounted for in another way. By a trial and error method it was established that multiplying the activation energy values in Boroson's equation by 0.85 yielded final gas tar contents that were in a similar range to observed values, so this approach was used in the model. In addition, the tar cracking temperature in eq. 4 was taken as:

$$T_C = \frac{T_s + T_g}{2} \quad \text{for} \quad (T_s > T_g)$$

$$T_C = T_g \quad \text{for} \quad (T_s \le T_g)$$

Product yields for primary and secondary pyrolysis were chosen on the basis of published data [26]. By carrying out an elemental mass balance using the data provided by Boroson *et al.* [26] and lumping the results, a formula for the tars (primary tar) which are the product of primary pyrolysis (also known as oxygenates) was derived: $C_6H_{10.71}O_{3.264}$. In addition the composition of secondary tar (hydrocarbon) was taken to be 100% benzene (C_6H_6), since benzene was the major component found in tar from downdraft gasifiers when this was analysed by direct mass spectrometry [30]. The assumption that secondary tar can be represented by benzene (C_6H_6) has been previously used in a wood particle decomposition model [31] and a coal particle combustion model [32].

Some uncertainty exists concerning the energetics of biomass pyrolysis as over the temperature range concerned both endothermic as well as exothermic processes have been observed to occur [33]. In this model, the heat of the primary pyrolysis process was assumed to be zero [34] and a value of the heat of reaction of 50 kJ/kg was assumed for the secondary pyrolysis process (tar cracking) [35].

Heterogeneous Char Combustion and Gasification

The heterogeneous char oxidation and gasification reactions are shown as Eq. (5)-(8) in Table 1. The unreacted shrinking core model, in which reaction rate depends on film diffusion and intrinsic chemical reaction rate, was used to represent these processes in the model. The heterogeneous reaction rates were estimated using eq. 9: this is adapted from Hobbs *et al.* [10].

$$R_i = \left(\frac{M_C}{\nu_j M_j}\right) \frac{A_v \rho_j}{\left[\left(\frac{1}{k_m}\right) + \left(\frac{1}{k_i}\right)\right]} \tag{9}$$

where $k_i = A_i T_s \exp\left(\frac{-E_i}{RT_s}\right)$, $i = 1\text{-}4,\ j = O_2, CO_2, H_2O, H_2$

A_v, the specific surface area or the reaction surface area, is one of the important fuel characteristics where heat and mass transfer to the particles in the bed are concerned. This reaction surface area is related to the bed void fraction and the particle diameter as follows:

$$A_v = \left(\frac{6}{d_p}\right)(1-\varepsilon) \tag{10}$$

The main products of char oxidation are CO and CO_2 (Eq. 5) and the ratio of CO/CO_2 increases with increases in the reaction temperature. These ratios have been correlated by various researchers for several types of char using first order Arrhenius-type relationships. In this study, the ratio of CO/CO_2 was calculated using a correlation developed by Evans and Emmons [36] from their experimental data for wood char.

$$\gamma = \frac{2 + A_C \exp\left(\frac{-E_C}{RT_s}\right)}{2 + 2\left(A_C \exp\left(\frac{-E_C}{RT_s}\right)\right)} \tag{11}$$

Homogeneous gas reactions

Six homogeneous reactions (eq. 12-17 in Table.1) were considered when developing the model. The first of these is the well-known water-gas shift reaction while the rest involve the oxidation of volatile gases, including the secondary tar. These are significant reactions in the downdraft gasification process.

An expression for the forward water-gas shift reaction rate was proposed by Grebenshchikova [38].

$$R_{5F} = \varepsilon A_{5F} \exp\left(\frac{-E_{5F}}{RT_g}\right) C_{CO} C_{H_2O} \tag{18}$$

The reverse water-gas shift reaction rate was determined using the equilibrium expression as

$$R_{5R} = \varepsilon \frac{A_{5F} \exp\left(\frac{-E_{5F}}{RT_g}\right)}{K_C} C_{CO_2} C_{H_2} \tag{19}$$

K_C is the equilibrium constant, a function of gas phase temperature and can be found from the following correlation [39]:

$$K_C = \exp\left(\frac{1}{RT_g}\left(11321 - 31.08T_g + 3T_g \ln T_g - 2.8E-4T_g^{\,2} - \frac{91500}{T_g}\right)\right) \tag{20}$$

For the partial combustion of this hydrocarbon group the reaction products were assumed to be CO and H_2 [40] and reaction rate was determined using eq. 21:

$$R_6 = \varepsilon A_6 \exp\left(\frac{-E_6}{RT_g}\right)\left(C_{C_6H_6}\right)^{0.5}\left(C_{O2}\right)T_g P^{0.3} \tag{21}$$

CO is a major gas component released during the secondary pyrolysis process. The mechanism for oxidation of carbon monoxide (see below), proposed by Howard *et al.* [41], was used in this model.

$$R_7 = \varepsilon A_7 \exp\left(\frac{-E_7}{RT_g}\right)\left(C_{CO}\right)\left(C_{O2}\right)^{0.5}\left(C_{H_2O}\right)^{0.5} \tag{22}$$

The mechanism for oxidation of methane, in which the reaction products are CO and H_2O [42], was used in this model:

$$R_8 = \varepsilon A_8 \exp\left(\frac{-E_8}{RT_g}\right)\left(C_{CH_4}\right)^{0.7}\left(C_{O2}\right)^{0.8} \tag{23}$$

The oxidation of ethylene was considered in the study to occur by a mechanism similar to that followed by other hydrocarbons [40]:

$$R_9 = \varepsilon A_9 \exp\left(\frac{-E_9}{RT_g}\right)\left(C_{C_2H_4}\right)^{0.5}\left(C_{O2}\right)T_g P^{0.3} \tag{24}$$

The last homogenous gas reaction considered in this model is oxidation of hydrogen which is formed during partial oxidation of hydrocarbons (C_6H_6 and C_2H_4). The global kinetic expression (see below) proposed by Varma *et al.* [43] was used in this study.

$$R_{10} = \varepsilon A_{10} \exp\left(\frac{-E_{10}}{RT_g}\right)\left(C_{H_2}\right)^{1.1}\left(C_{O2}\right)^{1.1} \tag{25}$$

All of the reactions and rate constants used in this model are summarized in Table 1.

Heat and mass transfer coefficient

The solid-gas heat transfer process is characterized by a convective heat transfer coefficient, which has been experimentally determined by a number of researchers using various experimental techniques, under steady state and unsteady-state conditions. Recently, the quite extensive collection of data on forced convection for the case of gases flowing through packed beds has been critically analysed and correlated. The resulting correlation (see below) was the one used in the model [44].

$$h_{sg} = Cp_g m''_g \left(2.19(\mathrm{Re})^{-2/3} + 0.78(\mathrm{Re})^{-0.381} \right)(\mathrm{Pr})^{-2/3} \tag{26}$$

Mass flux from a fluid to a solid surface is given by the product of the mass transfer coefficient and the concentration driving force. By applying an empirical analogy, known as the Chilton-Colburn analogy, a correlation for mass transfer in packed beds can be derived [44]. This correlation (shown below) was used in this study:

$$k_m = \frac{\rho_g}{m''_g} \left(2.19(\mathrm{Re})^{-2/3} + 0.78(\mathrm{Re})^{-0.381} \right)(Sc)^{-2/3} \tag{27}$$

Effective thermal conductivity

Two modes of heat transfer, conduction and radiation, were considered in developing the effective thermal conductivity term, since at the temperatures normally found in gasifiers, radiation becomes the most significant process, especially for a low conductivity solid material such as wood. The correlation proposed by Kunii and Smith [45] and shown below was used in the model.

$$k_{eff} = k_g \frac{\beta(1-\varepsilon)}{\gamma\left(\dfrac{k_g}{k_s}\right) + \dfrac{1}{\left(\dfrac{1}{\varphi} + \dfrac{d_p h_{rs}}{k_g}\right)}} + \varepsilon\left(k_g + d_p h_{rs}(\beta)\right) + \psi \,\mathrm{Re}\,\mathrm{Pr}\, k_g \tag{28}$$

where $\gamma = 2/3$, $\beta = 0.9 \sim 1$ and $\varphi = 0.1\varepsilon$,

$$h_{rv} = \text{Void-to-void radiation coefficient} = \frac{4\sigma T_g^{\,3}}{\left[1 + \dfrac{\varepsilon}{2(1-\varepsilon)}\left(\dfrac{1-\varepsilon'}{\varepsilon'}\right)\right]} \tag{29}$$

$$h_{rs} = \text{Solid radiation coefficient} = 4\sigma T_s^{\,3}\left(\frac{\varepsilon'}{2-\varepsilon'}\right) \tag{30}$$

and ψ depends on the ratio of particle diameter to effective reactor diameter [46]:

$$\psi = \frac{0.14}{1 + 46\left(\frac{d_p}{D}\right)^2} \quad (31)$$

Bed-to-wall heat transfer coefficient

The effective bed-to-wall heat transfer coefficient used in this model is calculated using a correlation developed by Froment and Bischoff [47]. Heat losses to the gasifier wall from the solid and gas phases were determined using equations initially proposed by DeWasch and Froment [46] and modified later by Hobbs *et al.* [10].

Conservation of mass

Mass balance equations were developed for several components, as described below.

Solid Phase:

Moisture:
$$(1-\varepsilon)\frac{\partial \rho_m}{\partial t} + (1-\varepsilon)\frac{\partial \rho_m V_S}{\partial x} = -R_m \quad (32)$$

Volatile matter:
$$(1-\varepsilon)\frac{\partial \rho_V}{\partial t} + (1-\varepsilon)\frac{\partial \rho_V V_S}{\partial x} = -R_{p1} \quad (33)$$

Char:
$$(1-\varepsilon)\frac{\partial \rho_C}{\partial t} + (1-\varepsilon)\frac{\partial \rho_C V_S}{\partial x} = -\sum_{i=1}^{4} R_i \quad (34)$$

Char density along the reactor was assumed constant and equal to 0.205 times the biomass density (this is derived from the product yield data given in Fig. 1). This assumption causes the time dependent char density term in Eq. 34 to disappear and enables the solid velocity along the gasifier (V_S) to be determined.

The rates of mass loss from particles are important when it comes to determining local heat and mass transfer coefficients. These are a function of char particle diameter, which diminishes along the combustion and gasification zones of the reactor as char is consumed. Assuming that no fragmentation, attrition or agglomeration of particles takes place, the reduction in the particle diameter can be related to the solid velocity, V_S, [48] and can be calculated as follows:

$$\frac{\partial}{\partial t}\left(\frac{(1-\varepsilon)\rho_C}{m_P}\right) + \frac{\partial}{\partial x}\left(\frac{(1-\varepsilon)V_s\rho_C}{m_P}\right) = 0 \qquad \text{where } m_P = \rho_C \frac{\pi}{6} d_p^{\ 3} \quad (35)$$

Gas Phase Species:

A generalised mass balance equation for the gas phase species is shown below. Equation numbers (36) to (44) refer respectively to the equations for the individual components listed below the generalised equation.

$$\varepsilon\frac{\partial \rho_i}{\partial t}+\frac{\partial m_i''}{\partial x}=\sum_{j=1}^{4}\nu_i\frac{M_i}{M_C}R_j+\sum_{j=5}^{10}\nu_i M_i R_j+X1_i R_{p1}+X2_i R_{p2}+C_i R_m \quad (36\text{-}44)$$

where, $i = O_2, CO, CO_2, H_2, H_2O, CH_4, C_2H_4, C_6H_{10.71}O_{3.264}, C_6H_6$

(Note that $C_{H2O}=1$, and that for $i \neq H_2O$ the term C_i is zero. This is because the rate of drying, R_m, is only applicable for the H_2O mass balance equation.)

$X1_i$ = (Yield of species i in primary pyrolysis) / (Total yield of volatiles)

$X2_i$ = (Yield of species i in secondary pyrolysis) / (Total yield of primary tar)

[i.e. $X1_{CO}$ = (3.2/79.5), $X2_{CO}$ = (27.62/52.8)]

The overall gas mass balance equation is:

$$\varepsilon\frac{\partial \rho_g}{\partial t}+\frac{\partial m_g''}{\partial x}=R_{p1}+R_m+\sum_{i=1}^{4}R_i \quad (45)$$

Conservation of energy

The energy balance relationships include individual equations for the solid phase and for the gas phase as described below.

Solid Phase:

$$(1-\varepsilon)\frac{\partial}{\partial t}(\rho_s h_s)+\frac{\partial}{\partial x}(m_s'' h_s)=\frac{\partial}{\partial x}\left(k_{eff}\frac{dT_s}{dx}\right)-h_{sg}A_V(T_s-T_g)-\frac{4h_{sw}}{D}(T_s-T_w)+R_m\Delta H_m+R_{p1}\Delta H_{p1}+\sum_{j=1}^{4}R_j\Delta H_j \quad (46)$$

Gas Phase:

$$(\varepsilon)\frac{\partial}{\partial t}(\rho_g h_g)+\frac{\partial}{\partial x}(m_g'' h_g)=\frac{\partial}{\partial x}\left(k_g\frac{dT_g}{dx}\right)+h_{sg}A_V(T_s-T_g)-\frac{4h_{gw}}{D}(T_g-T_w)+R_{p2}\Delta H_{p2}+\sum_{j=5}^{10}R_j\Delta H_j \quad (47)$$

The boundary conditions

Operator determined boundary conditions include the inlet air composition, flow rate and temperature, and the fuel (wood) composition and temperature at the inlet to the reactor. Both the fuel and air inlet temperature were set to be the same as the ambient temperature; since the fuel bed is relatively thick the amount of heat transferred upwards from the combustion zone to the unreacted wood at the top of the gasifier is negligible. The gas flow rate at the top of the fuel bed was set equal to the air supply rate and the inlet gas composition is that of air.

Even at the very bottom of the char bed, on the grate, temperatures are still high. This means that radiative heat transfer will occur between the base of the char bed and the ash pit since heat losses to the surroundings will cause ash pit temperatures to be lower than those at the base of the char bed. Accurately determining the ash pit temperature is difficult so in the model this temperature was defined as a function of the temperature at the base of the reactor bed. Since reaction rates are negligible at the base of the reactor, temperature gradients at the lower boundary of the bed can be represented as follows:

for the solid phase,
$$k_{eff}\frac{dT_s}{dx} = h_{rs}(T_s - T_{ashpit}) \tag{48}$$

for the gas phase,
$$k_g\frac{dT_g}{dx} = h_{rv}(T_g - T_{ashpit}) \tag{49}$$

Zero char loss through the grate is assumed so that V_S, the solid fuel velocity, can be set to zero at the bottom of the reactor.

The initial conditions

Before the model could be run, the initial conditions in the system needed to be specified. The initial conditions used in the model simulations were selected so as to match as closely as possible those employed in an experimental investigation on a stratified downdraft gasifier conducted as part of this study [15].

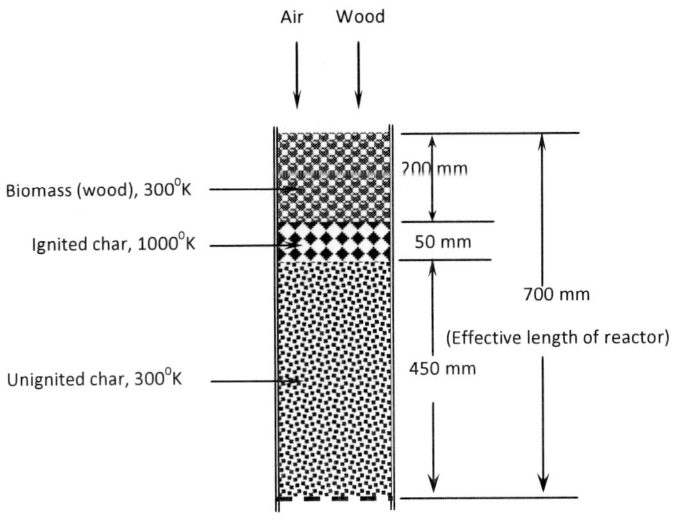

Figure 2. Initial conditions in the model reactor.

Gas temperatures were set equal to the inlet air temperature. It was assumed that a 500 mm depth of char was present in the reactor. The top 50 mm of char was assumed to be the combustion initiating fuel layer. This was a necessary assumption, since the chemical reactions are temperature dependent and some fraction of the fuel must be assumed to be at the ignition temperature. Suitable ignition temperatures range from 900 K to 1200 K. In this model an intermediate ignition temperature value of 1000 K was specified. It was assumed that wood is only added to the reactor once ignition has occurred and that, as suggested by Bhattacharya *et al.* [49], it is supplied continuously together with air, so that the release of large amounts of tar and smoke formed as the bed is warming up can be avoided. Other parameters, such as mole fractions and particle diameter, were assumed to be uniform along the reactor at the start of the run. The initial conditions in the reactor are illustrated in Fig. 2.

Additional assumptions

Other assumptions made in developing the model are:

1) The gas mixture behaves as an ideal gas.
2) Gas and solid phases move through the reactor in plug flow (i.e. temperatures can be assumed uniform in the radial direction).
3) Void fraction is assumed to remain constant at its initial value.
4) No pressure difference exists along the reactor.
5) Particles are of uniform diameter and spherical in shape.

Numerical method

The sixteen partial differential equations governing the mass and energy balances (eqs. 32-47) were solved for gas and solid temperatures, gas species mole fraction, solid density and velocity, and particle diameter by using the control volume method proposed by Patankar [50]. The discretisation equations were derived from the partial differential equations by applying the fully implicit scheme for time and the upwind scheme for space (i.e. a central differencing scheme was used for diffusion terms, and backward differencing was used for convective terms). Since the reaction source terms are highly nonlinear, these were linearised to achieve a converged solution using a linearization method recommended by Patankar [50]. In the iterative solution of strongly nonlinear equations such as those in use here, unacceptably large changes in parameter values can occur in a single iteration. To avoid the problems this introduces, an underrelaxation process, which slows down the changes between successive iterations, was applied [50]. In the oxidation zone, where very steep temperature and mole fraction gradients occur, an adaptive grid method (static rezone method) was used to produce computational grids with nodes that are highly non-uniformly distributed in space [51].

When solving the numerical model, the sequence of steps followed during each time interval was as follows:

For all points along the gasifier:

1) Calculate the values of temperature dependent properties.
2) Determine solid velocity values (starting from the bottom of the reactor) (Eq. 34).
3) Calculate solid densities and particle sizes (Eqs. 32, 33, 35).

4) Calculate the mole fractions of major gas species using mass balance equations (Eqs. 36-45).
5) Estimate gas and solid temperatures from the energy balance equations (Eq. 46 & 47) using a *TDMA* (TriDiagonal-Matrix Algorithm) method.
6) Repeat steps 1-5 until both solid and gas temperatures converge.

The size of the time step used in this model was determined by trial and error. A small time step improves the accuracy of the solution but increases the computing time. A large time step enables the program to be run faster but could result in a failure to converge. The range of suitable time steps determined for this model was 1 to 4 seconds. Tests showed that a 3 second time step was the most suitable for use in the computer program developed.

The expressions used in this study for predicting temperature dependent thermophysical and transport properties of solid fuel are listed in Table 2. The selected operating parameter values were chosen to be as representative as possible of experimental conditions. These parameter values are shown in Table 3.

Table 2. Thermophysical data and transport properties of solid fuel

Quantity	Expression (or) value	Source	Remark
Thermal conductivity (solid particle) (W/m K)	$= 0.13 + (3.0 \times 10^{-4})\, T$	[52]	- -
Thermal conductivity (gas mixture) (W/m K)	$= (4.77 \times 10^{-4})\, T^{0.717}$	[53]	- -
Specific heat (solid) (kJ/kg K)	$= 0.1031 + 0.003867 * T$	[54]	
Specific heat (gas mixture)	$= 9.41079 \times 10^{2} + 1.66918 \times 10^{-1}\, T_g + 6.50858 \times 10^{-5}\, T_g^{2} - 3.63930 \times 10^{-8} T_g^{3}$	[55]	Polynomials curves fitting for air
Viscosity (gas mixture) [kg/(ms)]	$= (1.98 \times 10^{-5}) \times (T/T_0)^{2/3}$	[56]	-
Absolute enthalpy (gas species)	-	[57]	Estimated from absolute enthalpy
Heat of reactions	-	[57]	Fuller *et al.* equation
Molecular diffusivity (gas species)	-	[58]	-
Char emissivity	= 0.85	[35]	

Table 3. Baseline operating parameter values used in the model

Quantity	Value
Particle diameter (mm)	20
Apparent density of fuel (kg/m³)	950
Moisture content of the fuel (%db)	10
Inner diameter of reactor (mm)	206
Effective length of the reactor (mm)	700
Reactor pressure (kPa)	101.325
Inlet air temperature (K)	300
Ignition temperature (K)	1000
Ignition depth (mm)	50
Air supply rate (kg/hr)	15
Fuel bed initial void fraction (-)	0.46

3. Results and Discussions

Temperature changes over time

At the beginning of a simulation, ignition was assumed to occur across a 50 mm deep section of the char bed and char oxidation within this layer was therefore the dominant heat source in the gasifier. Most of the heat released from this zone of combustion is initially utilized to heat up the lower part of the char bed but some initiates devolatilisation in those wood particles immediately above the combustion zone. While the char below the combustion zone is still heating up, rates of gasification are very low. However, above the ignition zone, pyrolysis of the lower part of the wood bed is by this time taking off quite rapidly. The volatiles released in this process burn readily and, as a result, the combustion zone moves upwards. This phenomenon is discussed in detail in Section 3.6. After sufficient time has elapsed for the entire char bed to heat up, temperature profiles stabilise and the gasifier starts producing combustible gas. This latter condition is known as the stable operating mode of the reactor. Based on model predictions, for all the combinations of model and operating parameter inputs employed, 30 minutes of operation was sufficient for the gasifier to achieve stable operation. Therefore the results presented in this paper as representing stable operating conditions are those predicted for a time 30 minutes after ignition.

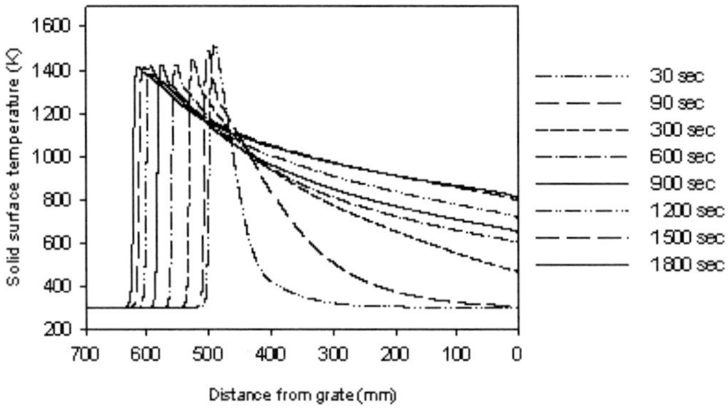

Figure 3. Transient behaviour of solid surface temperature profile.

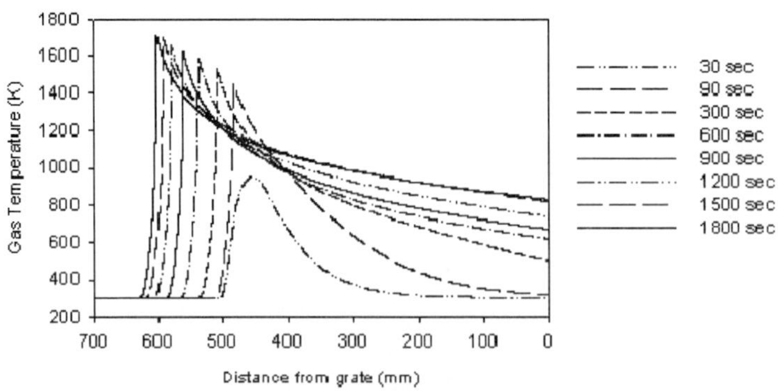

Figure 4. Transient behaviour of gas temperature profile.

The predicted changes over time (from ignition to stable operating condition) of temperatures in the solid phase are shown in Fig. 3. The gas temperature was initially assumed to be the same as the inlet air temperature. It then increases as the gas absorbs heat released during the char oxidation reaction. Since there are no gas phase oxidation reactions at the beginning of the start up period, the gas temperature is initially lower than the solid temperature. Once gas phase oxidation of pyrolytic decomposition products (volatile gases and secondary tar) begins, the gas temperature increases very rapidly to a peak level. Heat released from oxidation of volatiles is utilized to produce more volatile matter from the pyrolysis zone and this continues until pyrolysis is complete. This phenomenon is also known as *flaming combustion* in the case of a combustor and *flaming pyrolysis* in the case of a downdraft gasifier. An important aspect of flaming pyrolysis is the rapid evolution of large volumes of a highly combustible gas having a potential flame temperature > 1500°C for air and > 2500°C for oxygen [22].

Before the reactor reaches a stable operating condition, the heat released during gas phase oxidation also contributes to heating up the lower part of the char bed by convective gas-to-solid heat transfer. Fig. 4 shows the predicted changes of gas phase temperature with time.

Gas composition changes over time

Fig. 5 shows the predicted changes in gas composition along the axis of the gasifier during the period from start up to achievement of a stable condition. Initially, while the char bed below the ignition level is being heated up by heat released in the oxidation zone, heterogeneous gasification reaction rates are not high enough to produce gas with a significant heating value; this is because the concentrations of the two main combustible components, carbon monoxide and hydrogen, are very low.

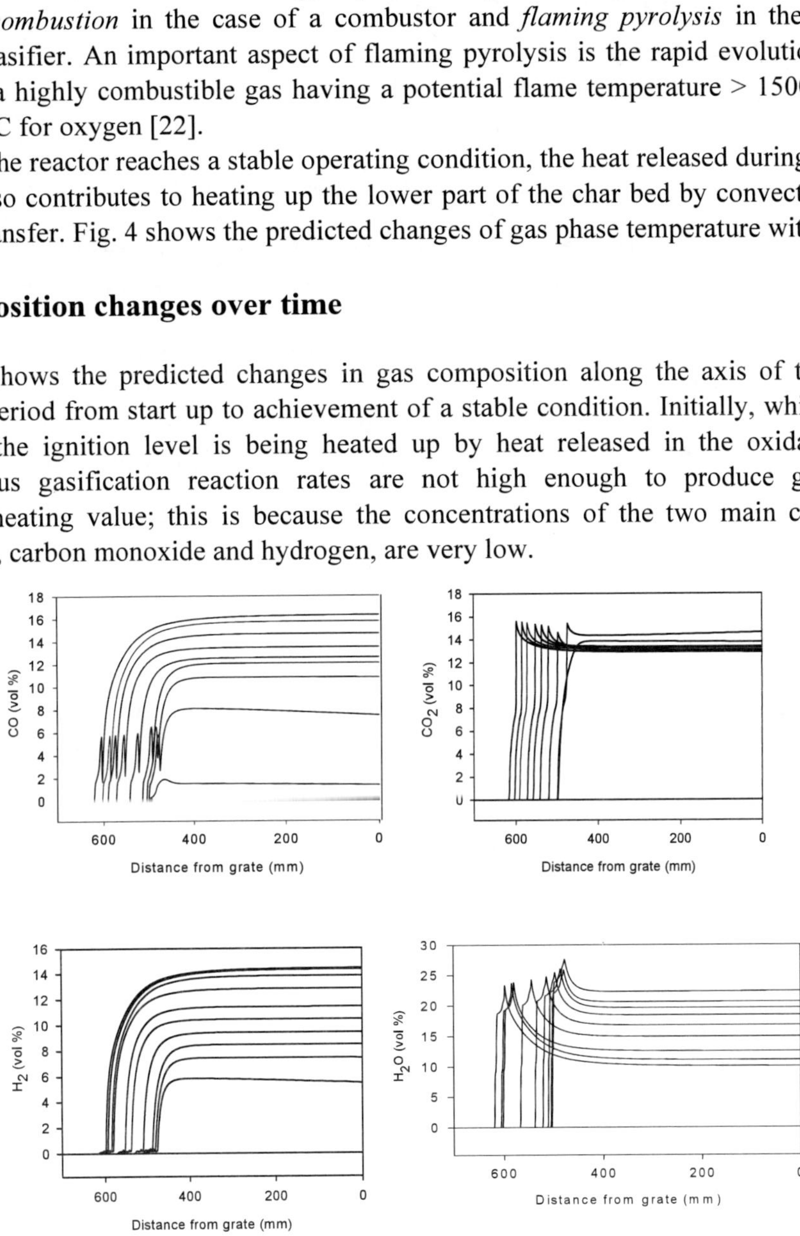

Figure 5. Transient behaviour of gas composition profile (a) CO, (b) CO_2, (c) H_2, (d) H_2O.

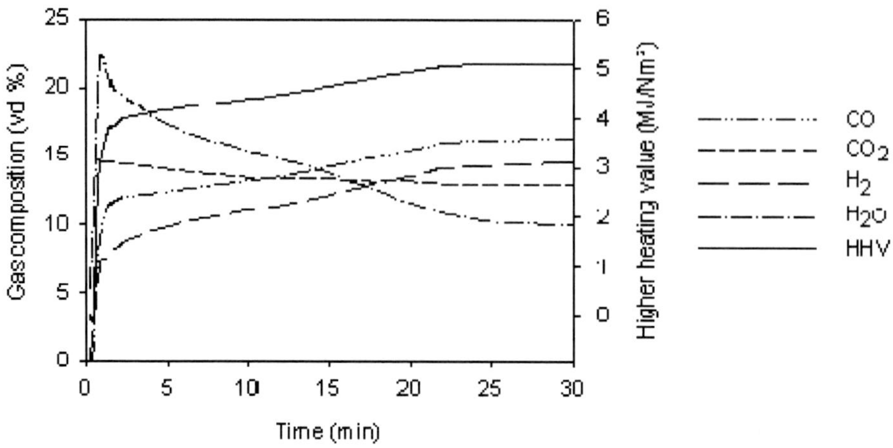

Figure 6. Transient behaviour of outlet producer gas properties.

As temperatures in the gasification zone increase over time, then, as would be expected, the amounts of CO and H_2 along the gasification zone increase while the amounts of CO_2 and H_2O decrease correspondingly.

In Fig. 6, outlet gas composition changes over time are shown together with the changes in the corresponding higher heating value of the producer gas (dry basis). It can be seen that, not unexpectedly, the heating value of the produced gas is very poor at the beginning but increases over time till it reaches a steady value. Model results show that it takes about 20-25 min to reach a stable condition.

Temperature profiles

Fig. 7 shows the predicted steady state temperature profiles for the set of baseline operating parameters. These results show that as one moves down the gasifier, the solid surface temperature increases very sharply at the point where the highly exothermic char oxidation reaction starts. The increase in solid temperature causes an increase of gas temperature through convective solid-to-gas heat transfer. When the gas temperature is hot enough to initiate oxidation of volatile gases and secondary tars, which are the products of primary and secondary pyrolysis, the gas temperature increases very rapidly. Then, once all the volatiles have been oxidised, the gas temperature ceases to rise any further. This peak in the gas temperature occurs slightly further down the gasifier than the peak in the solid temperature.

Beyond its peak, the solid temperature starts to decrease. This happens mainly because exothermic surface oxidation reactions cease, due to utilization of the remaining oxygen in gas phase oxidation reactions. Also, the predominantly endothermic reactions of char with steam and carbon dioxide that are characteristic of the gasification zone start to increase in importance. Gas temperatures, having peaked at the end of the oxidation zone, now decrease, mainly due to heat loss to the solid phase (where the endothermic gasification reactions take place) and to the reactor walls.

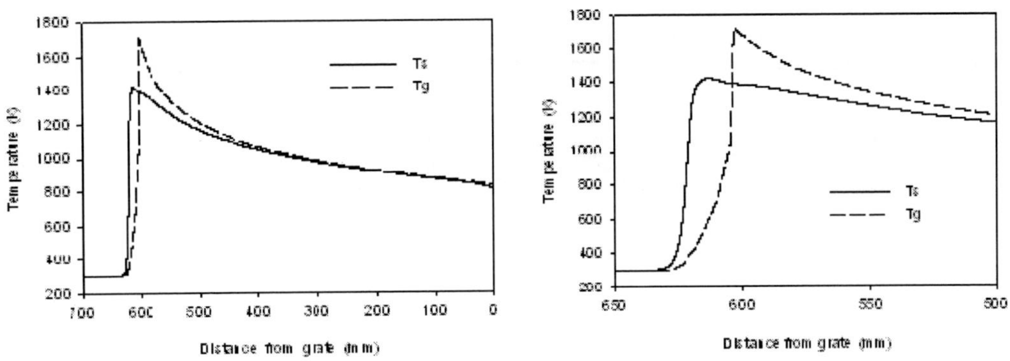

Figure 7. The predicted solid (Ts) and gas (Tg) temperature profiles along the axis of the gasifier (a) over the total reactor length (b) in the domain where rates of change are greatest.

The increase in solid temperature from ambient levels to a peak of around 1450 K is predicted to occur within a zone only about 20 mm long (Fig. 7b). The length of the region over which the gas phase temperature increases from ambient to its maximum (about 1700 K) is found to be similar to the value of 20 mm computed for the solid phase. Solid temperatures lead gas temperatures over the first few millimetres of this zone of rapid change and then gas temperatures lead over the rest of the reactor length. Differences between the solid and gas temperatures are substantial only in the combustion zone and tend to become smaller in the reduction zone.

Velocity profiles

The predicted solid and gas velocity profiles along the gasifier are shown in Fig. 8. The solid velocity at the top of the reactor shows as constant; this is because one of the assumptions made in developing the model was that particle shrinkage due to moisture evaporation and primary pyrolysis processes could be neglected. As mentioned earlier, this was not expected to have any significant impact on what was predicted to happen lower down in the gasifier. Once the char oxidation and gasification zones are reached, however, the particle size begins to reduce, causing the fuel velocity to decrease. This velocity drops to zero at the grate since there was assumed to be no char loss through the grate.

The predicted superficial gas velocity profile along the gasifier shows a very different pattern. In the upper part of the gasifier, that contains unreacted biomass, the gas velocity is constant. It then increases rapidly due partly to the large amount of gas released in drying and volatilization processes and in heterogeneous gas-char oxidation reactions, and partly due to expansion caused by increases in the gas phase temperature. Its peak

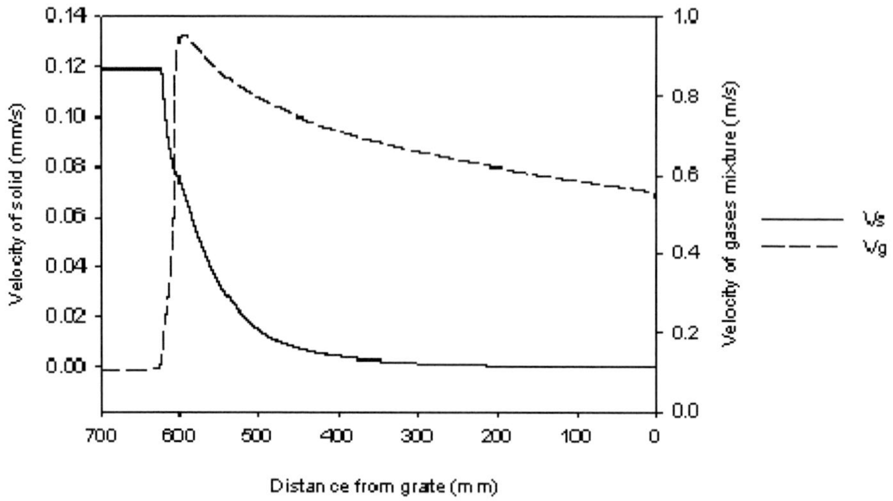

Figure 8. The solid and gas velocity profiles (Vs and Vg respectively) along the axis of the gasifier.

occurs close to the maximum gas temperature point and thereafter the gas velocity gradually decreases as gas phase temperatures fall along the rest of the gasifier.

Gas composition profiles

Fig. 9 and Fig. 10 show the predicted gas composition profile along the reactor under steady state conditions. Fig. 9 shows that the volume percent of oxygen drops rapidly as the air enters the drying and volatilization zone. This does not mean that oxygen is being consumed in these zones. In fact the amount of oxygen remains almost the same, but due to dilution by water vapour and primary pyrolysis volatiles, the concentration of oxygen in the total gas is reduced. When the solid surface temperature reaches a temperature of around 900 K, oxygen begins to be consumed in heterogeneous char oxidation reactions, causing the solid temperature to increase further. At this stage, heat produced during solid phase combustion reactions is transferred to the gas phase by convection, increasing the gas temperature. Once the gas temperature is high enough for homogenous oxidation of combustible volatiles to begin, the rest of the oxygen is consumed very rapidly. Fig. 9b illustrates how rapidly the oxygen volume percent decreases from 21 to zero. This plot can be divided into 3 parts, each with a different gradient: the first drop is largely attributable to mass transfer from the solid phase to the gas phase as a result of drying and primary pyrolysis processes; the second is largely attributable to oxygen consumption during char oxidation reactions; and the last is associated with gas oxidation reactions. Since gas phase oxidations occur very rapidly and are completed over just a few millimetres of the oxidation zone, the gradient of the last segment of the oxygen content plot is very steep.

Carbon monoxide is the main combustible component in the producer gas. The model predicts that the volume percentage of CO initially increases along the reactor because of CO evolution during pyrolysis processes, especially secondary pyrolysis. Whilst free oxygen is present, most of the CO produced in the pyrolysis zone is subsequently lost by oxidation to CO_2. This accounts for the small peak of CO within the combustion zone shown in Fig. 9b. A similar result has been reported for a char combustor model [48]. Once the char gasification zone is reached, CO levels build up again as a result of the reaction of char with H_2O and CO_2.

As would be expected, hydrogen concentrations are predicted to be very low in the oxidation zone. Although hydrogen is formed during oxidation of C_6H_6 and C_2H_4,

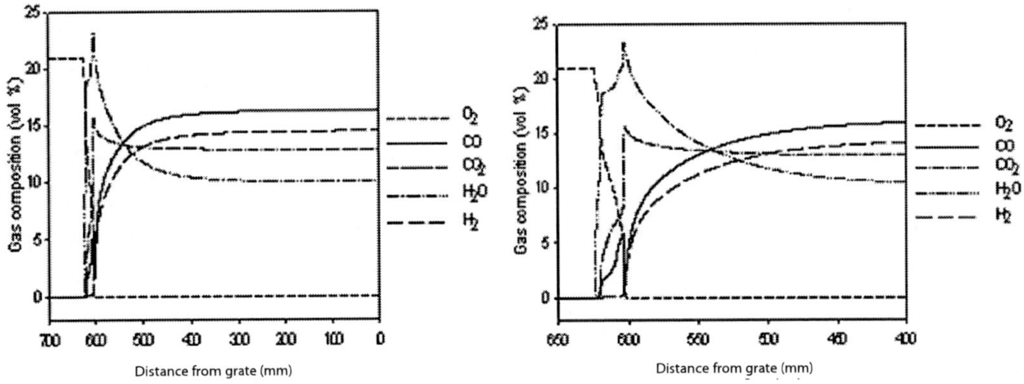

Figure 9. Major gas species profiles along the axis of the gasifier (a) over the total reactor length (b) in the domain where rates of change are greatest.

temperatures are so high that in the presence of free oxygen it is instantaneously oxidized to water. The content of hydrogen in the gas mixture, therefore, only increases once oxygen levels are depleted. The final hydrogen content in the producer gas is determined by the extent of char reactions with H_2O in the char gasification zone.

Initially CO_2 is formed by combustion of char. Further amounts of CO_2 are then produced by combustion of CO and, as shown in Fig. 9, CO_2 levels reach a maximum quite close to the point where the gas temperature is at its maximum. Then, in the gasification zone, CO_2 concentrations decrease monotonically due to its reaction with char to produce CO, finally levelling out at a more-or-less constant value.

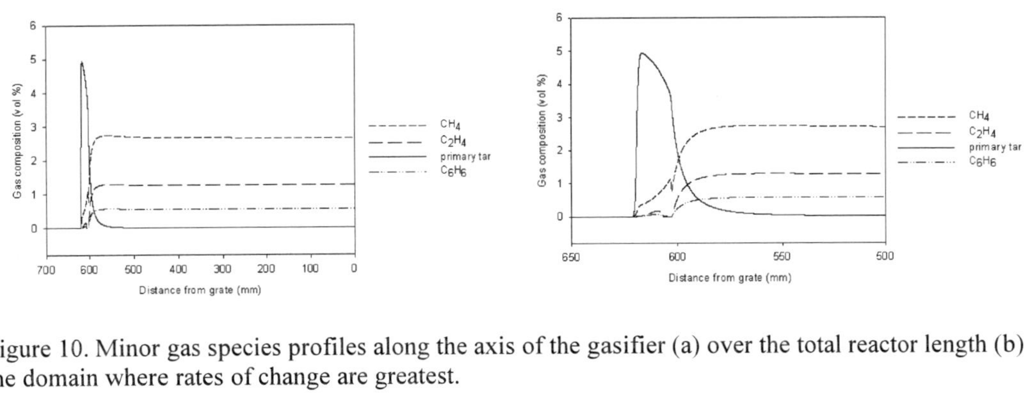

Figure 10. Minor gas species profiles along the axis of the gasifier (a) over the total reactor length (b) in the domain where rates of change are greatest.

Initial H_2O concentrations in the gas phase depend upon the moisture content of the biomass fuel. Fig. 9 shows H_2O levels building up due to the evaporation of moisture and being augmented by the water formed when hydrogen formed in primary pyrolysis processes is oxidised. Concentrations of H_2O, like those of CO_2, decrease in the gasification zone; this occurs as a result of the heterogeneous reaction of carbon with water vapour.

Other, minor, gaseous components include methane and ethylene, products of tar cracking; Fig. 10 shows that these are partially oxidized in the combustion zone with only small amounts being found in the final producer gas. Also shown in Fig. 10 are the predicted levels of primary tars (oxygenates), the major products of the primary pyrolysis reaction. The initial mass of these tars is more than half that of the initial biomass fuel but most primary tars are rapidly cracked into lower molecular weight gases and secondary tars (C_6H_6). These products are mostly oxidized in the combustion zone. However, as shown, not all primary tars are cracked before the gases leave the combustion zone. Much of this residual tar decomposes below the combustion zone, producing secondary tars that cannot be oxidized since no oxygen is available. This behaviour is more clearly shown in Fig. 11, which shows that cracking of primary tars still continues in the oxygen depleted zone beyond the point where the peak gas temperature occurs. This helps explain why small amounts of tar are still present in the gas produced from the conventional stratified downdraft gasifier.

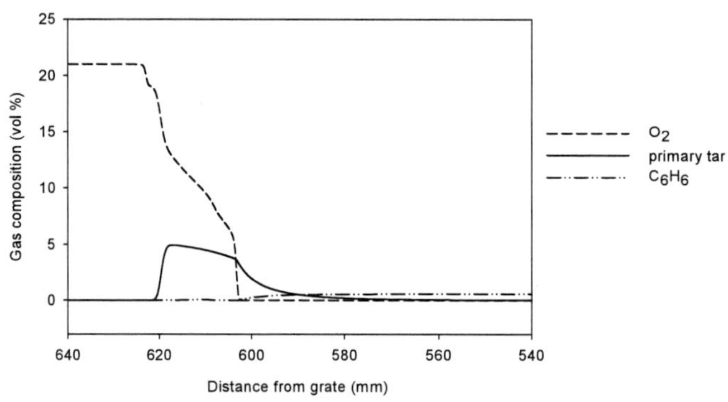

Figure 11. Primary tar, secondary tar, and oxygen profiles along the reactor axis.

Stability of reaction zones

In the downdraft stratified wood gasifier, the reaction front moves slightly upwards while the reactor is heating up; this occurs because of heat losses to the reactor wall and the low gasification rate. However, once the gasifier achieves full gas production, three operating modes are possible [59]:

1) Gasification dominant, where the rate of char formation by pyrolysis is slower than the rate of char consumption and as a result the reaction zones move downwards.
2) Pyrolysis dominant, where the rate of char formation by pyrolysis is faster than the rate of char consumption and as a result the reaction zones move upwards.
3) Steady state operation, where the rate of char formation by pyrolysis is equal to the rate of char consumption and, as a result, the reaction zone does not move.

This is the optimum mode of operation since the gasifier can be run for long periods with a relatively constant output.

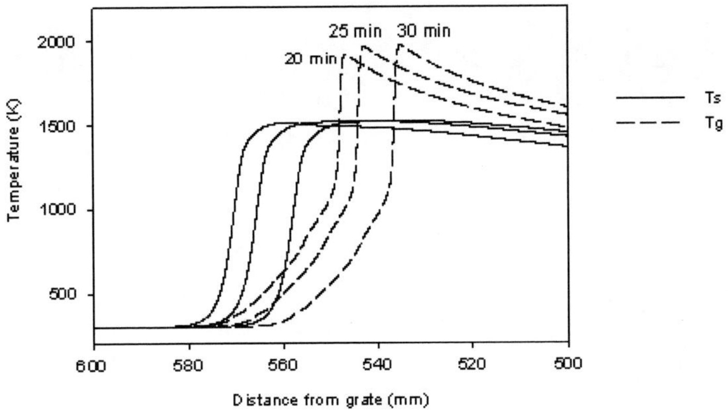

Figure 12. Illustration showing downward movement of the reaction zone as a result of an increase in the air supply rate (24 kg/hr).

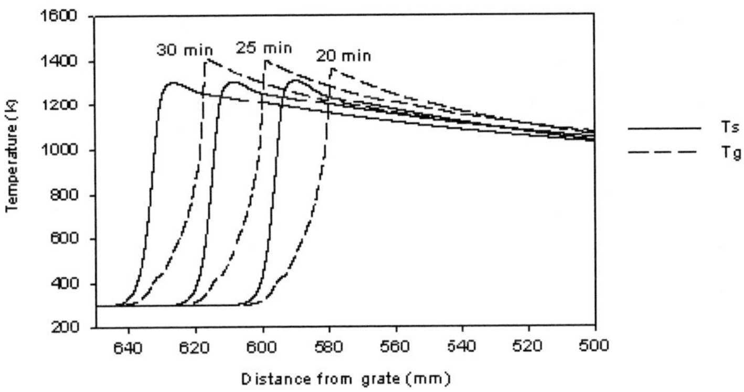

Figure 13. Illustration showing upward movement of the reaction zone as a result of a decrease in the air supply rate (8 kg/hr).

For a given gasifier operating under a specific set of conditions there is only one air to fuel ratio for which the reaction zones will be stationary [60]; this stability can easily be lost since even small changes in any one of the operating parameters will cause the reaction zones to start moving in one direction or the other [61]. Where upward movement of the front occurs there are similarities to what happens to the flame front in a flame speed experiment involving a premixed gas mixture. In both cases the front moves into an unburnt zone; in the latter case this contains a premixed gaseous fuel-air mixture, while in the gasifier it contains unburnt char and air [62].

The stability of the reaction zone was found to be particularly sensitive to changes in the air supply rate to the reactor. An increase in air supply rate increases the oxygen supply and hence the bed temperature and char consumption rates. As a result, reaction zones move downwards towards the grate. If on the other hand the air supply rate is reduced, the oxidation and gasification zone temperatures decrease, the rate of char consumption falls below the char production rate, and, as a result, the reaction zone moves upwards. This is illustrated by the model outputs shown in Fig. 12 and Fig. 13.

The range of air supply rates for which a steady state mode of operation can be achieved was found to be between 19 and 20 kg/hr for the base case (parameters defined in Table 3) giving a specific gasification rate of 290-296 kg/m^{2}hr (dry basis). This result is close to the value found in experimental work on an open core gasifier in which the specific capacity for stable operation was found to be 271 kg/m^{2}hr (DAF basis) [63].

Conclusion

A dynamic model has been developed for a stratified downdraft wood gasifier by applying a two-step pyrolysis mechanism in which primary tars (oxygenates) are initially formed and then cracked into secondary tars (hydrocarbons) and other combustible gases. The model is therefore capable of predicting the primary and secondary tar profiles along the gasifier as well as the gas composition profile and temperature profiles along the axis of the gasifier.

At the start of the combustion zone, solid surface temperatures are higher than gas temperatures but gas temperatures rise rapidly and exceed solid surface temperatures through the rest of the gasifier. Therefore the rates of the gas phase oxidation reactions peak at the end of the oxidation zone, where the gas temperature increases sharply.

In the outlet gas the concentrations of the two main combustible components, carbon monoxide and hydrogen, are very low when the gasifier is started up but gradually increase to a stable value. The model shows that 20-25 min is required to reach a stable operating condition. Within the gasifier, under such conditions, some CO is present in the gas leaving the oxidation zone but the major increase in CO concentration takes place in the gasification zone. The hydrogen content of the outlet gas is mainly determined by steam-carbon reactions in the char gasification zone.

Drying and the primary pyrolysis process are predicted to start and be completed over a relatively short distance (a few millimetres) just above the oxidation zone. Within this latter zone the solid surface temperature rises very steeply. Since most of the primary tars are cracked into secondary tars at the high temperatures present in the oxidation zone, more secondary tars (hydrocarbons) than primary tars (oxygenates) are found in the outlet gas from the gasifier. Model results also predict an average range of tar content between 20 and 200 mg/Nm^3 for the range of input parameters used in the model simulation, assuming that the fuel bed below the ignition ports was initially full of charcoal. The stability of the reaction zones was also investigated and for stable operation it was found that the specific gasification rate needed to be 290-296 kg/m^{2}hr (dry basis) which is close to experimental results obtained on an open core biomass gasifier.

It is appreciated that in developing the model described above some simplifying assumptions have been made. For example, in modelling the kinetics of primary pyrolysis processes it was not practicable to allow for the temperature gradients that exist within an individual particle and an equivalent temperature for the whole particle was used. In addition, no tractable way was found of incorporating into the model a means of accurately representing the anisotropic particle shrinkage and cracking processes that occur during pyrolysis. However, as shown in subsequent papers in which the sensitivity of the model to variations in operating and other parameters is examined [14] and the predictions of the model are compared with measurements made on a pilot scale gasifier constructed to validate the numerical model [15], model predictions are well supported by experimental results.

ACKNOWLEDGMENT

Financial support from the University of Melbourne for one of the authors, San Shwe Hla, is gratefully acknowledged.

REFERENCES

[1] International Energy Agency. World energy outlook 2000. IEA, Paris; 2000.

[2] Maniatis K. Progress in biomass gasification: An overview. In: Progress in thermochemical biomass conversion 2001. p. 1-31.

[3] Knoef HAM. Status and development of fixed bed gasification. Report EWAB 9929, NOVEM, Utrecht; 2000.

[4] Knoef HAM. Inventory of biomass gasifier manufacturers & installations. Final report to European commission, Contract DIS/1734/98-NL, Biomass Technology Group B.V, University of Twente, Enschede; 2000.

[5] Henriksen U, Christensen O. Gasification of straw in a two stage 50 kW gasifier. Proceedings of the 8th European conference of biomass for energy, environment, agriculture and industry, Chartier, P.H., Beenackers, A.A.C.M, Grassi, G (eds), Pergamon, Elsevier Science Ltd, Oxford, U.K; 1994. p. 1568-78.

[6] Bui T, Loof R, Bhattacharya SC. Multi-stage reactor for thermal gasification of wood. Energy 1994;19:397-404.

[7] Susanto H, Beenackers AAC. A moving-bed gasifier with internal recycle of pyrolysis gas. Fuel 1996;75:1339-47.

[8] Brandt P, Larsen E, Henriksen U. High tar reduction in a two-stage gasifier. Energy & Fuels 2000;14:816-9.

[9] Bhattacharya SC, San Shwe Hla, Pham H-L. A study on a multi-stage hybrid gasifier-engine system. Biomass & Bioenergy 2001;21:445-60.

[10] Hobbs ML, Radulovic FT, Smoot LD. Modelling of fixed-bed coal gasifiers. AIChE Journal 1992;38:681-702.

[11] Groeneveld HJ. The co-current moving bed gasifier. PhD Dissertation. Twente University of Technology, Enschede, The Netherlands; 1980.

[12] Chen J, Gunkel W. Modeling and simulation of Co-current moving bed gasification reactors- Part II: A detailed gasifier Model. Biomass 1988;14:75-98.

[13] Di Blasi C. Dynamic behaviour of stratified downdraft gasifiers. Chemical Engineering Science 2000;55:2931-44.

[14] San Shwe Hla, Lu Aye, Connor MA, Stewart DF. Chapter 6: A dynamic model for predicting tar content in gas from a stratified downdraft gasifier: Part II – Model

sensitivity analysis, in: Gasification: Chemistry, Processes and Applications (Ed: Michael D. Baker), Nova Science Publishers, Inc. 2011.
[15] San Shwe Hla, Lu Aye, Connor MA, Stewart DF, Fung P, Ng S. Chapter 7: A dynamic model for predicting tar content in gas from a stratified downdraft gasifier: Part III –Experimental validation, in: Gasification: Chemistry, Processes and Applications (Ed: Michael D. Baker), Nova Science Publishers, Inc. 2011.
[16] Salazar CM. The influence of particle size and shape on the mechanisms of decomposition of wood during pyrolysis. PhD Dissertation. The University of Melbourne, Australia; 1987.
[17] Basak AK, Bhattacharya SC. Performance of a down-draft charcoal gasifier. Applied Energy 1987; 26:193-216.
[18] Purnomo, Aerts DJ, Ragland KW. Pressurized downdraft combustion of wood chips. In: Twenty-third International Symposium on Combustion. The combustion Institute; 1990. p.1025-32.
[19] Bryden KM. Computational modeling of wood combustion. PhD Dissertation. University of Wisconsin-Madison; 1998.
[20] Chan W-CR, Kelbon M, Krieger BB. Modelling and experimental verification of physical and chemical processes during pyrolysis of a large biomass particle. Fuel 1985;64:1505-13.
[21] Chan W-CR, Kelbon M, Krieger, BB. Product formation in the pyrolysis of large wood particles. In: Fundamentals of Thermochemical Biomass Conversion 1985. p.219-36.
[22] Reed TB, Markson M. A predictive model for stratified downdraft gasification of biomass. Progress in biomass conversion, vol.4, edited by Tillman DA & Jahn EC; 1983. p. 217-54.
[23] Morf P, Hasler P, Hugener M, Nussbaumer T. Characterization of products from biomass tar conversion. In: Progress in Thermochemical Biomass Conversion 2001. p. 150-61.
[24] Fagbemi L, Khezami L, Capart R. Pyrolysis products from different biomasses: application to the thermal cracking of tar. Applied Energy 2001;69:293-306.
[25] Rath J, Staudinger G. Cracking reactions of tar from pyrolysis of spruce wood. Fuel 2001;80: 1379-89.
[26] Boroson ML, Howard JB, Longwell JP, Peters WA. Product yields and kinetics from the vapor phase cracking of wood pyrolysis tars. AIChE Journal 1989;35:120-8.
[27] Liden AG, Berruti F, Scott DS. A Kinetics model for the production of liquids from the flash pyrolysis of biomass. Chemical Engineering Communications 1988;65:207-21.
[28] Diebold JP. The cracking kinetics of depolymerized biomass vapours in a continuous, Tubular reactor. M.S Thesis. Dept. Chem. Petroleum-Refining Eng., Colorado Sch. Mines, Golden; 1985.
[29] Kosstrin H. Direct formation of pyrolysis oil from biomass. In: Specialists workshop on fast pyrolysis of biomass, Copper Mountain, Colorado: 19-22 October 1980. p. 105-121.
[30] Milne TA, Evans RJ. Biomass gasifier "tars": Their nature, formation and conversion. Golden, CO: National Renewable Energy Laboratory, the U.S Department of Energy; 1998.

[31] Peters B, Bruch C. A flexible and stable numerical method for simulating the thermal decomposition of wood particles. Chemosphere 2001;42:481-90.

[32] Veras CAG, Saastamoinen J, Carvalho JA Jr., Aho M. Overlapping of the devolatilization and char combustion stages in the burning of coal particles. Combustion and Flame 1999;116:567-79.

[33] Roberts AF. A Review of Kinetics Data for the Pyrolysis of wood and related substances. Combustion and Flame 1970;14:261-72.

[34] Bryden KM, Ragland KW. Numerical modeling of a deep, fixed bed combustor. Energy and Fuel 1996;10:269-75.

[35] Gronli MG, Mellaaen MC. Mathematical model for wood pyrolysis- Comparison of experimental measurements with model predictions. Energy & Fuels 2000;14:791-800.

[36] Evans DD, Emmons HW. Combustion of wood charcoal. Fire Research 1977;1:57-66.

[37] Yoon H, Wei J, Denn MM. A model for moving-bed coal gasification reactors. AIChE Journal 1978;24:885-903.

[38] Grebenshchikova GB. Study of the kinetics of the conversion of carbon monoxide by steam in the pressure of ash from Lisichansk Coal. Podzemnaya Gazifikatsiya Uglei 1957:2:54-7: (Lawrence Livermore Laboratory Translation UCRL-TRANS-10900).

[39] Benson HE. Chemistry of coal utilisation: 2nd supplementary volume. Ed. Elliott MA, John Wiley; 1981.

[40] Smooth LD, Smith, PJ. Volatiles combustion. Pulverized-coal combustion and gasification, edited by Smoot, L.D and Pratt, D.T; 1979. p.169-82.

[41] Howard JB, Williams GC, Fine DH. Kinetics of carbon monoxide oxidationin postflame gases. In: 14th Symposium (International) on Combustion; 1973. p. 975-86.

[42] Dryer FL, Glassman I. High-temperature oxidation of CO and CH_4. In: 14th Symposium (International) on Combustion; 1973. p. 987-1003.

[43] Varma AK, Chatwani AU, Bracco FV. Studies of premixed laminar hydrogen-air flames using elementary and global kinetics models. Combustion and Flame 1986;64:233-6.

[44] Bird RB, Stewart WE, Lightfoot EN. Transport phenomena. New York: (2nd eds) J. Wiley; 2002.

[45] Kunii D, Smith JM. Heat transfer characteristic of porous rocks. AIChE Journal 1960;6:71-8.

[46] DeWasch AP, Froment GF. A two-dimensional heterogneous model for fixed bed catalytic reactors. Chemical Engineering Science 1971;26:629-34.

[47] Froment GF, Bischoff KB. Chemical reactor analysis and design. New York: J. Wiley; 1979.

[48] Cooper J, Hallett WLH. A numerical model for packed-bed combustion of char particles. Chemical Engineering Science 2000;55:4451-60.

[49] Bhattacharya SC, Siddique AHM, Pham H-L. A study on wood gasification for low-tar gas production. Energy- The International Journal 1999;24:285-96.

[50] Patankar SV. Numerical heat transfer and fluid flow. New York: McGraw-Hill; 1980.

[51] Hyman JM, Naughton MJ. Static rezone methods for tensor-product grids. Report LA-UR-83-3245, Los Alamos National Laboratory, Los Alamos, NM; 1983.

[52] Koufopanos CA, Panayannakos N. Modelling of the pyrolysis of biomass particles: Studies on kinetics, thermal and heat transfer effects. The Canadian Journal of Chemical Engineering 1991;69:907-15.

[53] Wayne WS. Analysis of single particle wood combustion in convective flow. PhD Thesis. Department of mechanical engineering, University of Wisconsin- Madison; 1983.

[54] Ragland KW, Aerts DJ, Baker AJ. Properties of wood for combustion Analysis. Bioresources Technology 1991;37:161-8.

[55] Rogers GFC, Mayhew YR. Thermodynamic and transport properties of fluids: SI Units. (5th eds) Black Well, Oxford, UK; 1995.

[56] White FM. Fluid Mechanics. New York: McGraw-Hill; 1979.

[57] Turns SR. An introduction to combustion: Concepts and application. New York: (2nd eds) McGraw-Hill; 2000.

[58] Reid RC, Prausnitz JM, Poling RE. The properties of gases and liquids. New York: (4th eds) McGraw-Hill; 1987.

[59] Earp DM, Reyes-Nunez LR, Evans GD, Bridgwater AV. Mass and energy balances over an open-core downdraft gasifier. In: Biomass for Energy and Industry vol.2, EC Conference, Lisbon; 9-13 October 1989.

[60] Earp DM. Gasification of biomass in a downdraft reactor. PhD Thesis. Aston University, Birmingham, UK; 1988.

[61] Reed TB, Markson M. Biomass gasification reaction velocities. In: Fundamentals of Thermochemical Biomass Conversion 1985. p. 951-65.

[62] Dasappa S, Paul PJ. Gasification of char particles in packed beds: analysis and results. International Journal of Energy Research 2001;25:1053-72.

[63] Milligan JB. Downdraft gasification of biomass. PhD Dissertation. The University of Aston in Birmingham, UK; 1994.

NOMENCLATURE

A pre-exponential factor
A_v particle surface area per unit volume, m^2/m^3
C molar concentration of gas species
C carbon
C_p specific heat, J/kg.K
d_p particle diameter, m
D diffusivity, m^2/s
D reactor diameter, m
E activation energy, J/mol
ΔH enthalpy of reaction, J/kg
h specific enthalpy, J/kg
h heat transfer coefficient, $W/m^2.K$
h_{rs} solid radiation coefficient, $W/m^2.K$

h_{rv} void-to-void radiation coefficient, W/m².K
H hydrogen
k intrinsic reaction rate
k thermal conductivity, W/m.K
k_{eff} effective thermal conductivity, W/m.K
k_m mass transfer coefficient, m/s
K_C equilibrium constant for water gas shift reaction
m'' mass flux, $kg/m^2.s$
M molecular weight, g/mol
m_p mass of single particle, kg
N nitrogen
O oxygen
P gas mixture pressure, kPa
Pr Prandtl number (= $Cp.\mu/k$)
R reaction rate, $kg/m^3.s$ (or) $kmol/m^3.s$
R universal gas constant, J/kmol.K
Re Reynolds number (= $d_p m''/\mu$)
Sc Schmidt number (=$\mu/\rho D_i$)
t time, s
T temperature, K
V velocity, m/s
x vertical coordinate, m
$X1_i$ yield of species i in primary pyrolysis process (mass fraction)
$X2_i$ yield of species i in secondary pyrolysis process (mass fraction)

Greek symbols

$\acute{\varepsilon}$ emissivity
ε bed void fraction, void volume/ bed volume
ξ ratio of reacting to non-reacting heat transfer coefficient
γ stoichiometric coefficient of char oxidation reaction, mol/mol
v stoichiometric coefficient, mol oxidant /mol carbon
μ viscosity, kg/m.s
ρ density, (kg/m^3)
σ Stephan-Boltzmann constant (=5.6703E-8, W/m^2K^4)

Subscripts

avg average value
c char
C cracking of primary tar
g gas phase
gw gas-to-reactor wall
m moisture

s	solid phase
sg	solid-to-gas
sw	solid-to-wall
tar1	primary tar (condensable tar, oxygenates)
tar2	secondary tar (taken as benzene)
v	volatiles
w	reactor wall

Subscripts for reactions

m	drying process
p1	primary pyrolysis process
p2	secondary pyrolysis process
1	char combustion reaction
2	char-carbon dioxide reaction
3	char-water vapour reaction
4	char-hydrogen reaction
5	water-gas shift reaction
5F	forward water-gas shift reaction
5R	reverse water-gas shift reaction
6	oxidation of secondary tar
7	oxidation of carbon monoxide
8	oxidation of methane
9	oxidation of ethylene
10	oxidation of hydrogen

In: Gasification: Chemistry, Processes and Applications
Editor: Michael D. Baker
ISBN: 978-1-61209-681-0

Chapter 6

A Dynamic Model for Predicting Tar Content in Gases from a Stratified Downdraft Gasifier: Part II –Model Sensitivity Analysis

San Shwe Hla[a], Lu Aye[a,*], M. A. Connor[b] and D. F. Stewart[a]

[a]International Technologies Centre (IDTC), Department of Civil and Environmental Engineering, Australia

[b]Department of Chemical and Biomolecular Engineering

The University of Melbourne, Victoria, Australia

Abstract

An analysis was carried out to determine the sensitivity of a dynamic model developed for a stratified downdraft wood gasifier to changes in input parameter values. The sensitivity of the model to small increments and decrements in three operating parameters and eight model parameters was investigated. The results show that a higher air supply rate increases the combustion zone temperature and reduces outlet tar levels. Changes of ±20% in air supply rate, initial particle diameter and solid to gas heat transfer coefficient led to changes in predicted reaction zone velocity of more than 50%. The final mass of condensable tars in the producer gas was predicted to change by more than 300% as a result of 20% changes in the values of the following model parameters: solid-to-gas heat transfer coefficient, mass transfer coefficient and tar cracking reaction rates.

Keywords: Gasification; Tar content; Modelling, Sensitivity analysis

* Corresponding author. e.mail: lua@unimelb.edu.au, Tel.: +613-8344-6879, fax: +613-8344-6868

1. INTRODUCTION

Details are provided in a previous paper [2] of the development of a dynamic model capable of predicting how the performance of a stratified downdraft gasifier changes over time under different operating conditions. The model can predict how gas composition, primary and secondary tar content, and gas and solid surface temperature profiles along the gasifier axis develop over the period from ignition to achievement of stable gas production. A set of baseline values for parameters used in the numerical model was established using empirical correlations taken from the work of a number of other researchers. The diversity of sources used and the fact that many correlations were not derived specifically for conditions existing within a gasifier made it essential to determine the sensitivity of the model outputs to changes in these baseline parameter values.

The model outputs used to characterise the sensitivity of the model to changes in input parameter values were temperature profiles, gas composition, outlet gas tar content and reaction zone stability. The sensitivity analysis considered the effects of changes in both operating parameters and model parameters. Three operating parameters were investigated, namely airflow rate, initial particle diameter and moisture content. Eight model parameters were also investigated, namely the solid-gas heat transfer coefficient, the mass transfer coefficient, the particle specific surface area, primary pyrolysis rate coefficients, secondary pyrolysis (tar cracking) rate coefficients, secondary tar oxidation rate coefficients, effective thermal conductivity and heat losses.

The sensitivity studies were conducted by changing the value of one specific parameter whilst all the other operating and model parameters were kept constant at the baseline values given in Table 1. An airflow rate of 15 kg/hr was selected as the reference value for the sensitivity analysis. As discussed in a previous paper [2], the rate and even the direction of reaction zone movements alter during the reactor warm up period. Therefore, to ensure that results relating to reaction zone stability are comparable across the entire sensitivity analysis, the reaction front movement characteristics presented here were all calculated for the case where the gasifier is producing a stable gas output (this condition is reached between 25 minutes and 30 minutes after ignition).

Table 1. Baseline operating and model parameter values used in the model

Quantity	Value
Air supply rate (kg/hr)	15
Initial particle diameter (mm)	20
Moisture content of the fuel (%db)	10
Reacting to non-reacting heat transfer coefficient ratio (-)	0.60
Maximum mass transfer coefficient (m/s)	0.045
Ratio to effective and actual transport surface area (-)	1.0
Primary pyrolysis rate multiplication factor (-)	1.0
Secondary pyrolysis rate multiplication factor (-)	1.0
Tar oxidation rate multiplication factor (-)	1.0
Effective conductivity of solid phase multiplication factor (-)	1.0
Bed to wall heat transfer coefficient multiplication factor (-)	1.0

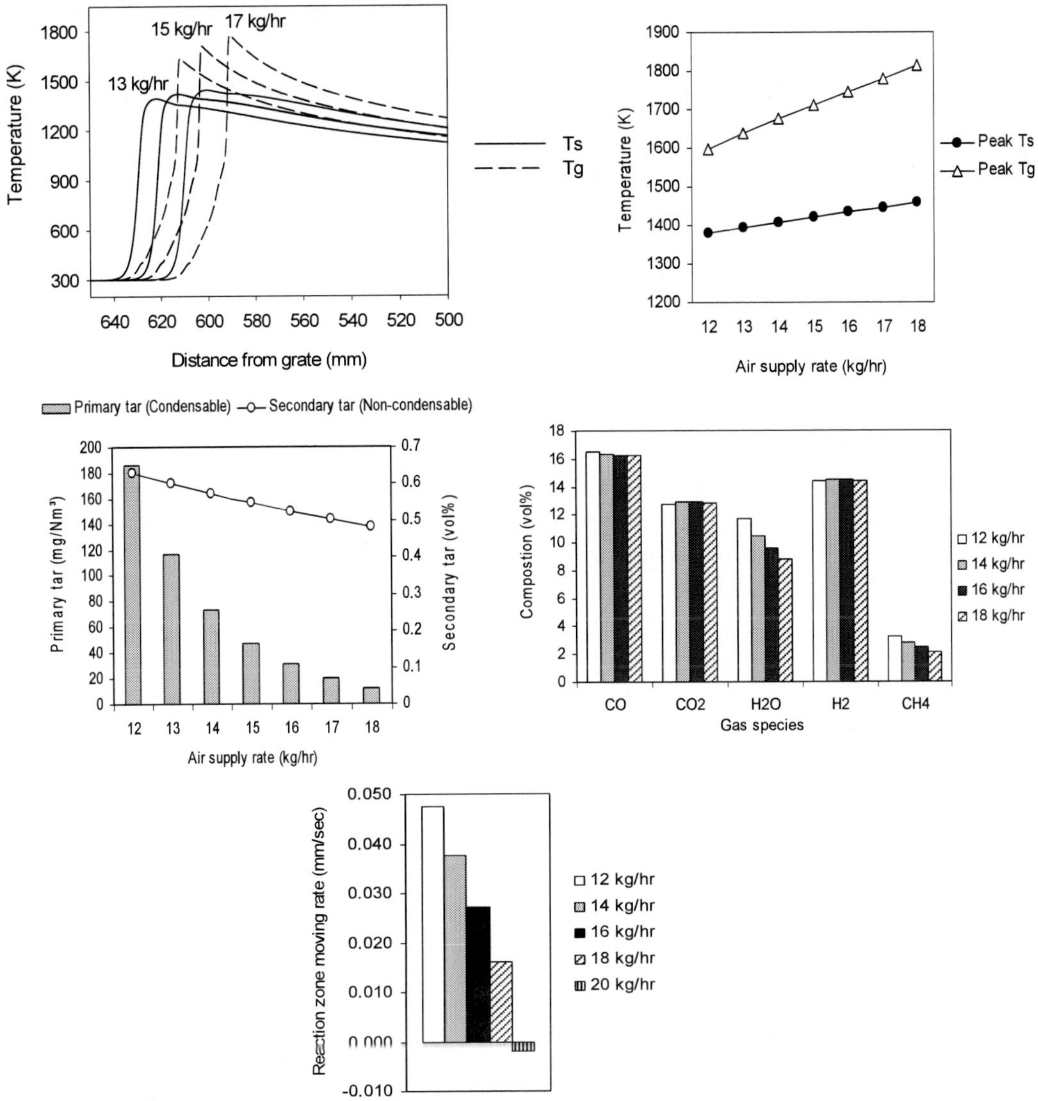

Fig. 1. Effects of air supply rate (a) on solid (Ts) and gaseous (Tg) temperature profiles (b) on peak temperatures (c) on outlet primary and secondary tar levels (d) on outlet gas composition (e) on reaction zone movement (+ve for upward, -ve for downward).

2. Sensitivity Analysis on Operating Parameters

Airflow rate

The rate at which air is supplied to the gasifier is one of the most important operating parameters, strongly influencing the temperature profiles, tar content, and stability of the reaction zones. Fig. 1 shows the effects of variations in airflow rate on gasifier behaviour.

Increasing the air supply rate means a greater oxygen supply to the system, a higher rate of heat release from both the heterogeneous and homogenous oxidation reactions, and an increased peak temperature for both the solid and gas phases in the combustion zone. Fig.1b shows that peak gas phase temperatures increased faster than peak solid phase temperatures as the air supply rate was increased. This is attributed to a more rapid increase in gas phase oxidation rates than in solid phase oxidation rates as temperatures rose. As a result, both primary and secondary tar outlet levels drop at higher air supply rates (see Fig. 1c); this is attributed to higher thermal cracking rates at higher gas phase temperatures.

The composition of the outlet producer gas as a function of air supply rate is presented in Fig. 1d. Of particular interest is that changes in the concentration of CO in the outlet gas at different airflow rates are not significant. Increases in the solid temperature (Ts) would be expected to increase the rate of heterogeneous gasification reactions and therefore the CO level. However, this is offset by the higher gas temperature, which increases the rate of oxidation of CO released in the pyrolysis processes. Higher solid phase temperatures also increase the heterogeneous water-gas reaction rates; as a result the quantity of water vapour found in the final gas outlet would be expected to decrease and the concentration of H_2 to increase. However, as Fig. 1d shows, the predicted drop in H_2O level appears more significant than the increase in H_2 level as airflow rates rise. The main reason for this is that increasing the airflow rate also increases nitrogen concentrations in the gas; this has the effect of making the decrease in the amount of H_2O present seem more important than it is while making the corresponding increase in the amount of hydrogen seem less important. The other gaseous component that changes significantly is methane. At higher air supply rates, lower CH_4 levels are observed because of the higher CH_4 oxidation rates at higher temperatures.

As discussed in a previous paper [2], air supply rate controls the stability of the reaction zones. At higher air supply rates, the char consumption rate increases, which leads to a higher solid velocity and causes the reaction zone to move downwards towards the grate (see Fig.1e). At lower air supply rates the opposite happens and the reaction zones move upwards.

Initial particle size

The predicted sensitivity of system output parameters to changes in initial wood particle diameter is shown in Fig. 2. Varying the particle diameter affects both the heat and mass transfer coefficients and therefore has an influence both on primary pyrolysis processes and on the heterogeneous gas-char reactions. In the pyrolysis zone, the rate at which heat from the particle surface penetrates to the particle core is faster in the case of a smaller particle. Thus the rate of volatiles production is greater for a small particle than for a larger particle. In the combustion and gasification zones, a smaller char particle diameter increases the total surface area per unit volume and decreases mass transfer resistances, thereby promoting both char oxidation and gasification. As a result, smaller particles enhance char combustion rates and lead to higher solid peak temperatures. At larger particle sizes, char oxidation rates fall and more oxygen is utilized for gas phase combustion, leading to increases in gas phase

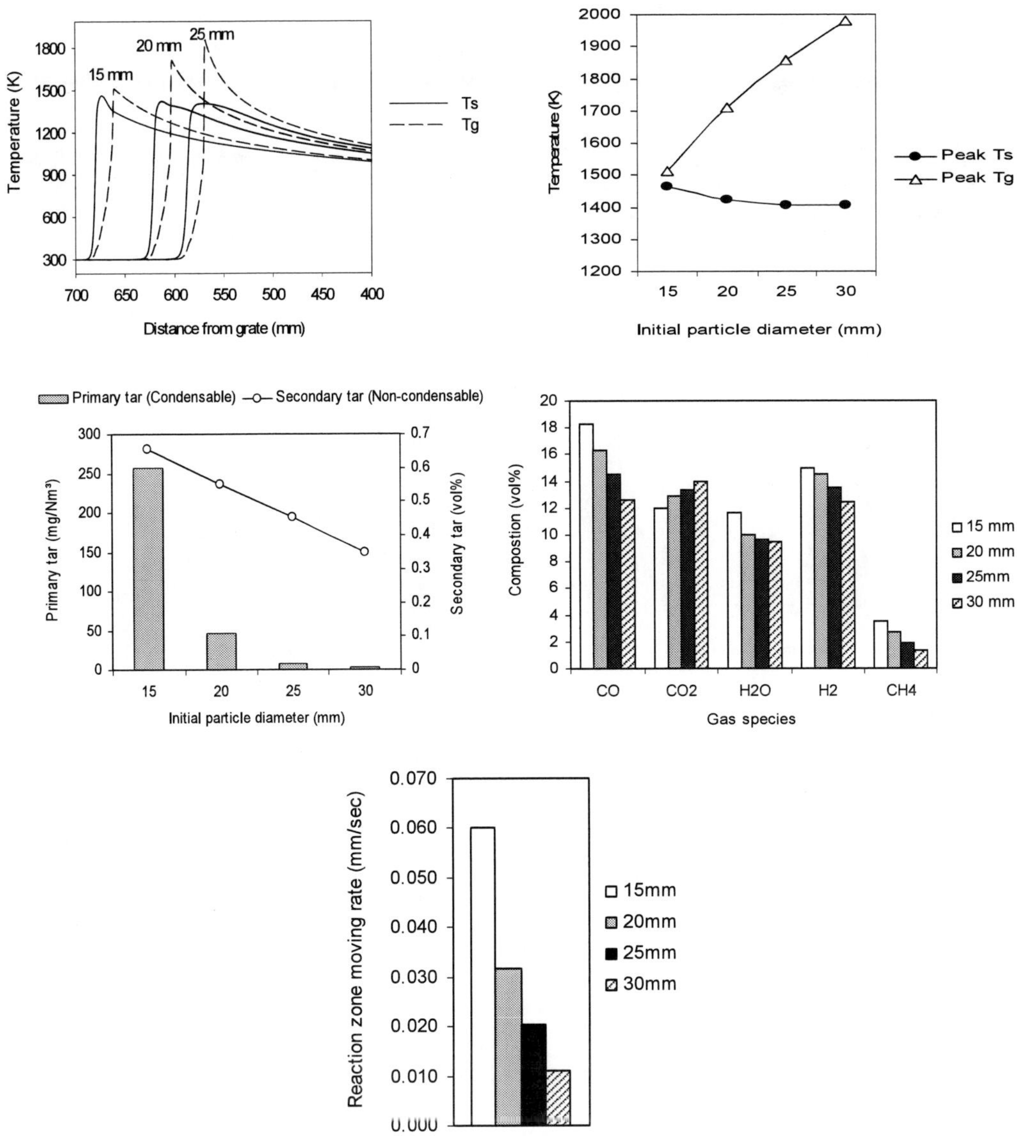

Fig. 2. Effects of initial wood particle size (a) on solid (Ts) and gaseous (Tg) temperature profiles (b) on peak temperatures (c) on outlet primary and secondary tar levels (d) on outlet gas composition (e) on reaction zone movement (+ve for upward, -ve for downward).

temperatures. As a result, lower primary and secondary tar outputs are observed in the case of large wood particles. These effects are illustrated in Fig. 2a-2c.

Predicted exit gas compositions as a function of initial wood particle size are presented in Fig. 2d. Because of the lower specific surface areas in beds of larger particles, higher solid temperatures are needed to attain particular levels of carbon conversion in the gasification zone than are required in the case of smaller particles. This results in more combustible gas

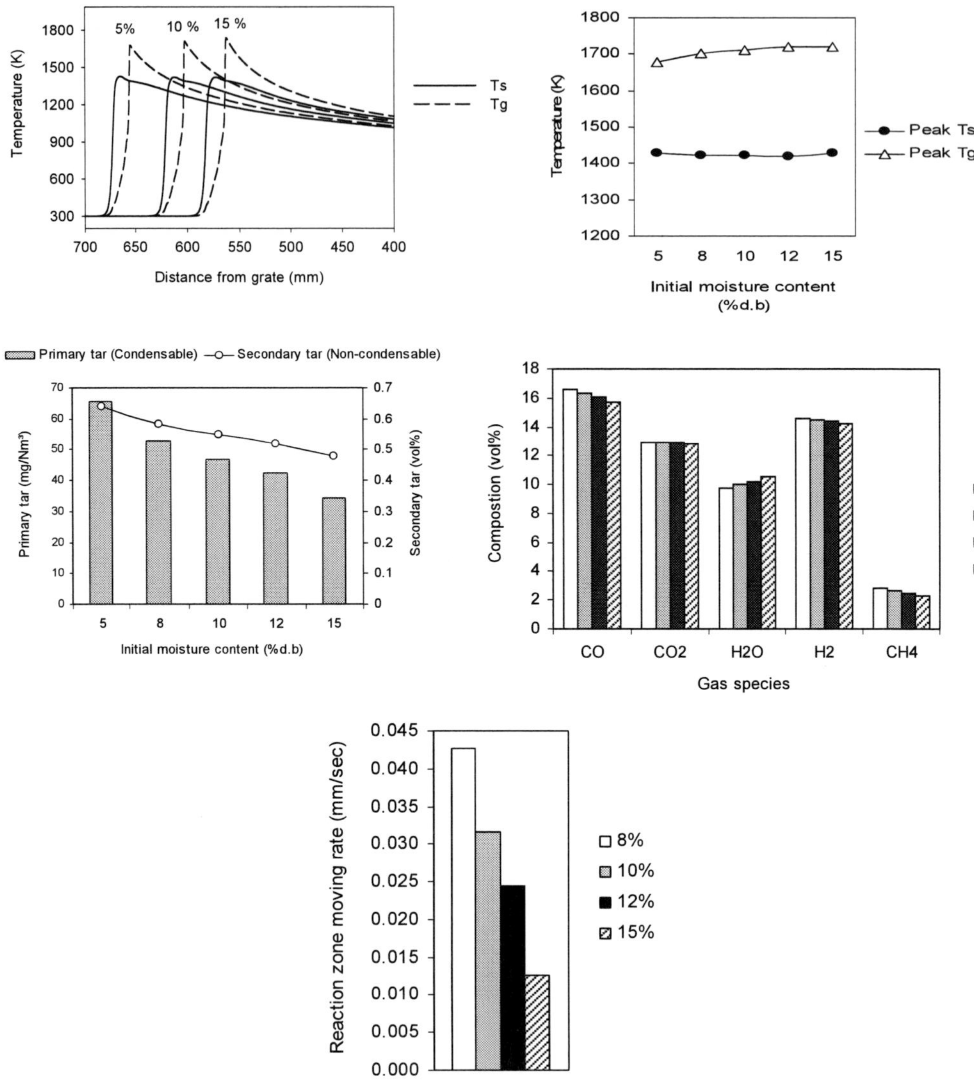

Fig. 3. Effects of initial moisture content (a) on solid (Ts) and gaseous (Tg) temperature profiles (b) on peak temperatures (c) on outlet primary and secondary tar levels (d) on outlet gas composition (e) on reaction zone movement (+ve for upward, -ve for downward).

species being consumed within the oxidation zone when larger char particles are present. As a result, lower exit CO and H_2 concentrations, and a higher exit CO_2 concentration are predicted for larger solid fuel particles (see Fig. 2d). CH_4 is also present in decreasing amounts as particle sizes increase.

As particle sizes decrease, reaction zones show a tendency to move upwards; this can be attributed to the faster pyrolysis rate (Fig. 2e).

Initial moisture content

The effect of increasing fuel moisture content on the phenomena occurring within the reaction region is shown in Fig. 3 for five fuel moisture contents: 5, 8, 10, 12 and 15% (dry basis).

As discussed in a previous paper [2], drying of the wood is expected to be complete before temperatures high enough for char combustion to begin are reached. Therefore, as shown in Fig. 3a, increases in wood moisture content favour the char gasification processes involving water vapour, and the reaction zones move downwards. Fig. 3b shows that changing initial moisture content does not affect the peak solid temperature but that a slight increase in peak gas temperature occurs at higher moisture content levels. This rather unexpected result seems to be linked to CO oxidation processes. The CO oxidation reaction rate depends in part on the water vapour concentration [2] and a higher water vapour concentration increases the CO combustion rate, the associated heat release rate, and the local gas temperature. It also increases the hydrocarbon (secondary tar) combustion rate and, as a result, a slight decrease in both the primary and secondary tar content of the gas is found at higher initial moisture contents (Fig. 3c).

Fig. 3d shows that changing the moisture content also causes slight changes in the levels of all gas phase components. As discussed above, higher moisture contents lead to an increase in CO oxidation rates, and a slight decrease in gas phase CO level results. Fig. 3d also shows a slight decrease in hydrogen levels at higher moisture content levels, due to the higher gas temperatures and consequent higher hydrogen oxidation rate. This presumably contributes to the observed increase in gas phase H_2O content as the wood moisture content increases. The CO_2 concentration appears not to be affected by changes in fuel moisture content even though the CO oxidation rate increases. This is attributed to the increased amount of H_2O in the gas phase, which affects volume percentage calculations. At higher wood moisture contents, a lower CH_4 level is observed; this can be attributed to the higher peak gas temperatures and the higher associated CH_4 oxidation rates. The effect of fuel moisture content increases is to slow the upward movement of the reaction zones in the gasifier, as shown in Fig. 3e.

Sensitivity Analysis on Model Parameters

Solid-gas heat transfer coefficient

Heat transfer coefficients derived for a reacting surface are scarce. Therefore the basic solid-gas heat transfer coefficient used in this model was one formulated for a non-reacting surface. As reported by Dzhapbyev *et al.* [65], however, the solid-gas heat transfer coefficient for a nonreacting system exceeds that for a reacting system. Therefore the initially calculated heat transfer coefficient needs to be modified for use in the gasifier model. The ratio of reacting to non-reacting heat–transfer coefficient (ξ) has been introduced as a means of doing this [66], [67]. According to Dzhapbyev *et al.* [65] the factor ξ that accounts for the reactive nature of the fixed bed can have a value ranging from 0.02 to 1.0. For the present study a base case value of 0.6 was selected for ξ; this value was determined using the experimental results

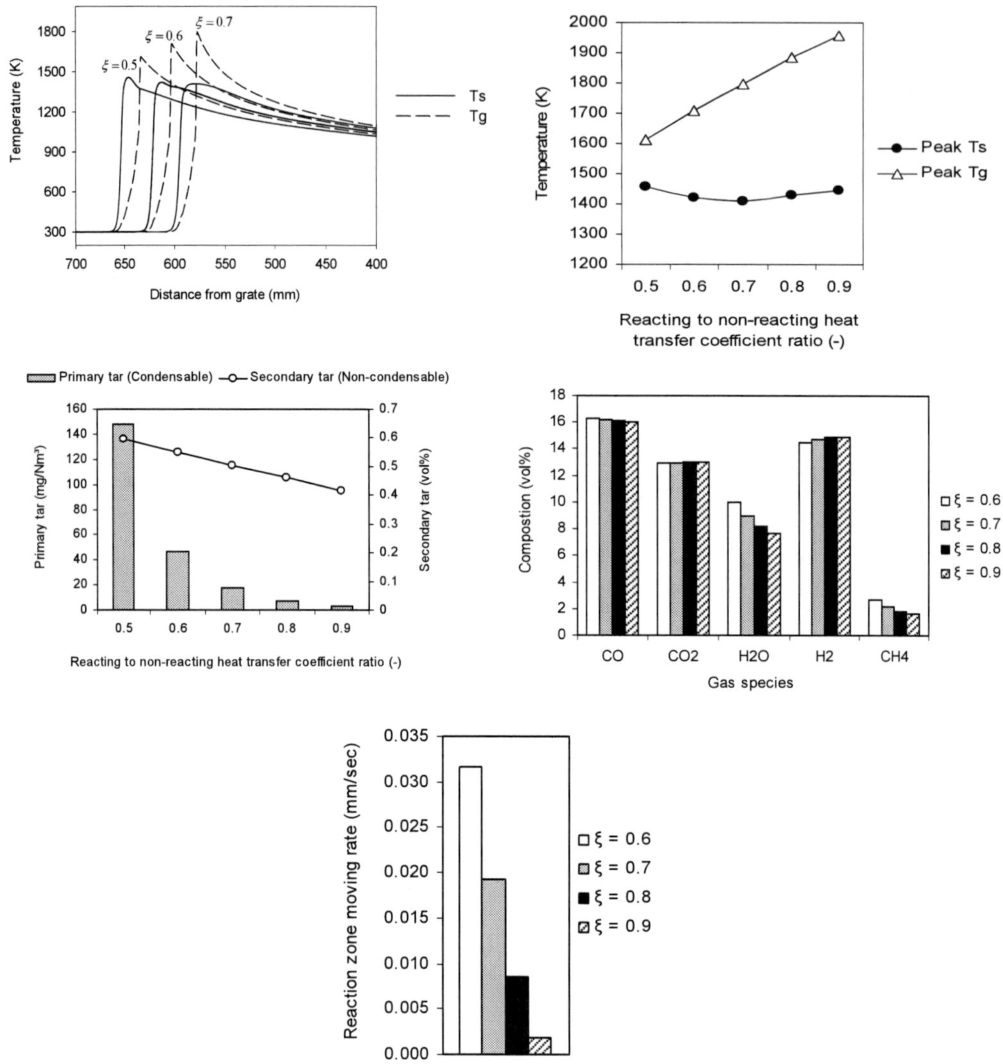

Fig. 4. Effects of solid-to-gas heat transfer coefficient (a) on solid (Ts) and gaseous (Tg) temperature profiles (b) on peak temperatures (c) on outlet primary and secondary tar levels (d) on outlet gas composition (e) on reaction zone movement (+ve for upward, -ve for downward).

obtained on a pilot scale gasifier [68]. The effects of changes in ξ on the model outputs are shown in Fig. 4.

A reduction in the value of ξ leads to a decrease in the heat transfer rate from the solid phase to the gas phase. Since oxidation of volatile gases depends on gas phase temperature, a smaller ξ results in a more prolonged gas combustion process and consequently a lower peak gas phase temperature (Fig. 4a & 4b). More oxygen is consumed in char oxidation reactions and so less is available for hydrocarbon oxidation reactions. Because of this, a higher primary and secondary tar content is observed at smaller values of ξ (Fig. 4c). By increasing ξ, a higher heat transfer rate and a higher peak gas phase temperature are achieved and increases occur in the solid temperature along the gasification zone (though not in the peak solid temperature) (Fig. 4a) and hence in heterogeneous gas-char reaction rates. As a result, solid velocities increase, leading to a decrease in the rate of upward movement of reaction zones (Fig. 4e).

A higher value of ξ leads to higher gas temperatures and more rapid gas phase combustion reactions in the oxidation zone as a result. The trend of increasing H_2O and CO_2 and decreasing H_2, CO and CH_4 with increasing ξ is a direct consequence of this increase in gas phase reaction activity (Fig. 4d)

Mass transfer coefficient

There is always some uncertainty associated with the use of empirical mass transfer correlations derived from experimental data obtained in equipment different from one's own. This is especially true at low values of the Reynolds number (as occur in the case of packed beds of small particles at high temperature). Kato *et al.* [69] describe the wide divergence between different correlations at low *Re* and also show that the actual mass transfer coefficient in solid-gas fixed beds becomes considerably smaller than the theoretical coefficient at low *Re* values. In this study, direct application of existing empirical correlations to determine the rate of the diffusion-controlled combustion reactions in the gasifier resulted in unrealistic temperature values. To overcome this problem, use of a limiting maximum value of the mass-transfer coefficient has been recommended [70],[71]. A value of 0.045 m/s, which was found to give results closely matching those obtained experimentally, was therefore selected as a maximum mass transfer coefficient value to use in the simulation for the base case. The effects on the system outputs of varying this mass transfer coefficient value are presented in Fig. 5.

The mass transfer coefficient strongly affects the rates at which gaseous reactants diffuse to the solid fuel surface and react with the fuel. Thus a lower mass transfer coefficient leads to decreased heterogeneous gas-char reaction rates. A reduced mass transfer rate results in heterogeneous char oxidation reactions slowing down in comparison with gas phase combustion reactions. This leads to a greater proportion of the oxygen being consumed in reactions with volatile products, with the result that the gas temperature increases (Fig. 5b), and both primary and secondary tar levels decrease (Fig. 5c).

Conversely, if mass transfer rates increase (due to an increase in the mass transfer coefficient value), gas temperatures drop, the amounts of combustible components consumed in the oxidation zone fall, and larger quantities of CO and CH_4 are present in the exit gas. A slight increase in H_2 levels is also observed (Fig. 5d).

Since gas temperatures are lower in the case where mass transfer rates increase, solid temperatures along the gasification zone (but not peak temperatures) also decrease. As a

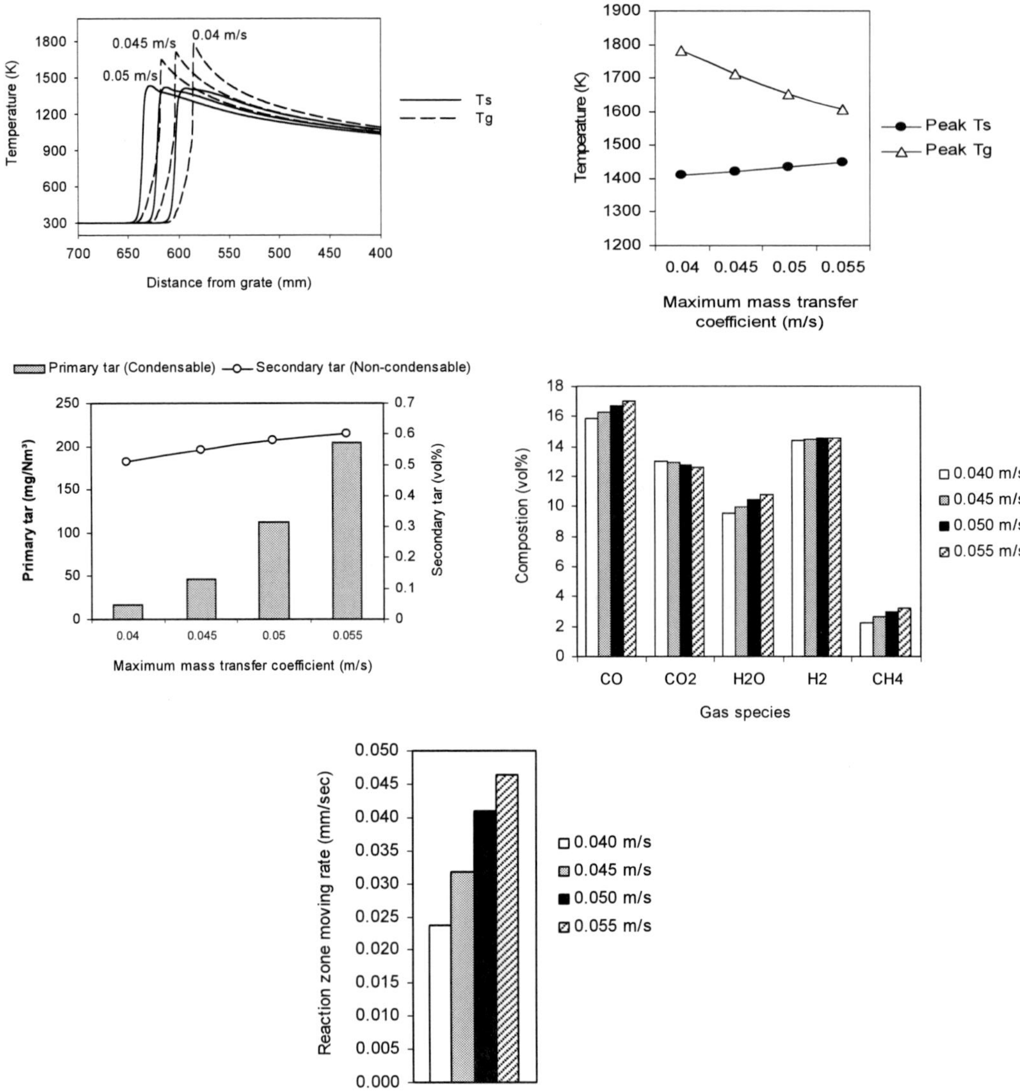

Fig. 5. Effects of mass transfer coefficient (a) on solid (Ts) and gaseous (Tg) temperature profiles (b) on peak temperatures (c) on outlet primary and secondary tar levels (d) on outlet gas composition (e) on reaction zone movement (+ve for upward, -ve for downward).

result, despite the higher value of the mass transfer coefficient, there is nevertheless a decrease in overall gas-char reaction rates and reaction zones tend to move upwards (Fig. 5e).

Particle specific surface area

In this model, the biomass particles were taken to be spherical in shape and the effect of char particles breaking up was not accounted for. In reality, however, particles are not spherical and char particles do break up into smaller pieces. The effect of this is to increase the particle surface area per unit volume above the value assumed in the model. The model

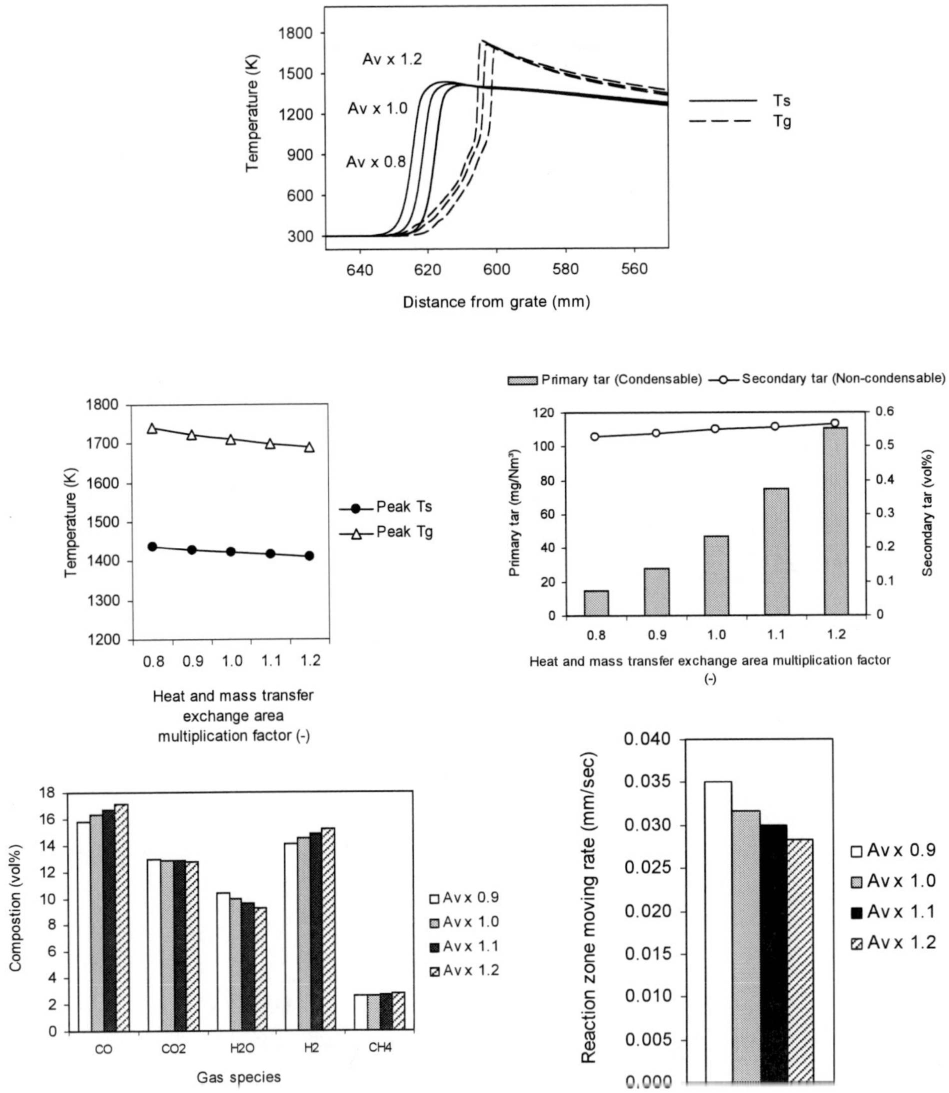

Fig. 6. Effects of heat and mass transfer exchange area (a) on solid (Ts) and gaseous (Tg) temperature profiles (b) on peak temperatures (c) on outlet primary and secondary tar levels (d) on outlet gas composition (e) on reaction zone movement (+ve for upward, -ve for downward).

also neglects the loss of effective surface area when particle surfaces are in contact. Therefore a sensitivity analysis was carried out to assess the effect of changes in heat and mass exchange area. This was done by multiplying the base case area by factors ranging from 0.8 to 1.2.

The effects of varying the heat and mass transfer exchange area (A_V) are presented in Fig. 6. A reduction in heat and mass transfer exchange area reduces the overall heterogeneous reaction rate. This results in more oxygen being consumed in volatiles combustion processes, an increase in gas temperature (Fig. 6a and Fig. 6b), and a decrease in primary and secondary tar levels in the outlet gas (Fig. 6c).

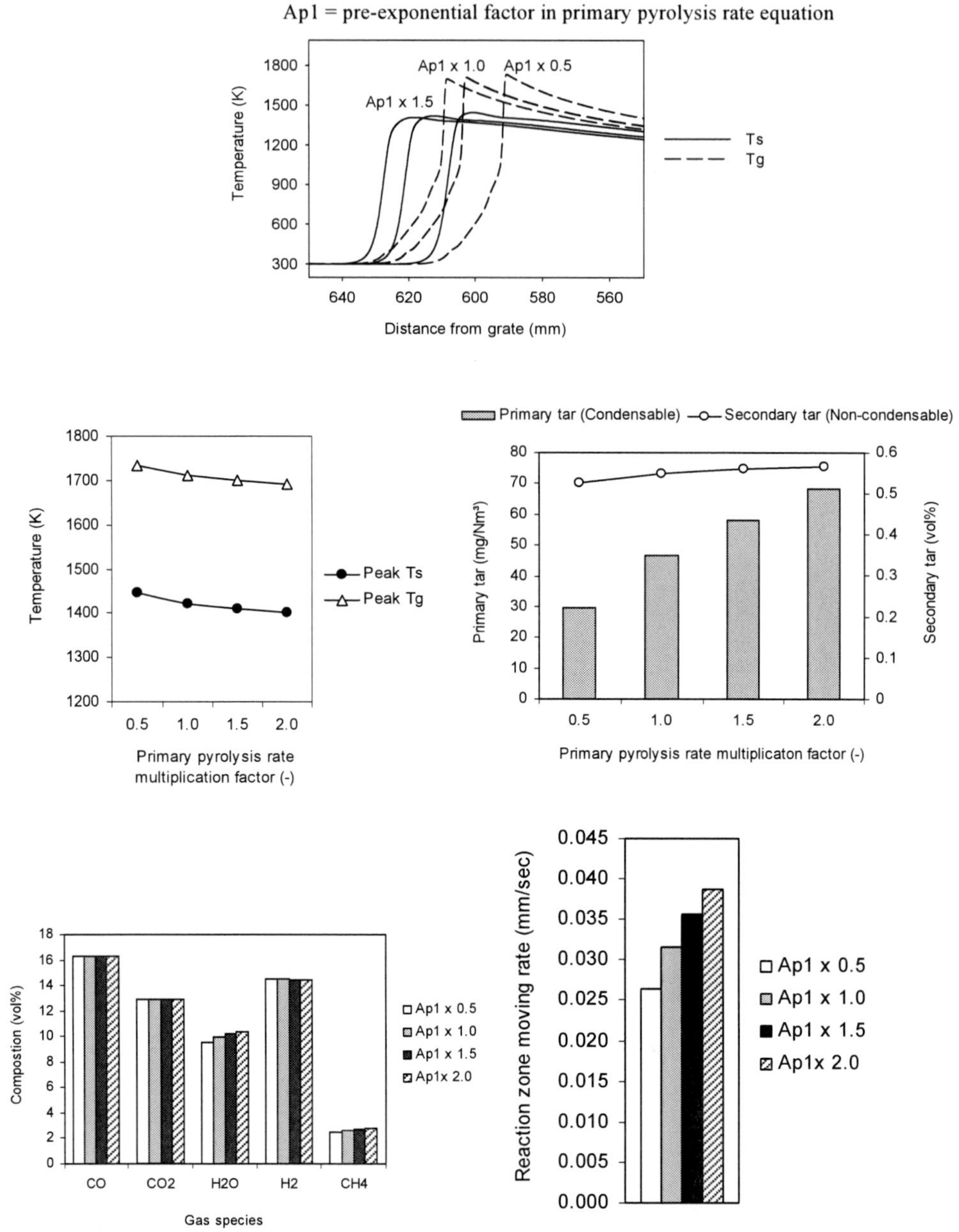

Fig. 7. Effects of changes in primary pyrolysis rate (a) on solid (Ts) and gaseous (Tg) temperature profiles (b) on peak temperatures (c) on outlet primary and secondary tar levels (d) on outlet gas composition (e) on reaction zone movement (+ve for upward, -ve for downward).

As also happens when the mass transfer coefficient is reduced, decreasing the heat and mass exchange area decreases the proportions of CO, H_2, and CH_4 in the outlet gas due to the higher combustion rates at higher gas phase temperatures (Fig. 6d). Reducing the heat and mass transfer exchange area also decreases char consumption rates and, as a result, increases the rate of the upward movement of reaction zones (Fig. 6e).

Kinetics of primary pyrolysis reactions

As described in the paper describing the development of the model [2], rate coefficients for the primary pyrolysis reactions were derived from experimental results obtained by other researchers for the rapid pyrolysis of large wood particles. The effect of varying primary pyrolysis reaction rates was explored in a similar way to that in which the sensitivity analysis for the heat and mass exchange area was carried out, except that in this case the factor ranged in value from 0.5 to 2.0.

Primary pyrolysis is the process in which primary tars (oxygenates) are released from virgin wood. An increase in primary pyrolysis rates produces more primary tar within a given time and completion of tar cracking is hindered. Thus a higher tar content in the gas results at higher primary pyrolysis rates (Fig. 7c). A slight decrease in both gas and solid temperatures is observed at higher primary pyrolysis rates (Fig. 7a), probably because of the lower secondary tar oxidation rate reducing the peak gas temperature (Fig. 7b). No significant changes in gas composition result from changing the value of the primary pyrolysis reaction rate (Fig. 7d). As discussed in a previous paper [2], the primary pyrolysis process is one of the main factors affecting the stability of the reaction zone; a faster pyrolysis rate leads to the reactor running in a pyrolysis dominant mode and, as a consequence, reaction zones move upwards (Fig. 7e).

Kinetics of secondary pyrolysis reactions

The effect of changes in the rate coefficients for the secondary pyrolysis process on the model output parameters are shown in Fig. 8. Changes in the rate coefficients were made by multiplying the pre-exponential factor in the rate equation by values ranging between 0.5 and 2.0. Secondary pyrolysis involves the cracking of primary tar into secondary tar and other combustible gases. Thus, increasing the secondary pyrolysis reaction rate directly reduces the amount of primary tar, as Fig. 8c shows.

Increasing the secondary pyrolysis reaction rate also leads to the formation of more combustible gases, the consumption of more oxygen in gas combustion reactions and increases in gas phase temperatures (Fig. 8a and Fig. 8b). The effect of changes in secondary pyrolysis rates on gas composition was insignificant (Fig. 8d).

Increases in secondary pyrolysis rates lead to higher gas phase oxidation rates, reduced char combustion rates and a tendency for reaction zones to move upwards faster (Fig. 8e).

Kinetics of secondary tar oxidation reactions

In the model, the amount of secondary tar in the outlet gas is determined mainly by the rate of secondary tar oxidation. An analysis of the sensitivity of the model to changes in this rate was carried out by multiplying the pre-exponential factor in the rate equation by values ranging from 0.8 to 1.2. The results of the analysis are presented in Fig. 9.

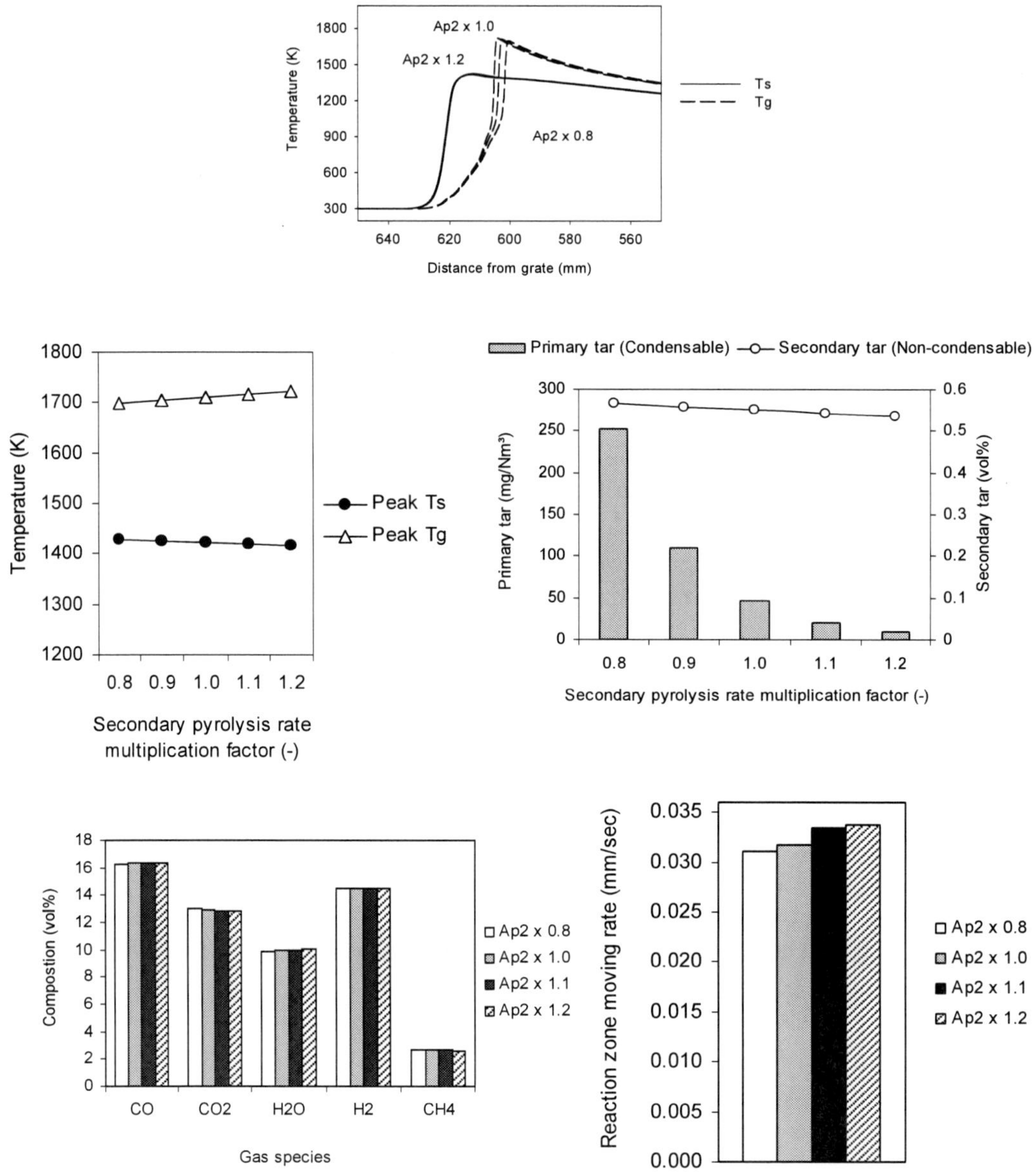

Fig. 8. Effects of changes in secondary pyrolysis rates (a) on solid (Ts) and gaseous (Tg) temperature profiles (b) on peak temperatures (c) on outlet primary and secondary tar levels (d) on outlet gas composition (e) on reaction zone movement (+ve for upward, -ve for downward).

As Fig. 9 shows, changing secondary tar oxidation rates seems to have no significant effect on model outputs. This suggests that secondary tar oxidation is not a rate limiting step in the process.

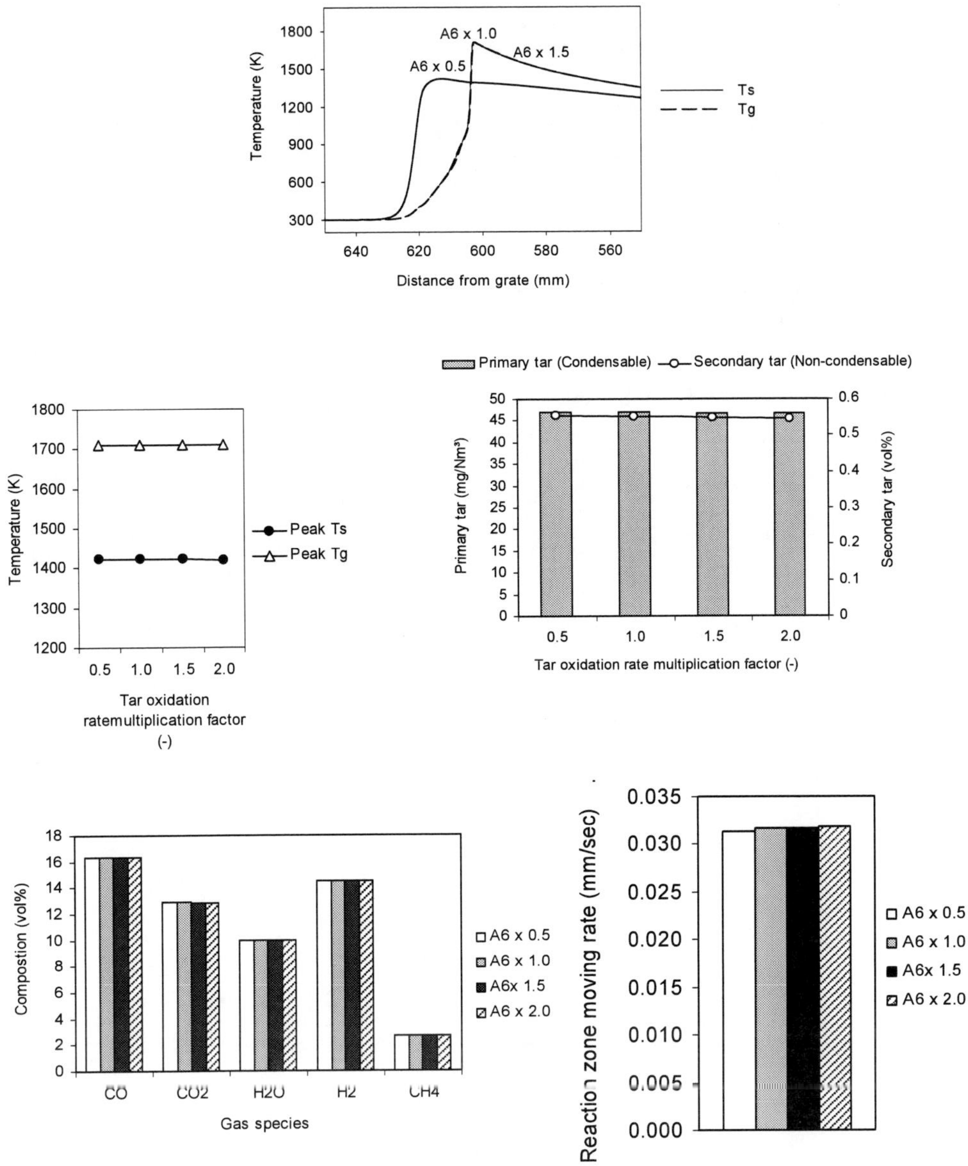

Fig. 9. Effects of changes in secondary tar oxidation rate (a) on solid(Ts) and gaseous(Tg) temperature profiles (b) on peak temperatures (c) on outlet primary and secondary tar levels (d) on outlet gas composition (e) on reaction zone movement (+ve for upward, -ve for downward).

Effective thermal conductivity

The effective thermal conductivity for the gasifier fuel bed was derived from a complex model [72]. An analysis of the sensitivity of the model outputs to changes in effective thermal conductivity (K_{eff}) is presented in Fig. 10.

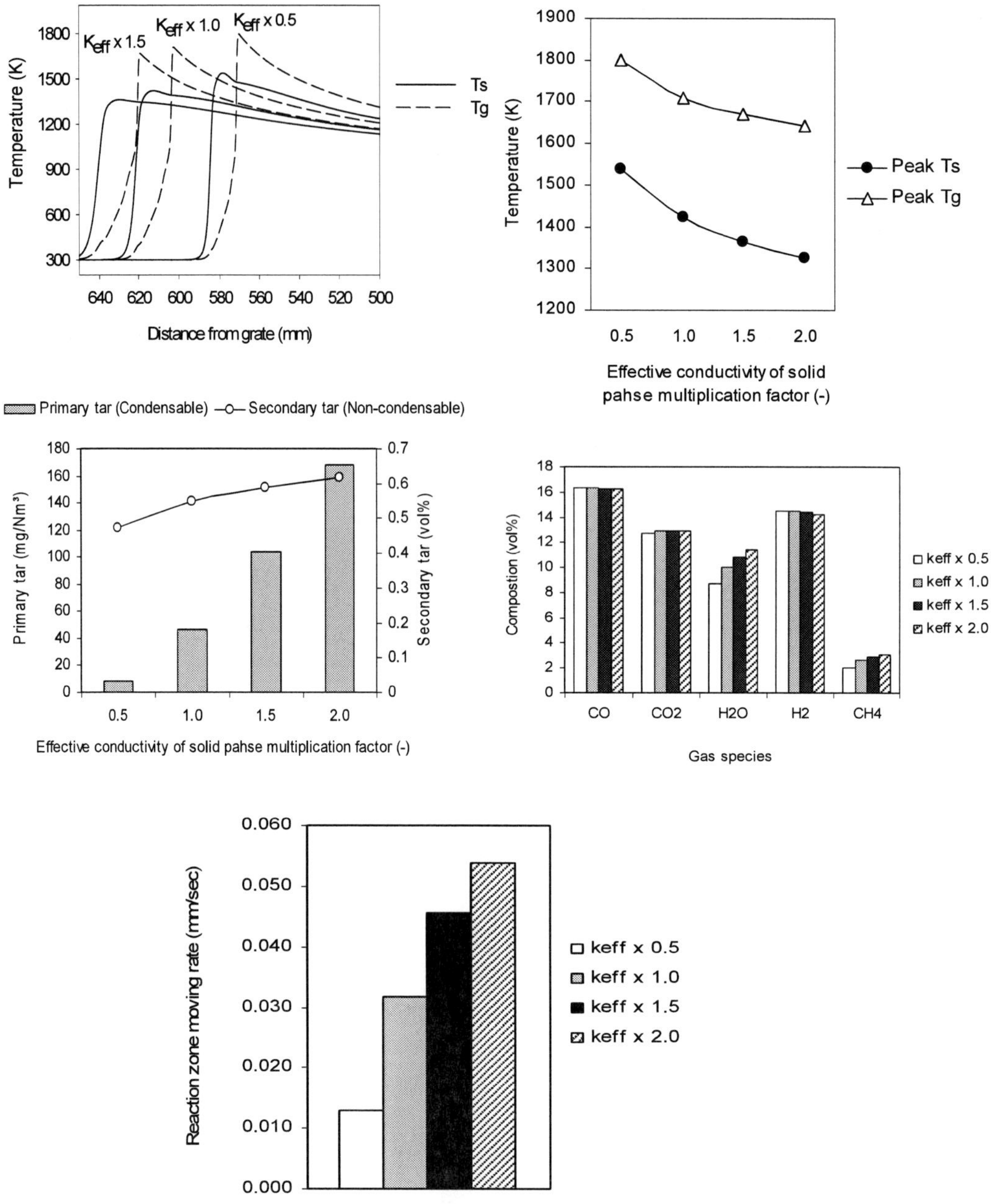

Fig. 10. Effects of changes in effective thermal conductivity (a) on solid (Ts) and gaseous temperature profiles (b) on peak temperatures (c) on outlet primary and secondary tar levels (d) on outlet gas composition (e) on reaction zone movement (+ve for upward, -ve for downward).

All model outputs except for gas composition were found to be sensitive to variations in the effective thermal conductivity. At a higher effective thermal conductivity, more heat from the oxidation zone is transferred to the upper part of the reactor bed while the solid and gas phase temperatures in the combustion zone fall (Fig. 10a & 10b). As a result, rates of tar cracking decrease and the outlet gas has a higher tar content (Fig. 10c).

Since the solid temperature falls in the case of a higher effective thermal conductivity, the rates of the heterogeneous gasification reactions are reduced and, as a result, a lower amount of CO and H_2, and a higher amount of H_2O is found in the outlet gas. Also, a larger amount of CH_4 is present due to less CH_4 combustion in the oxidation zone (Fig. 10d).

At higher effective thermal conductivity levels, conduction of heat upwards in the bed increases and pyrolysis occurs at a faster rate (Fig. 10e). Under such conditions the reaction zones rise up the gasifier. To restore stable operation under such conditions the input airflow rate would have to be increased.

Heat losses

In this study, an empirical correlation suggested in the literature [73] was used to estimate the bed-to-wall heat loss coefficient. Total heat loss through the reactor wall is directly controlled by the bed-to-wall heat transfer coefficient and by the reactor inner wall temperature, which depends on the thickness and type of materials in the reactor wall and insulation. In this model, the inner wall temperature was assumed to vary linearly with solid and gas temperatures over time.

The effect of varying the heat loss through the reactor wall was explored by running the model using bed-to-wall heat transfer coefficient values (h_{bw}) ranging from 0.5 to 2.0 times the base case value. The results obtained are presented in Fig. 11. This shows that, as might be expected, a lower rate of heat loss from the fuel bed to the wall increases the gas phase temperature while a higher heat loss rate has the opposite effect (Fig. 11a & 11b). When the heat loss is small, a low primary tar content is predicted because of the higher gas temperature and the almost complete cracking of the primary tar that this causes. Under these conditions a cleaner gas is produced (Fig. 11c).

In the case where heat losses are reduced, elevated H_2 and CO levels and lower H_2O and CO_2 levels are predicted due to the higher steam and CO_2 gasification rates that accompany higher solid temperatures along the gasification zone (Fig. 11d). As shown in Fig. 11e the effect of a reduction in heat loss is to slow the upward movement of reaction zones.

Sensitivity of model outputs to changes in parameter values

To compare the level of sensitivity of key model outputs to changes in the values of operating and model parameters, the effects of similar (±20%) changes in each baseline parameter value on outlet gas tar (both primary and secondary) content and reaction zone stability were determined. These are shown in Table 2.

For both outlet gas tar content and reaction zone stability, airflow rate is found to be the most important operating parameter while the reacting to non-reacting heat transfer coefficient ratio is the most important of the model parameters. Decreases of 20% in either the air supply rate or the solid-to-gas heat transfer coefficient value caused a 300% increase in the primary tar content of the outlet gas and a 50% change in reaction zone velocity.

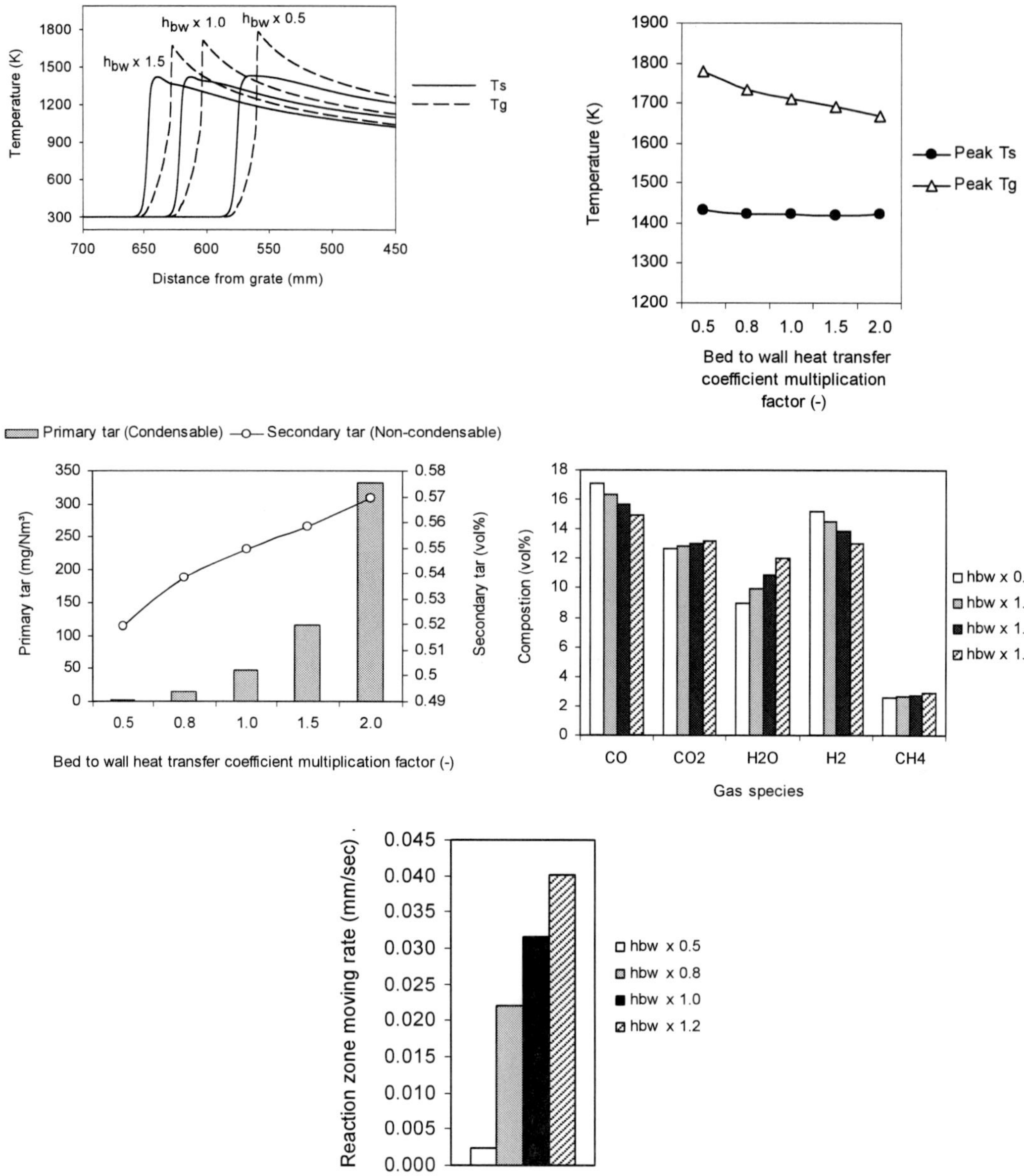

Fig. 11. Effects of changes in the extent of heat loss through the reactor wall (a) on solid (Ts) and gaseous (Tg) temperature profiles (b) on peak temperatures (c) on outlet primary and secondary tar levels (d) on outlet gas composition (e) on reaction zone movement (+ve for upward, -ve for downward).

As expected, the primary tar level in the outlet gas was found to be strongly controlled by the primary tar cracking rate (secondary pyrolysis reaction rate). A greater than 400% increase in outlet gas primary tar content was observed following a 20% decrease in the secondary pyrolysis rate. When reaction zone velocity alone is considered, initial particle size is observed to be the most important of the operating parameters and solid-to-gas heat transfer rate the most important of the model parameters.

Table 2. Effects of operating and model parameters on outlet primary tar level, outlet secondary tar level and reaction zone movement

Parameter	% changes in input baseline parameter	% changes in predicted output parameter		
		Primary tar	Secondary tar	Reaction zone velocity*
Air flow rate	-20	298.4	14.5	50.1
	+20	-73.2	-12.4	-49.0
Initial particle diameter	-20	-48.0	-2.9	-68.7
	+20	53.2	0.5	55.2
Initial moisture content (d.b)	-20	13.0	6.1	34.6
	+20	-9.6	-5.7	-22.5
Reacting to non reacting heat transfer coefficient ratio	-20	306.4	9.9	75.1
	+20	-70.7	-9.3	-49.3
Maximum mass transfer coefficient	-20	-87.7	-14.7	-41.1
	+20	297.1	8.7	44.7
Particle specific surface area	-20	-68.1	-4.0	17.3
	+20	136.8	3.0	-11.0
Primary pyrolysis reactions rate	-20	-12.4	-1.4	-0.3
	+20	12.3	0.9	8.2
Secondary pyrolysis reactions rate	-20	438.5	2.8	-1.8
	+20	-81.5	-2.5	6.6
Secondary tar oxidation rate	-20	0.2	0.2	-0.2
	+20	-0.1	-0.2	0.1
Effective thermal conductivity	-20	-38.4	-4.3	-16.1
	+20	43.4	3.3	17.9
Heat loss through reactor wall	-20	-67.2	-2.0	-30.5
	+20	147.0	1.6	27.0

* +ve = increase in upward moving rate
-ve = decrease in upward moving rate (or) moving downward

As discussed in section 0, the effects of changes in secondary tar oxidation rate on observed output values are very small, even negligible. Apart from the secondary tar oxidation rate, the primary pyrolysis rate is found to be the least important of the model parameters as far as both outlet gas tar content and reaction zone stability is concerned.

CONCLUSION

As described in Chapter 5 [2] a dynamic model of a stratified downdraft gasifier has been developed. In order to obtain a detailed understanding of this model's sensitivity to changes in key parameters, a systematic analysis of the effect of varying individual operating and model parameters on predictions of temperature profiles, outlet gas tar content, gas composition and reaction zone stability was carried out.

A higher air supply rate was found to increase both solid and gas phase temperatures and to reduce the outlet gas tar content. Over the wide range of air inlet rates investigated, an increase in the inlet airflow rate was found to lead to a decreased rate of upward movement of the reaction zone, and vice versa. At a certain stage, a further increase in the inlet air rate leads to the reaction zone changing direction and moving downwards. This makes it important for the long-term operability of the reactor to fix the airflow rate at a value ensuring stable operation.

Using a larger particle size was found to significantly increase the gas phase temperature and to reduce both primary and secondary tar outputs. However, for a smaller char particle size, due to the higher rates of mass transfer achievable, gasification was possible at lower temperatures than those necessary for a bed of larger particles to achieve the same extent of carbon conversion in the gasification zone. Increasing the initial moisture content of the fuel was shown to have a negative effect on the quality of the outlet producer gas.

As the solid-to-gas heat-transfer coefficient is increased, gas temperatures increase faster, a higher proportion of the oxygen present is used to oxidize the volatile gases, and the gas temperature increases. Both primary tar and secondary tar content were found to be sensitive to the ratio of the reacting to non-reacting solid-to-gas heat-transfer coefficient. Changing the mass transfer coefficient affected the location and maximum temperature of both the solid and gas phases. Decreasing either the mass transfer coefficient or the exchange area favoured gas oxidation and reduced the quantity of the outlet tar. An increase in the primary pyrolysis rate leads to the reactor operating in a "pyrolysis dominant" mode and the reaction zone moves upwards. Increasing the secondary pyrolysis rate has the effect of reducing the primary tar content of the outlet gas. The model was found to be insensitive to changes in the rate of tar oxidation. Temperatures, tar content and reaction zone stability were all significantly affected by variations in effective thermal conductivity. By reducing the heat loss through the wall, the tar content of the outlet gas can be reduced; however, the reaction zone moves downwards due to increases in heterogeneous reaction rates.

To check the sensitivity of model outputs to variation in operating and model parameter values, the effects of ±20% changes in each baseline parameter value on reactor performance were determined. Air supply rate and heat transfer coefficient value were found to be the most important operating and model parameters respectively. The initial moisture content and the secondary tar oxidation rate were the respective operating and model parameters that had the least influence on tar content and reaction zone stability.

The sensitivity analysis provided a good understanding of the effects of operating and model parameters on the performance of a stratified downdraft gasifier. However, the model developed still needed to be validated against experimental data. Therefore an experimental investigation was conducted, details of which are described in a subsequent paper [68], where details of the experimental set up, experimental procedures, experimental results and a comparison between model predictions and experimental results are presented.

Acknowledgment

Financial support from the University of Melbourne for one of the authors, San Shwe Hla, is gratefully acknowledged.

References

[1] Lu Aye, San Shwe Hla, Connor MA, Stewart DF. Chapter 5: A dynamic model for predicting tar content in gas from a stratified downdraft gasifier: Part I –Model development, in: Gasification: Chemistry, Processes and Applications (Ed: Michael D. Baker), Nova Science Publishers, Inc. 2011.

[2] Dzhapbyev K, Miropol'skii AL, Mal'kovskii, VJ. Investigation of unsteady heat transfer in a packed bed of spheres swept by gas. Thermal Engng. 1986;33: p.159.

[3] Cho YS, Joseph B. Heterogeneous model for moving-bed coal gasification reactors. Ind. Eng. Chem. Process Des. Dev. 1981;20:314-8.

[4] Ghani MU, Radulovic PT, Smoot LD. An improved model for fixed-bed coal combustion and gasification: sensitivity analysis and applications. Fuel 1996;75:1213-26.

[5] San Shwe Hla, Lu Aye, Connor MA, Stewart DF, Fung P, Ng S. Chapter 7: A dynamic model for predicting tar content in gas from a stratified downdraft gasifier: Part III – Experimental validation, in: Gasification: Chemistry, Processes and Applications (Ed: Michael D. Baker), Nova Science Publishers, Inc. 2011.

[6] Kato K, Kubota H, Wen CY. CEP Symp. Ser. 1970;66:p.87.

[7] Bhattacharya A, Salam L, Dudukovič B, Joseph B. Experimental and modelling studies in fixed-bed char gasification. Ind. Eng. Chem. Process Des. Dev. 1986;25:988-96.

[8] Di Blasi C. Dynamic behaviour of stratified downdraft gasifiers. Chemical Engineering Science 2000;55:2931-44.

[9] Kunii D, Smith JM. Heat transfer characteristics of porous rocks. AIChE Journal 1960;6:71-8.

[10] Froment GF, Bischoff KB. Chemical reactor analysis and design. New York: Wiley; 1979.

In: Gasification: Chemistry, Processes and Applications
Editor: Michael D. Baker
ISBN: 978-1-61209-681-0

Chapter 7

A Dynamic Model for Predicting Tar Content in Gases from a Stratified Downdraft Gasifier: Part III – Experimental Validation

San Shwe Hla[1], Lu Aye[1]*, M. A. Connor[2], D. F. Stewart[1], P. Fung[3] and S. Ng[3]

[1]International Technologies Centre (IDTC), Department of Civil and Environmental Engineering, Australia
[2]Department of Chemical and Biomolecular Engineering
The University of Melbourne, Victoria, Australia
[3]Forestry and Forest Products, CSIRO, Clayton South, Victoria, Australia

Abstract

A pilot-scale downdraft gasifier was designed, fabricated and tested. The experimental data obtained from it were compared with the predictions of a dynamic model the development and analysis of which are described in previous papers [1 & 2]. River red gum (*Eucalyptus camaldulensis*) wood blocks of two different sizes were used as gasifier feedstock. The gasifier was operated at inlet airflow rates of 7.5 to 27.6 kg/hr, with corresponding energy outputs ranging from 54 to 168 MJ/hr. An average cold gas efficiency of 70% was obtained. Two modes of operation were used. In the first mode, the section of the gasifier below the ignition port was filled with charcoal prior to start up and the remainder of the gasifier filled with wood; in the second mode, no initial charge of charcoal was used. Operating the reactor with wood fuel alone produced an outlet gas with a final tar content that was around 20 times higher than the average tar content values in gas produced when using an initial charge of charcoal. Transient temperature profiles, outlet gas compositions, and outlet tar contents predicted by the model were in generally good agreement with experimental values. However, the model predicted a more rapid increase in concentrations of H_2 and CO along the gasifier axis than was observed experimentally. There was good agreement between the predicted and

* Corresponding author. e.mail: lua@unimelb.edu.au, Tel.: +613-8344-6879, fax: +613-8344-6868

experimentally observed rates and directions of reaction zone movement within the gasifier over the range of inlet airflow rates studied.

Keywords: Biomass; Gasifier; Tar content; Experimental validation

INTRODUCTION

This paper describes the experimental component of a study undertaken to improve our understanding of stratified downdraft gasifiers. The first step in the study was to construct a numerical model, the detailed development of which was discussed in a previous paper [1]. The predictions of this model, together with details of a sensitivity analysis carried out to test the effect of changes in different operating and model parameters on model predictions, have been described in a second paper [2].

Unlike earlier stratified downdraft gasifier models that have been focused primarily on temperature profiles and outlet gas compositions, the model developed in this project was designed to also provide information on tar formation and reduction processes within the body of a gasifier. Validation of this model could not be done using previously published experimental data since there are no experimental data available on internal tar profiles along a downdraft gasifier either for a traditional (with throat) or a stratified (without throat) reactor. In addition, there is very limited experimental data on gas composition profiles along the reactor axis. It was therefore necessary to conduct an experimental investigation on a stratified downdraft gasifier constructed specifically for the purpose of model validation.

In this paper, details are provided of the experimental set up, experimental procedures, experimental results and model validation procedures.

EXPERIMENTAL APPARATUS

A schematic of the entire gasifier system is shown in Fig. 1. Detailed descriptions of the system components are given below.

Gasifier

A small scale stratified downdraft gasifier (40-75 kW thermal input), shown schematically in Fig. 2, was designed and fabricated for the purpose of the experimental work. The gasifier, which was cylindrical in shape, was fabricated from 6 mm thick mild steel tube with a 206 mm inner diameter. The reactor can be divided into three sections as illustrated in Fig. 2.

The top section, through which air and fuel were supplied to the gasifier, consisted of an inlet section shaped like an inverted funnel mounted above a 700 mm long cylindrical section. No significant reactions were expected to occur over the first 700 mm below the inlet so that the lower part of this top section acted in practice as a fuel storage bunker.

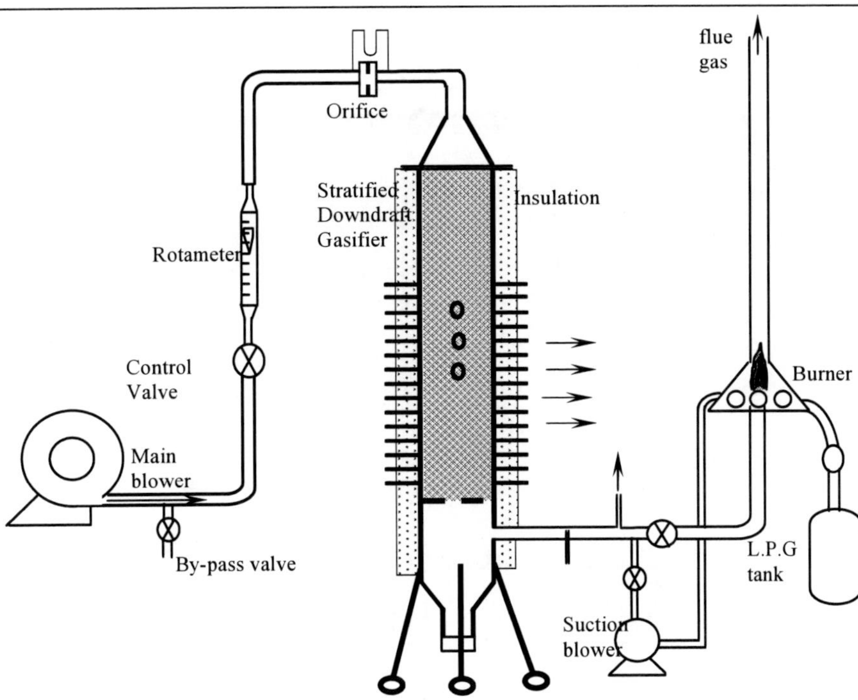

Figure 1. Experimental set-up.

In the middle section, groups of four 12 mm (OD) stainless steel tubes were inserted into the gasifier at points 50 mm apart along the gasifier body down to the level of the grate. These tubes were used for the purposes of temperature measurement and gas sampling and were located symmetrically around the reactor circumference. Twelve 25 mm (OD) stainless steel tubes were attached at three different levels in the middle part of the reactor for the purpose of ignition. It was in this middle section that all reactions took place and measurements were carried out. It was fabricated using mild steel tube (O.D 356 mm X I.D 340 mm). A 67 mm thick layer of bonded castable (Tufcast 1700QT) insulation was used to line the inside surface of the gasifier so that the effective internal diameter of this section was effectively the same as that of the top and bottom sections.

The bottom section contained a grate located at a distance of 1500 mm below the top of the gasifier. The grate was designed so that it could be rotated for the purpose of cleaning. The part immediately below the grate served as a gas solid separator, and below that an ashbin was fitted. A horizontal 50 mm (ID) diameter pipe attached to the gasifier below the grate served as the producer gas outlet.

Thermal insulation

To prevent excessive heat loss from the reactor wall, two layers of a ceramic fibre blanket of thickness 25 mm were wrapped around the reactor. To prevent the condensation of tar and moisture along the gas-sampling pipes, these were also wrapped with two layers of the same ceramic fibre blanket.

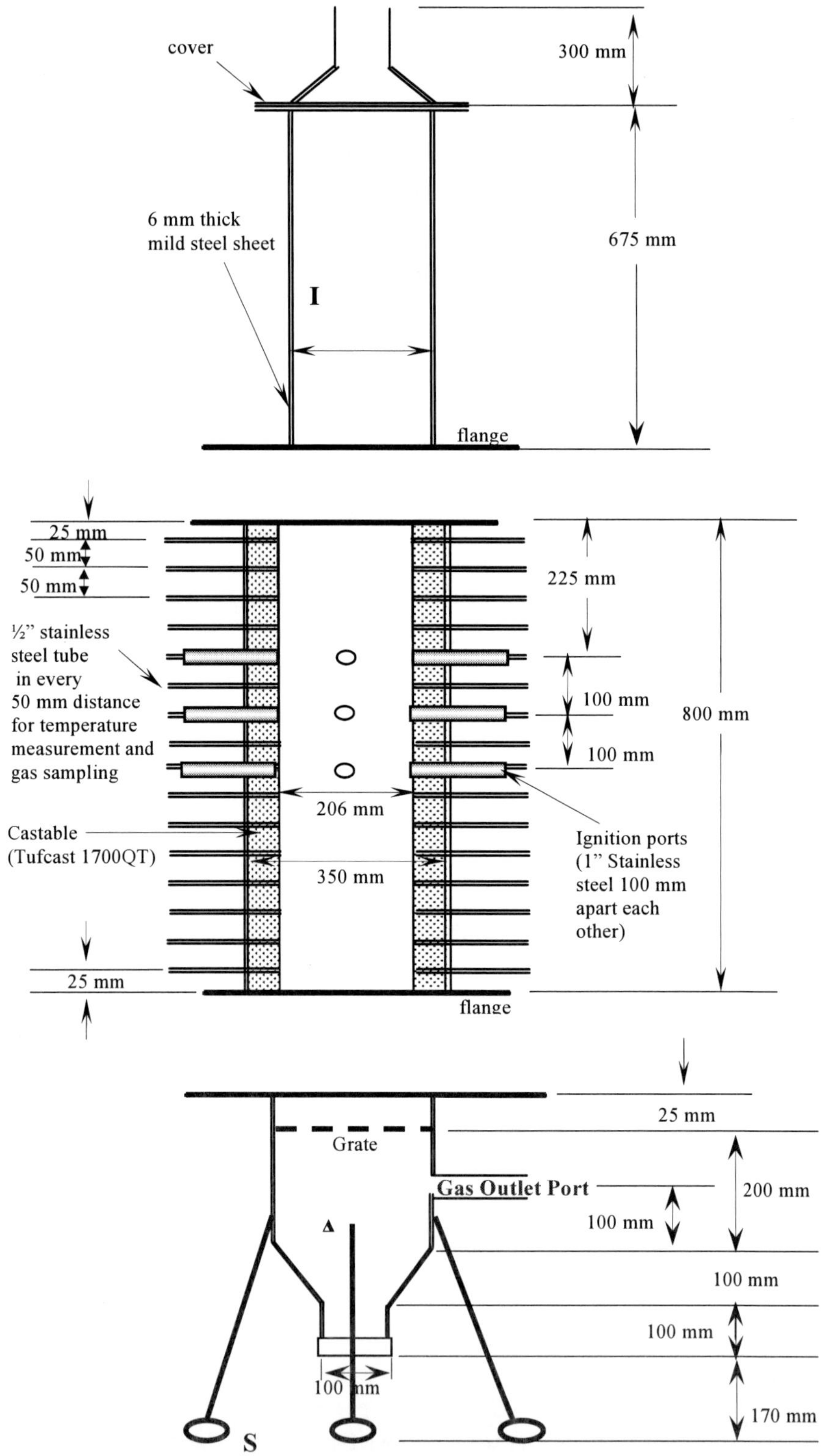

Figure 2. Schematic of experimental stratified downdraft gasifier.

Flare and Burner

A LPG torch was used both for ignition of fuel inside the gasifier and for ignition of producer gas downstream of the gasifier. A stainless steel burner, more details of which are shown in Fig. 3, was used to flare the producer gas. A pilot flame from the LPG torch was injected through one side of the burner, air was introduced through the other sides of the burner, while the producer gas entered through the bottom of the burner. The pilot flame from the LPG torch was turned on throughout the gasifier warm up process to ensure the complete combustion of the very low calorific value producer gas formed during the gasifier start-up stage.

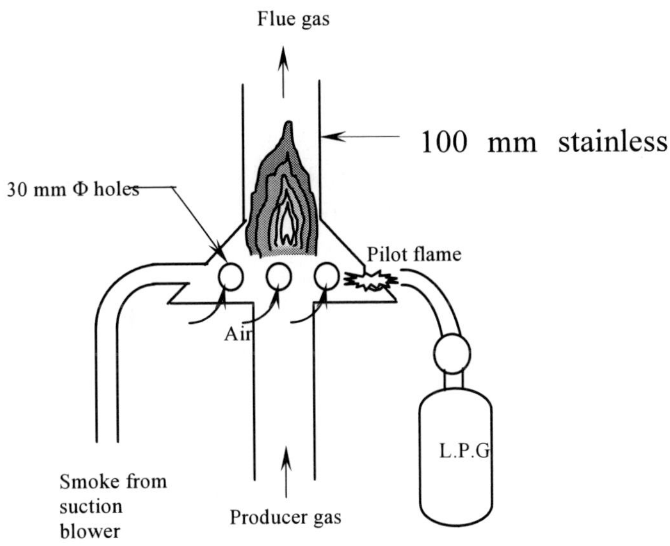

Figure 3. Schematic of the burner set-up.

Main air blower

In the gasification process, external power is required for moving the air and producer gas through the gasifier. This could be achieved by either applying vacuum to the bed outlet or blowing air into the bed inlet. In this experimental set-up the latter approach was followed and a rotary piston compact blower was used. The capacity of the air blower was much higher than the required air supply rate for some experimental runs. Therefore a by-pass line was fitted to the blower. This enabled excess air to be discharged to the atmosphere, thereby preventing pressure build up within the gasifier and possible overloading of the blower motor.

Suction blower

In this experiment, ignition of solid fuel inside the gasifier was achieved by holding a torch flame next to the ignition ports and using a small blower to apply vacuum to the bed outlet to draw the flame into the fuel bed of the gasifier. A rotary blower of capacity 0.5 kW was used for this purpose. After ignition of the solid fuel, this small blower was switched off and only the main blower was used to supply gasification air into the gasifier inlet.

MEASUREMENT SYSTEM AND INSTRUMENTATION

The parameters measured during the experiments included: the tar content of the producer gas at five different levels in the reactor; the gas composition at five different levels in the reactor; temperatures at each of 16 different levels in the reactor; inlet air pressure; inlet airflow rate; outlet gas flow rate; and wood consumption rate.

Gas sampling train

In this experiment, producer gas was sampled from five different levels in the gasifier; the method used was that set out in a recently proposed standard gas-sampling procedure guideline [3]. The modular sampling train was set up according to this guideline and consisted of 3 modules, namely, gas preconditioning and particle removal, tar collection and volume measurement. The sampling train described in the guideline can process gas from only one designated point in the reactor at any given time. In this study, gas samples needed to be collected from 5 different levels in the reactor and this necessitated modifying the original sampling train design. In the modified arrangement 5 trains were used, with a common water bath and a common vacuum pump. A direct drive high vacuum pump was used as the common vacuum pump. To transfer the gas from each sampling line to the sample containers another smaller vacuum pump was used. A schematic of the modified gas sampling train set-up is shown in Fig. 4.

Measurement of tar content

Gravimetric tar measurement enables quantitative measurement of the mass of the portion of a tar sample that is not amenable to GC analysis [4]. The tar content was determined by means of solvent distillation and evaporation according to the recent guideline [3]. Detailed measurement and calculation procedures for determining gravimetric tar content are provided in the guideline [3].

Measurement of gas composition

The H_2, CO, CH_4 and CO_2 content of discrete gas samples taken periodically from the gas-sampling stream was measured on a Varian 3700 gas chromatograph (GC) fitted with a thermal conductivity detector. 400 ml PVC gas sample containers were used to collect and transport these gas samples across to the GC. The sample containers were first thoroughly flushed with a small positive pressure of sampling gas and later filled with it so that air could not leak in before analysis. Next, 5-10 ml of gas sample was extracted through the rubber seal of the gas sample container with a hypodermic syringe; this was immediately injected into the gas chromatograph. The GC was calibrated using a calibration gas of the following composition: (H_2=71.77%, CO_2=16.40%, CO=10.70%, CH_4=1.13%).

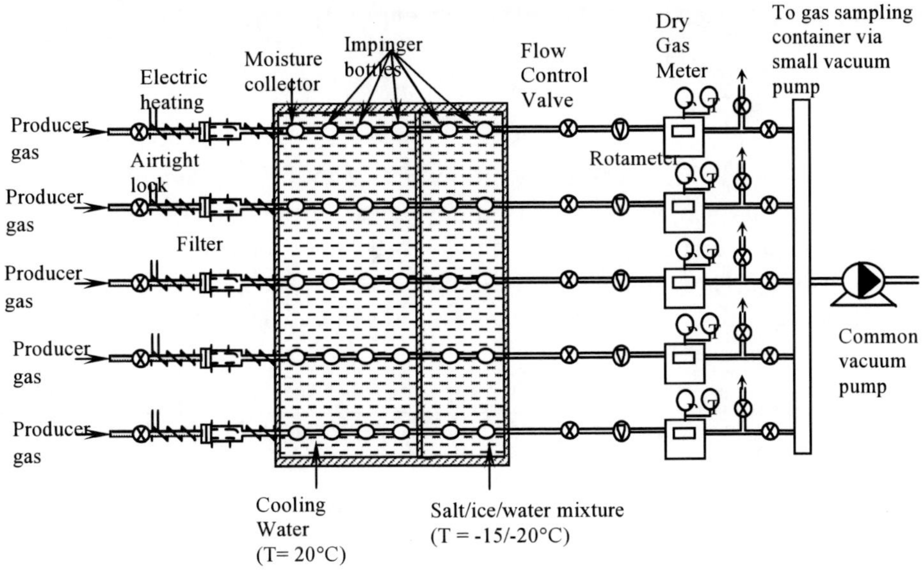

Figure 4. Modified gas sampling train.

Measurement of temperature

Chromel-Alumel (K-Type) thermocouples [these are capable of measuring to within ±8°C accuracy over the temperature range of the combustion zone (900-1200 °C)] were used to measure temperatures at 16 different levels along the experimental reactor and also the outlet gas temperature. Copper-Constantan (T-Type) thermocouples were used to measure the sampling gas temperature and inlet air temperature. These thermocouples were connected to a DataTaker 500 data logger which recorded the temperature values every 3 s.

Measurement of pressure

Inlet air pressure and the pressure drop across the orifice plate meter were measured by using Setra pressure transducers which were also connected to the data logger; this recorded the sensor outputs every 3 s. For the measurement of inlet air pressure, a Setra pressure transducer (Model 206, 25 PSIG, 0.15% accuracy), which can measure a range of gauge pressures between 0 and 25 psi (0 and 172 kPa) was used. Pressure drop through the orifice plate was measured by a Setra pressure transducer (Model 264, 25 WG, 1.0% accuracy), which can measure differential pressure in the range 0-25 in H_2O (0-6.2 kPa).

Measurement of airflow rate

The rate at which air is supplied to the gasifier is one of the most important operating parameters and it was necessary to measure this accurately and continuously. This airflow rate was measured by a standard orifice meter technique [5].

Measurement of gas flow rate

When a gas is cool and clean, its flow rate can be measured directly by flow measurement devices such as rotameters, Pitot tubes, orifice meters and venturi meters. However, such devices could not be used in this project since the hot producer gas was flared without prior cooling. Therefore, the total gas flow rate was estimated indirectly by means of a nitrogen balance across the gasifier, assuming the nitrogen content of wood and NO_X production in the gasifier to be negligible.

$$\text{Producer gas flow rate } (\text{Nm}^3/\text{hr}) = \frac{\text{Air supply rate}(\text{Nm}^3/\text{hr}) \times \text{N}_2\text{Vol fraction in air}}{\text{N}_2 \text{ Vol fraction in producer gas}} \quad (1)$$

By knowing the volume fraction of major gas species (H_2, CO, CH_4 and CO_2) in the final producer gas, the balance could be assumed to be N_2 since levels of O_2 and other gases such as hydrocarbons would be negligible at the reactor outlet.

Measurement of wood consumption rate

Average wood consumption rate was estimated by measuring the volume of fuel consumed during each experiment. The gasifier was normally filled to the top with wood at the start of an experiment. At the end of each run the distance of the upper fuel layer surface from the top of the reactor was measured and used to calculate the wood consumption rate as follows:

$$\text{Wood consumption rate (kg/hr)} = \frac{\text{Wood bulk density } (\text{kg/m}^3) \times \text{Cross-section area(m2)} \times \text{Lengthof wood column (m)}}{\text{Druration of Experiment (hr)}} \quad (2)$$

Cold gas efficiency

If the producer gas is used as a fuel in an internal combustion engine, the sensible heat of the gas is not utilized. The conversion efficiency based solely on the calorific energy content of the cold gas is called the cold gas efficiency and is defined as follows:

$$\text{Coal gas efficiency } (\%) = \frac{\text{Energy carried by producer gas}}{\text{Energy supplied to gasifier}} = \frac{VF_{gas} \times NHV_{gas}}{CR_{wood} \times NHV_{wood}} \times 100 \quad (3)$$

where VF_{gas} = volumetric flow rate of producer gas (Nm^3/hr)
NHV_{gas} = net heating value of producer gas (MJ/Nm^3)
CR_{wood} = consumption rate of the wood blocks (kg (dry basis)/hr)
NHV_{wood} = net heating value of the wood (MJ/kg (dry basis))

Table 1. Average properties of red gum wood blocks

Characteristic	Value	Test Method
Gross Heating Value (MJ/kg)	20.04	(*)
Apparent density (kg/m^3)	951	ASTM - D 2395-83
Small Blocks		
Size (mm)	20 x 20 x 20	
Bulk Density (kg/m^3) (20 mm cube)	527	ASTM - E 873-82
Void fraction, (-) (20 mm cube)	0.45	
Moisture content (% w.b)	9.7	ASTM - E 871-82
Larger Blocks		
Size (mm)	30 x 30 x 30	
Bulk Density (kg/m^3) (30 mm cube)	485	ASTM - E 873-82
Void fraction, (-) (30 mm cube)	0.49	
Moisture content (% w.b)	12.2	ASTM - E 871-82
Proximate Analysis (wt % d.b)		
Volatile Matter	75.95	ASTM - E 872-82
Fixed Carbon	23.92	ASTM - E 870-82
Ash	0.13	ASTM - D 1102-82
Ultimate Analysis (wt % d.b)		
C	52.90	**HRL – Method 1.4
H	5.30	HRL – Method 1.4
O	41.63	
N	0.01	HRL – Method 1.4
S	0.03	AS 1038.6.3.3
Ash	0.13	HRL – Method 1.6

* Calculated from values of ultimate analysis [6].
** HRL is the NATA registered laboratory that analysed the wood sample.

RAW MATERIALS

The raw materials used in the stratified downdraft gasification study were mainly wood blocks, with some charcoal used during the reactor warming up process.

Wood blocks

River red gum (*Eucalyptus camaldulensis*) wood stakes of square cross-section were purchased from a local wood supplier in two different sizes: 20mm x 20 mm x 1800 mm long and 30 mm x 30 mm x 1800 mm long. These stakes were only available undried and their moisture content was as high as 45% (db). They were therefore dried in a kiln for 10 days and their average moisture content decreased to 13% (db). The dried stakes were cut into cubes (20 x 20 x 20 mm and 30 x 30 x 30 mm) using a band saw. The properties of these wood blocks are presented in Table 1.

Red gum charcoal

Suitably sized charcoal was required during start-up of the gasifier if formation of large amounts of tar within the reactor was to be avoided. Bags of red gum wood charcoal were purchased from a local charcoal supplier. Larger pieces were broken manually into smaller pieces and the resulting mix sieved to obtain the 20–30 mm fraction. The properties of these charcoal pieces are presented in Table 2.

Table 2. Average properties of red gum charcoal blocks

Characteristic	Value	ASTM Standards Test Method No.
Size (mm)	20-30	
Bulk Density (kg/m^3) (20 mm cube)	244	E 873-82
Moisture content (% w.b)	5.06	E 871-82
Proximate Analysis (wt % d.b)		
Volatile Matter	13.58	E 872-82
Fixed Carbon	85.90	E 870-82
Ash	0.52	D 1102-82

EXPERIMENTAL PROCEDURE

Prior to each run, the gasifier was filled with sized charcoal up to the ignition port level. The remainder of the gasifier was then filled with wood blocks. After loading the wood, the top cover was closed and the suction blower started up in order to draw into the gasifier a flame from a torch held in front of the four air suction ports. After 1-2 minutes of ignition time, combustion was well developed inside the gasifier. The suction ports were closed and air introduced at the desired rate at the top of the gasifier. After 10-15 minutes of operation, the heating value of the producer gas was high enough for it to burn without assistance from a pilot flame. The experimental run continued until the wood blocks had burned down to the level of the ignition ports; this took 1 to 2 hours depending on the air supply rate.

EXPERIMENTAL RESULTS

Experimental runs were carried out for various air supply rates for both sizes of wood blocks. During experiments, the main parameters measured or calculated were: temperature profiles along the gasifier, the gas composition profile, the tar profile, fuel consumption, gas production rate, air to fuel ratio, reactor cold gas efficiency and velocity of reaction zone movement.

Material balance and elemental balance

The reliance that can be placed on experimental data is determined by the accuracy and completeness of the measurements obtained. Material and elemental balance closures are important indicators of the quality of experimental results since mass has to be conserved across the gasifier. Material balance closure over the entire period of experimental run was calculated by dividing the total mass output by the total mass input (and multiplying by 100 if it was to be expressed as a percentage). Elemental balances for oxygen, carbon and hydrogen were conducted in a similar way, i.e. by determining the ratio of the total outputs of each individual element to the total inputs of that element. The average closures are presented in Table 3.

It was found that for 14 out of the total 20 experimental runs, material balances closed to within 3%. Another 5 runs had closures within 5% and the last run closed to within 6%. The average value of mass balance closures for all experimental runs was found to be 99.3% and this indicates that considerable reliance can be placed on the accuracy of the experimental measurements.

Table 3. Mass and elemental balance closures, average values

	Average %	Standard deviation σ <95% mean
Mass	99.30	2.84
Carbon	91.77	8.24
Oxygen	102.34	6.36
Hydrogen	99.00	9.87

Temperature

At the beginning, as the reactor warms up, the heat from the burning char goes to heating up the bottom char bed and the reactor itself (mainly the thick refractory wall), both of which absorb a large proportion of the heat produced at this stage. Under these conditions rates of gasification are insignificant to small. It was observed that only after the temperature at the grate level reached 350-400 °C was the outlet producer gas able to burn by itself without the LPG flare. Fig. 5 shows a representative set of changes over time of temperatures at four different locations in the gasifier. The figure shows clearly how the rate of temperature change at the base of the reactor is high initially and then the temperature levels off at its steady state value about 15-20 minutes after ignition starts. The temperature at the fuel ignition level is at a high value except at the very beginning of the run, but fluctuates within the range of 950-1270 °C. These fluctuations are attributed to a combination of small reaction zone shifts and to changes in the micro-environment around the thermocouple as the solid bed moves down past it. Temperatures near the top of the gasifier (750 mm above the grate) were observed to stay about 10-20 °C higher than the ambient temperature throughout the experimental run and therefore no physical or chemical changes would be expected in the fuel bed at that level.

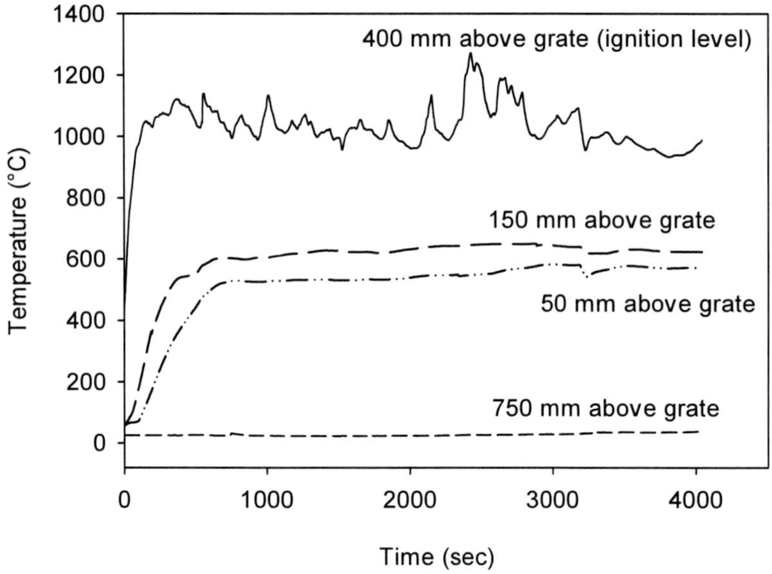

Figure 5. Temperature record for experimental run at an air flow rate of 16.92 kg/hr with 20 mm blocks.

Temperature profiles measured for a particular airflow rate (7.5 kg/hr) at three different levels above the ignition port are presented in Fig. 6. It is evident that in this run the reaction zone moved upwards in the reactor. From the distance between corresponding sections of the profiles at different times the rate of reaction zone movement can be estimated. From the profiles shown in Fig. 6, the rate of upward movement of the reaction zone was estimated to be 0.063 mm/s.

Fig. 7 shows how temperatures vary along the gasifier once stable gas producing conditions have been reached. Air supply rates for the three curves shown were 10.7, 14.8 & 19.6 kg/hr and the temperatures shown are averages over the period 30 to 40 minutes after ignition. The profiles show temperatures increasing rapidly at the beginning of the combustion zone and then gradually decreasing in the gasification zone. This pattern is similar to that predicted by the model (model validation is discussed in detail in Section 7). It can be seen that, as would be expected, the value of the peak temperature increases with increases in the airflow rate.

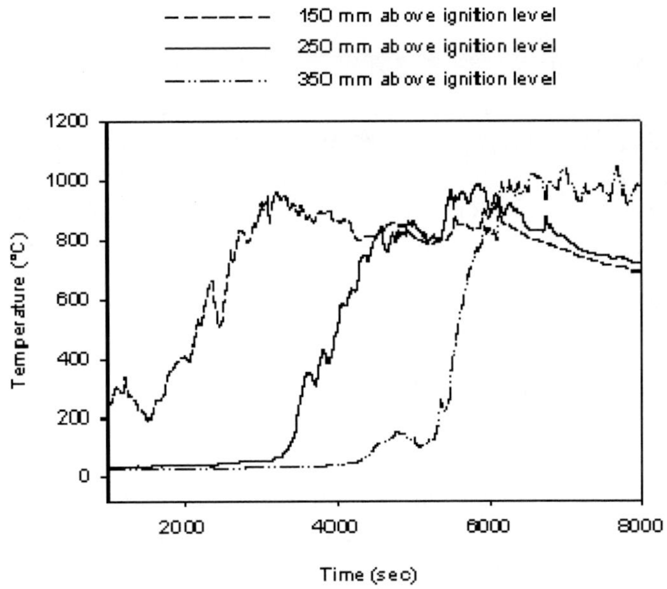

Figure 6. Temperatures measured at three different levels above the ignition port during an experimental run (Airflow 7.5 kg/hr, ignition started at 400 mm above the grate).

Fig. 7 also shows that peak temperatures for the three different airflow rates occur at different points along the reactor even though ignition was started at the same level (400 mm above ignition) in all cases. This illustrates the effect of airflow rate on movement of the reaction zone. As predicted by the numerical model, a higher airflow rate slows down the upward reaction front movement. This is discussed in more detail in Section 7.

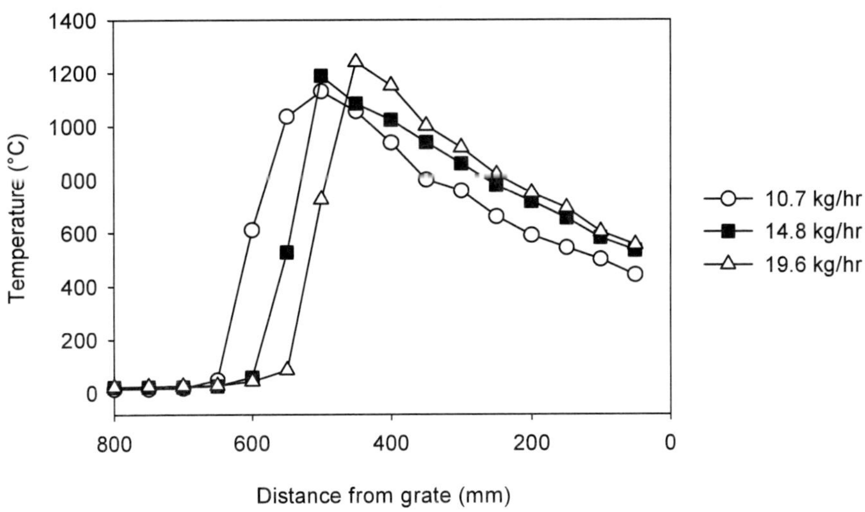

Figure 7. Temperature profiles at three different air flow rates (average temperature values between 30 min and 40 min after ignition).

Gas composition

Gas samples were taken from five different levels of the reactor at two different times during each experimental run and analysed by gas chromatography as discussed in Section 3.3. The outlet gas compositions were relatively consistent over the range of operating airflow rates used. The results show nitrogen to be the dominant constituent of the outlet gas (46-58 vol %). The concentration of CO_2 was in the range of 11-17 vol%. Of the combustible gas components, CO has the highest concentration (16-21 vol%) followed by H_2 (14-19 vol%). A small fraction of CH_4 (1.0-1.7 vol%) was also present. Net heating values of the outlet gas ranged from 4.3 to 5.2 MJ/Nm³.

Tar content

Tar contents of gas samples from five different levels in the gasifier were determined by following the procedures described in section 3.2. It was found that outlet gas tar contents for all experimental runs were less than 100 mg/Nm³, which is about 20 times less than the tar content found in gas from a conventional throatless downdraft wood gasifier [80]. This can be attributed to two factors. Firstly, the initial loading of charcoal rather than wood into the gasifier below the level of the ignition ports ensured no tar formation occurred below the combustion zone. Secondly, there was uniform distribution of air across the reaction zone, since air was introduced to the system at the top of the reactor rather than from the sides of the reactor. The positive effect of the charcoal bed was well illustrated by what happened in an experimental run in which wood fuel only was used and there was no initial char bed. For this run the tar content was almost 1000 mg/Nm³, which was about 15 times higher than that for an experimental run conducted at a similar air supply rate and for which an initial charcoal bed was used.

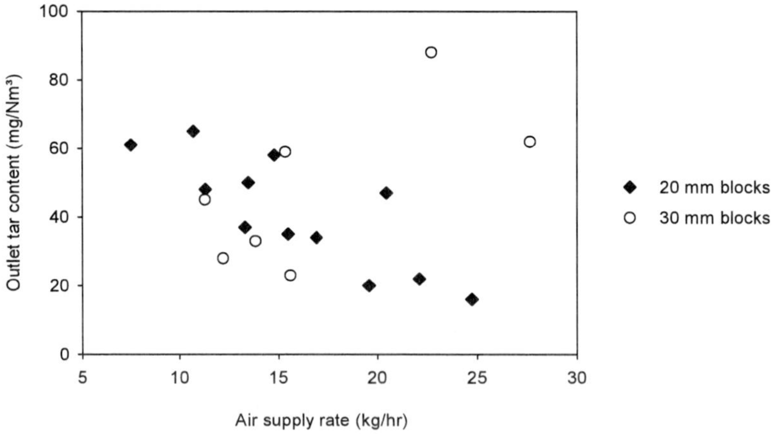

Figure 8. Outlet gas tar content as a function of air supply rate.

Tar content in the outlet producer gas is shown in Fig. 8 as a function of air supply rate for both wood block sizes. From Fig. 8, it can be seen that tar content generally decreased as the air supply rate increased, as predicted by the model.

Energy output

The total energy output in MJ/hr (determined from the outlet gas composition and flow rate, which only takes into account the calorific value of the gas) is plotted against dry wood consumption rate in Fig. 9. As can be seen from the figure, and not unexpectedly, total energy output increased linearly with the increase in fuel consumption. A similar result was also obtained in an experimental study in which a 600 mm internal diameter stratified downdraft gasifier was operated using wood chips and wood pellets [81].

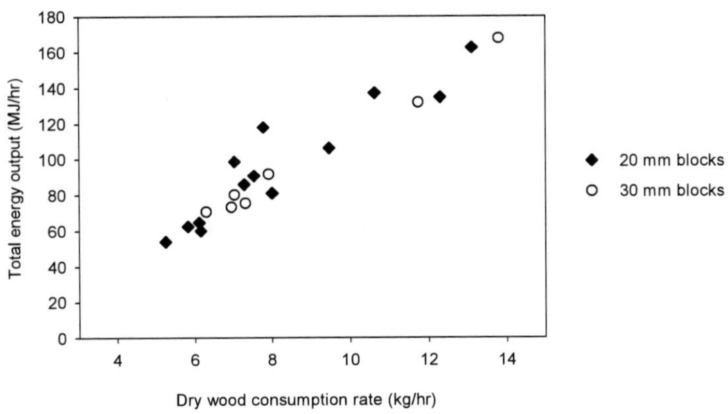

Figure 9. Energy output versus dry wood consumption rate.

Cold gas efficiency

Cold gas efficiency is plotted against the air supply rate in Fig. 10. Within the range of air supply rates (7.5 - 27.6 kg/hr) tested, average cold gas efficiency was found to be 70%. As shown in Fig. 10, similar values of cold gas efficiency were achieved for most of the runs. A slight increase in cold gas efficiency with increases in air supply rate is suggested by the results but, given the extent of scatter in the results, further work would be needed to confirm this apparent trend.

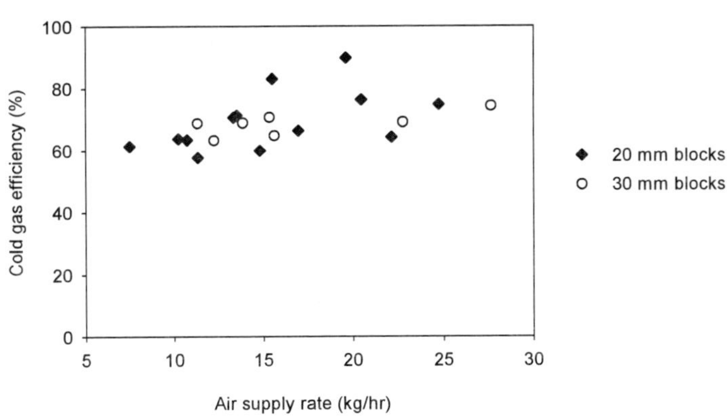

Figure 10. Cold gas efficiency versus air supply rate.

Comparison of Model Predictions with Experimental Results

To compare the model predictions with experimental results, the numerical model was run using the same input parameters as those used in the experimental runs. The predicted and experimentally measured values of various gasifier performance parameters (transient and steady temperature profiles, gas composition profiles, tar profile, wood consumption rate, air to fuel ratio, gas production rate, and reaction zone movement) were then compared.

Temperature profiles

In Fig. 11, measured transient temperature profiles are compared with the temperature profiles predicted by the model. The numerical model predicts that within the first ten minutes after initial ignition, heat released from the ignited char will cause the gasification zone temperature to increase. After 15-20 minutes of ignition, the rate at which the gasification zone heats up is predicted to slow and the temperature profiles to stabilise. Since gasification zone temperatures over the initial warm-up period are not high enough for gasification rates to be significant, the pyrolysis rate is predicted to be higher than the gasification rate, causing the reaction zone to move upwards. The trend of the experimental temperatures over time is quite similar to that predicted by the model.

The model predicts that temperatures in the gasifier will increase sharply in the oxidation zone, peak, and then decrease more gradually along the gasification zone. However, the values of the peak gas temperature predicted by the model are higher than the experimentally measured values and experimentally measured temperatures rise to their peak values more gradually than those predicted by the model. Two reasons can be put forward to explain this discrepancy. Firstly, in the model oxidation of volatiles released from the pyrolysis zone is described by a kinetic expression alone, since mass transfer effects are very difficult to include. The effect of this is that the model predicts complete oxidation of the volatiles within too short a length of the reactor, too rapid attainment of the adiabatic flame temperature, and therefore too sharp an increase in the gas temperature. The second reason relates to the difficulties involved in obtaining accurate measurements of true flame temperatures in the gaseous oxidation zone within a packed bed of particles. Temperature gradients within the gas spaces are large, and the bed of particles in the reaction zone is continually moving, especially at the time of reactor start up. In addition, temperatures measured inside the reactor using the thermocouples inserted through the reactor wall are more likely to reflect solid surface temperatures or intermediate temperatures rather than true gas phase flame temperatures [82]. Resolution of this problem will have to wait until temperature sensors better suited to measuring the temperatures in the gas spaces of a moving packed bed are available.

Fig. 12 shows the gas temperature at the reactor outlet (which was measured with thermocouple T_1 in Fig. 1) as a function of air supply rate for both wood block sizes. It can be seen that there were no significant differences in gas outlet temperature for the two different wood block sizes, while gas temperature tends to increase with an increase in air supply rate.

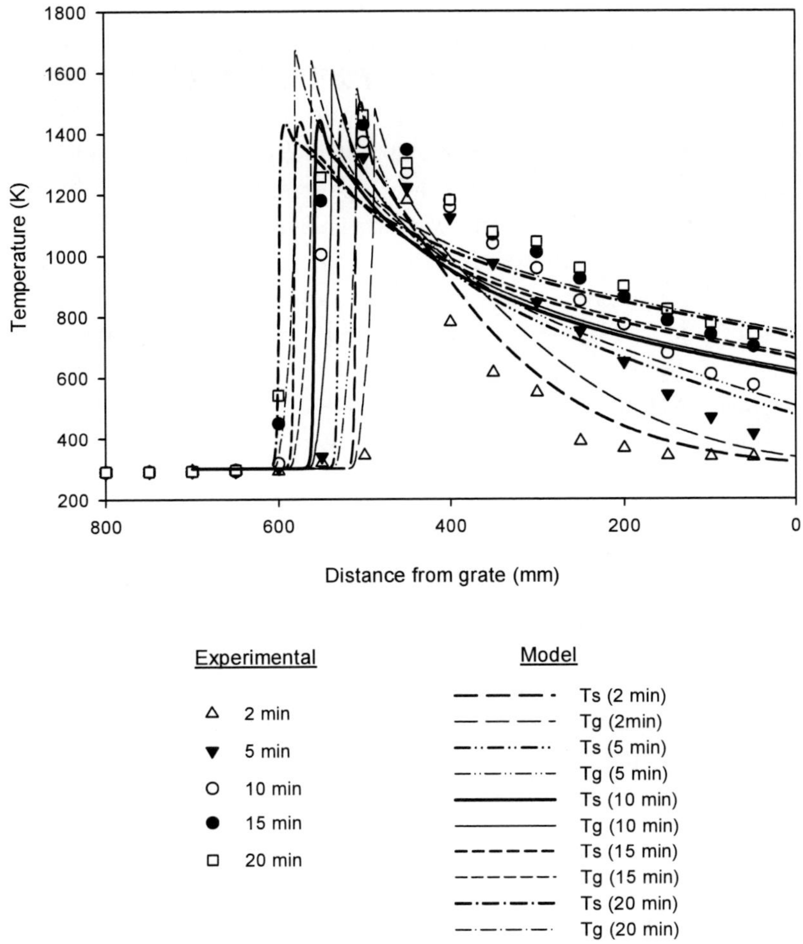

Figure 11. Comparison of predicted and experimentally measured temperature profiles along the axis of the gasifier (air flow rate: 15.5 kg/hr, wood block size 20x20x20 mm).

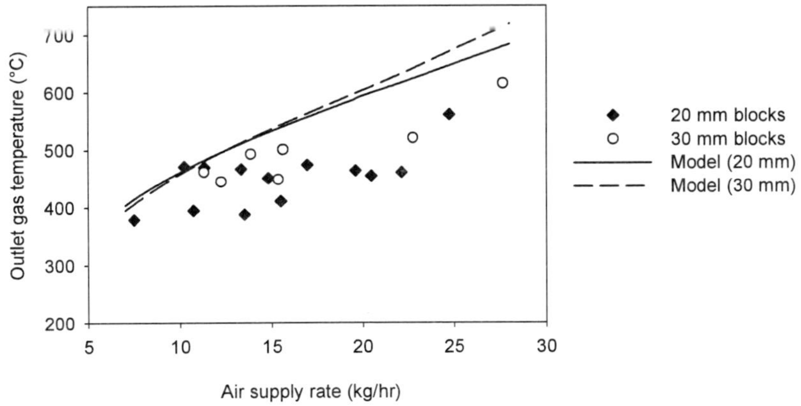

Figure 12. Predicted gas outlet temperatures compared with experimental results as a function of air supply rate (experimental points are average values over the 2 minute interval 29-31 minutes after ignition; predicted values are those at 30 min after ignition).

This is because a higher air supply rate leads to higher rates of combustion and higher gas temperatures in the combustion zone, which also raises the gas outlet temperature.

Fig. 12 also compares the experimental results with model predictions. The figure shows that the outlet temperature predicted by the model increased over the range of air supply rates shown by an amount similar to that by which the experimental results increased. The predicted gas outlet temperatures, however, were generally higher than the experimentally measured values, especially at higher airflow rates. This discrepancy is largely attributable to the fact that the gas outlet temperature predicted by the model is the temperature at the level of the grate while the experimental gas outlet temperature was that measured inside a gas outlet pipe located about 250 mm downstream from the level of the grate (see Fig. 1). It was not possible to insert a thermocouple at the level of the grate since clearance was required for rotation of the grate. Even though attempts to minimise heat losses downstream of the grate were made by wrapping the pipe in insulation, some loss of heat was inevitable.

Gas composition profiles

Comparisons between the model predictions and experimentally measured values of gas compositions (dry basis) along the gasifier axis are presented in Fig. 13. It can be seen that the model was able to predict gas composition profiles reasonably well. For CO_2, the experimental data confirmed the model prediction that CO_2 levels would increase to a peak in the combustion zone and then decrease slightly in the gasification zone. For CO, H_2 and CH_4, there was a greater difference between model predictions and experimental data.

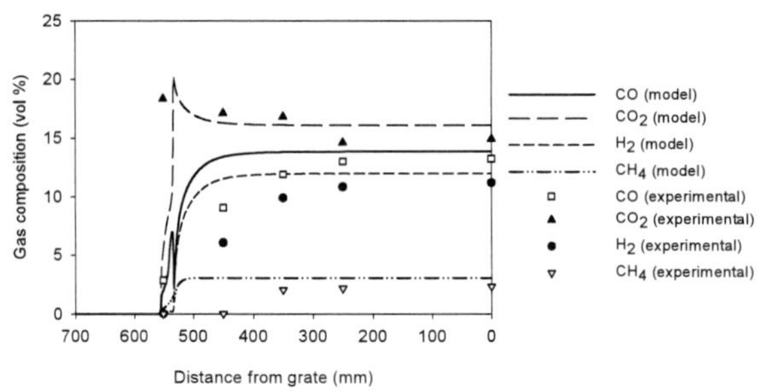

13(a)

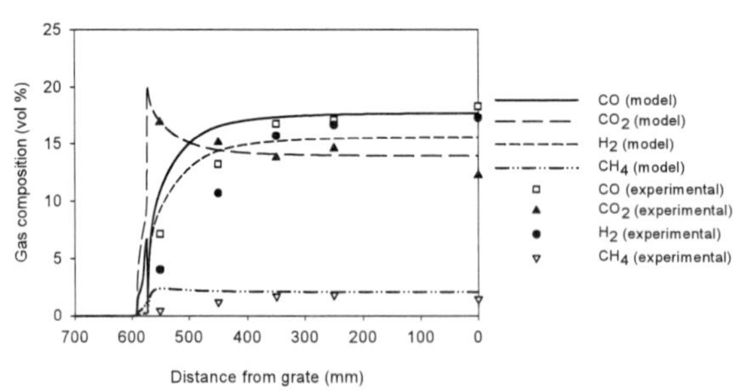

13(b)

Figure 13. Comparison of predicted and experimental dry gas composition profiles along the gasifier for two different time intervals (air flow rate 19.6 kg/hr, small wood blocks) (a) 10-15 min after ignition (b) 30-40 min after ignition.

The model predicted a much more rapid rate of increase in the levels of these gases and attainment of steady values well upstream of the point where this occurred in the experimental runs. This is attributed to the fact that predicted temperatures distributions rose more sharply than experimental ones, as discussed in section 7.1. Predicted and observed steady values were in good agreement, however.

Tar profile

As described in section 3, tar contents of gas samples drawn from five different levels in the gasifier were measured. The top sampling point was located either 50 mm or 150 mm above the ignition level depending on where the ignition was started; the amount of tar in the sample from this sampling point was taken to be representative of gas phase tar content in the pyrolysis zone. The gases sampled from the other three sampling ports (each 100 mm apart along the gasifier below the top sampling port) enabled the tar profile through the gasification zone to be determined. The last sampling port enabled the tar content of the producer gas at the gasifier outlet to be determined.

Fig. 14 shows predicted and experimentally measured tar profiles along the gasifier for four different airflow rates. Both sets of results show the tar content in the pyrolysis zone to

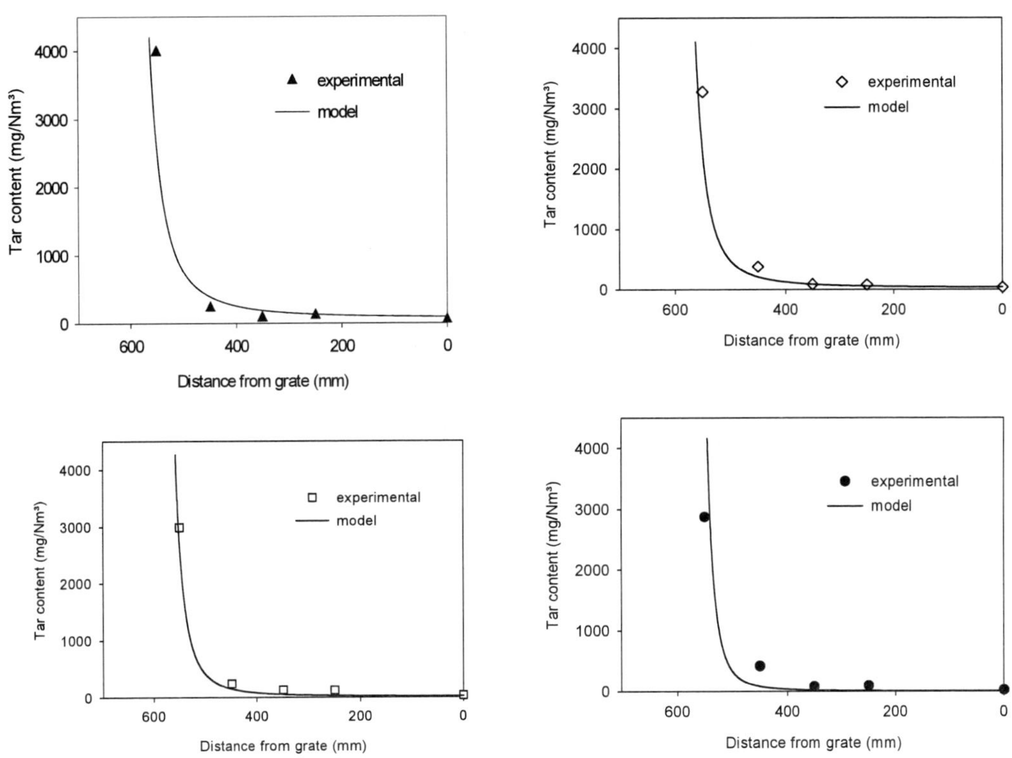

Figure 14. Comparison of predicted and experimentally measured condensable tar profiles along the gasifier for different airflow rates under stable operating conditions (a) air flow rate 13.50 kg/hr, (b) 15.48 kg/hr (c) 16.92 kg/hr (d) 19.60 kg/hr.

be much higher than that in the gasification zone and outlet gases. The highest level of tar (2500-4000mg/Nm³) occurs in the pyrolysis zone and this decreases rapidly as the gas passes through the oxidation zone. Tar levels in the gasification zone are orders of magnitude less than that in the pyrolysis zone. Not unexpectedly, the lowest tar level occurs in the outlet gas samples.

Fuel consumption

Predicted and experimentally observed wood fuel consumption rates are presented in Fig. 15 as a function of air supply rate. In line with expectations, it was found that the wood fuel consumption rate was directly dependent on the amount of air supplied to the gasifier. A higher air supply rate increases the rate of combustion because more oxygen is available to

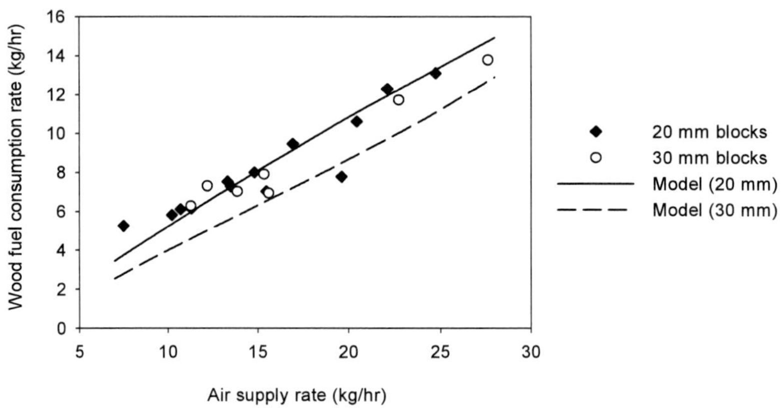

Figure 15. Wood fuel consumption versus air supply rate.

react with both char and volatile gases, and the gasification rate also increases since a higher amount of CO_2 and H_2O is released from the combustion zone. Fig. 15 shows that model predictions agree very well with the experimental data for the case of the 20 mm blocks. The predictions are less accurate for the case of the 30 mm blocks, for reasons discussed in the following section.

Air-fuel ratio

Fig. 16 shows the predicted effect of airflow rate on the air-fuel ratio (the ratio of mass of air supplied to the mass of fuel consumed) together with that calculated from experimental results. The figure shows that varying the air supply rate has no significant effect on the air-fuel ratio. This is consistent with the finding, discussed in previous section (7.4), that the rate of fuel consumption is a linear function of air supply rate. In other words, the air-fuel ratio is a parameter which cannot be controlled by the operator but which is controlled by the gasifier itself and by the properties of the fuel, as reported by other researchers [83]-[84]. In the case of the larger wood blocks the model predicts an air-fuel ratio higher than that for the smaller wood blocks. This is as expected given that the 30 mm cubes have a lower specific surface area than the 20 mm cubes. The fact that the observed air-fuel ratio for the 30 mm blocks lies below the predicted value and close to that for the 20 mm cubes suggests that the 30 mm blocks are cracking or breaking up into smaller pieces in the gasifier to a greater extent than allowed for in the model. This would be consistent with the observations of Salazar [85] who observed a greater incidence of cracking as the diameter of the cylinders of *Eucalyptus delegatensis* he was pyrolysing was increased above 20 mm. For the 20 mm cubes, however, the agreement between predicted and observed air-fuel ratios is very good.

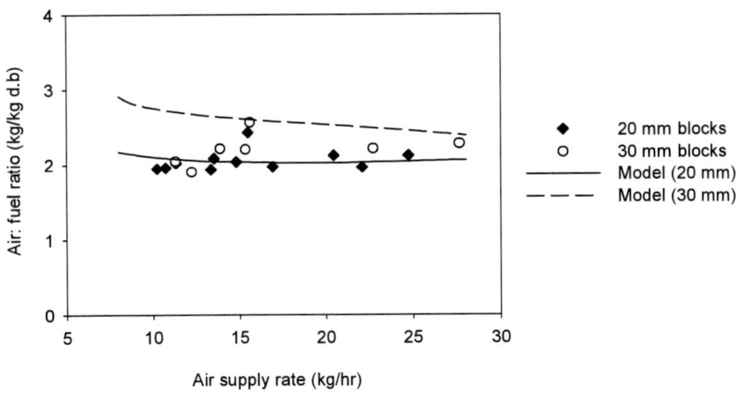

Figure 16. Air-fuel ratio (kg/kg d.b) versus air supply rate.

Gas production rate

Fig. 17 shows predicted and experimentally measured gas production rates as a function of air supply rate. The figure shows that there is a linear increase in gas production rate as the air supply rate increases. The model predicts a higher gas production rate when 20 mm blocks are used than when 30 mm blocks are used but there is no significant difference between the observed gas production rates for 20 mm and 30 mm blocks at the same airflow rate. The likely reasons for this divergence between model predictions and experimental results were discussed in Section 7.5.

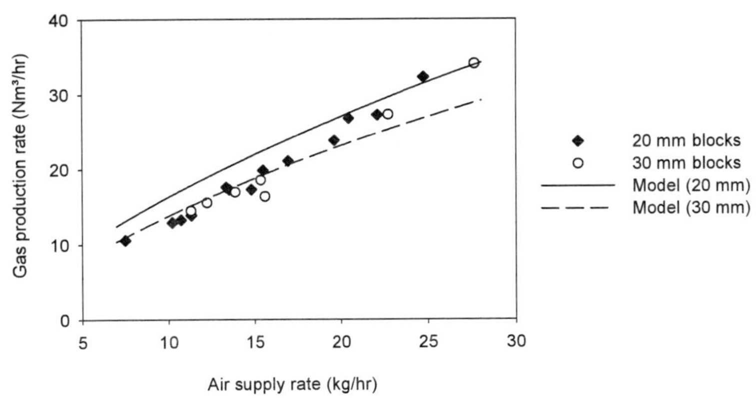

Figure 17. Gas production versus air supply rate.

Variation of reaction zone movement with airflow rate and wood block size

By measuring temperature profiles along the gasifier at different times after ignition it is easy to establish if the reaction zone is changing position within the gasifier. Fig. 18a shows temperature profiles measured at two different times for two different airflow rates but the same wood block sizes. Fig. 18b shows a similar pair of temperature profiles for two different wood block sizes and the same airflow rate. The profiles in Fig. 18a show that in the first case, where the airflow rate was 11.3 kg/hr, movement of the reaction front was in an upward direction. However, in the case where the airflow rate was considerably higher (24.7 kg/hr) the reaction zone moved in the other direction, i.e. downwards.

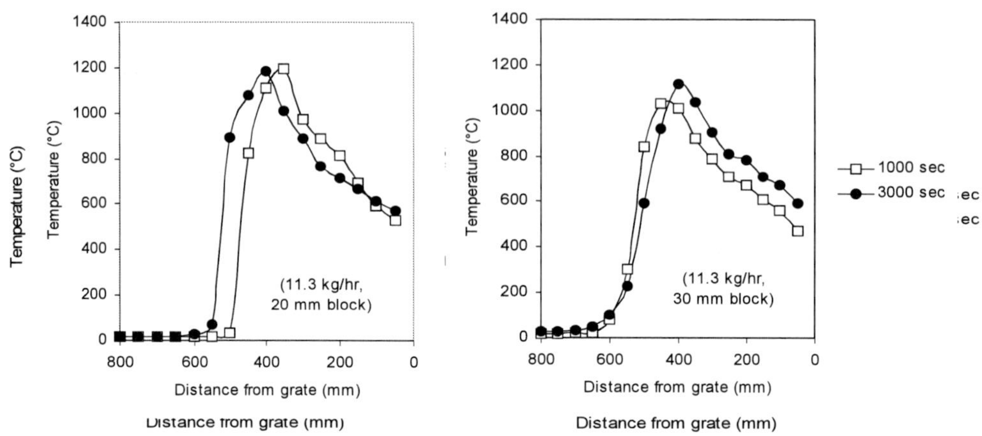

Figure 18. Temperature profiles at two different times (1000 seconds and 3000 seconds respectively after ignition of the charcoal bed) for two different air flow rates (same fuel size) (b) Temperature profiles at the same two times for two different fuel sizes (same air supply rate).

The same effect of air supply rate on reaction front movement was observed in model predictions. As discussed in the sensitivity analysis of the model [3], at lower air supply rates the predicted direction of reaction zone movement was upwards and, as the air supply rate increased, the rate of zone upward movement decreased. Beyond a certain limit (in this study this is about 20 kg/hr), further increases in air supply rate reverse the direction of movement, causing the reaction zone to move downwards towards the grate and the reactor operates in the gasification dominant mode.

The effect of wood block size on reaction front movement is shown in Fig. 18b. When using small wood blocks (20mm) at an airflow rate of 11.3 kg/hr, it is evident that the reaction front moves upwards. However, when larger wood blocks (30mm) are used at the same airflow rate the reaction front moved downwards. A similar effect of fuel particle size on zone movement rate and direction is predicted by the model, as discussed in the section on sensitivity analysis in a previous paper [3]. Relatively good agreement was obtained between model predictions and experimental results. This is well illustrated in Fig. 19 which compares predicted and observed rates of reaction zone movement for a range of airflow rates.

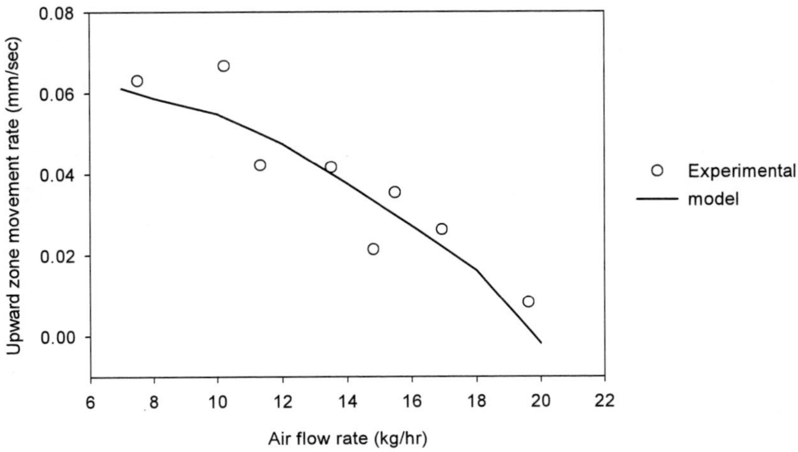

Figure 19. Variation of reaction zone movement rate with air supply rate (comparison between predicted and measured values).

Conclusion

To characterize the performance of a stratified downdraft gasifier in practice and to validate the developed computer model, a small scale stratified downdraft gasifier (206 mm ID) was run over a range of air supply rates using two different sizes of red gum wood blocks as fuel. Temperatures at 16 different levels in the reactor were measured, gas from five different levels in the reactor was sampled, and both the composition of the gas and its tar content were measured.

The major conclusions of this experimental investigation can be summarized as follows:

1) The overall material balance for the experimental stratified downdraft gasifier closed very well for all experimental runs. Reasonably good elemental mass balance closures were also achieved.
2) A substantial reduction in the tar content of the outlet gases could be achieved by initially filling the gasifier with charcoal up to the ignition port level. The average measured tar content of the outlet gas for all experimental runs using an initial charcoal bed was found to be 44 mg/Nm³, which is about 40 times less than the tar content found in gas from a traditional downdraft gasifier.
3) The experimental gasifier could be operated over a fairly wide range of air supply rates (7.5 to 27.6 kg/hr) with a corresponding energy output ranging from 54-168 MJ/hr. The producer gas composition did not vary significantly over the range of airflow rates used. However long term stable operation is only possible within a limited range of airflow rates.
4) The variation over time of the measured temperature profile along the axis of the gasifier agreed reasonably well with model predictions. It was found that 20 to 25 minutes was required to reach a stable temperature profile. A discrepancy between the predicted gas temperatures and measured values within the combustion zone was observed. This discrepancy between predicted and

measured temperatures is in part attributable to the difficulties in measuring gas temperatures in a moving packed bed and partly due to the nature of the kinetic expressions used to characterise gas phase combustion reactions.

5) The model was able to predict reasonably well the gas composition profiles at different times after start up. The rate of increase in the concentrations of two combustible components (H_2 and CO) along the gasifier was observed to occur more gradually during experimental runs than was predicted by the model; however, there was good agreement between exit gas compositions predicted by the model and those obtained experimentally.
6) As expected, the gravimetric tar content of the gas was found to be highest in the pyrolysis zone and reduced dramatically while passing through the oxidation zone. Good agreement between predicted and measured tar contents was obtained in the gasification region.
7) Wood fuel consumption rate and gas production rate were both observed to increase linearly with an increase in air supply rate. Air-fuel ratio was not affected by changes in air supply rate.
8) A higher airflow rate favoured the downward movement of reaction zones as did the use of larger wood block size. The value of the specific gasification rate for which stable operating conditions were achieved was predicted by the model to be in the range of 290-296 kg/hr/m^2 (dry basis) and this agreed very well with the range of 290-308 kg/hr/m^2 determined from experimental results.

Overall, there was an encouragingly good level of agreement between the results of model simulations and experimental measurements; however, deficiencies in our understanding of combined kinetic, heat transfer and mass transfer processes in the gas spaces of packed beds at high temperatures mean that some discrepancies remain between predicted and observed changes in the gasifier combustion zone.

ACKNOWLEDGMENT

The experimental work was partially funded by the Open Society Institute, and this assistance is gratefully acknowledged. One of the authors, San Shwe Hla, expresses his sincerest thanks to the University of Melbourne for financial support for carrying out this research.

REFERENCES

[1] Lu Aye, San Shwe Hla, Connor MA, Stewart DF. Chapter 5: A dynamic model for predicting tar content in gases from a stratified downdraft gasifier: Part I –Model development, in: Gasification: Chemistry, Processes and Applications (Ed: Michael D. Baker), Nova Science Publishers, Inc. 2011.

[2] San Shwe Hla, Lu Aye, Connor MA, Stewart DF. Chapter 6: A dynamic model for predicting tar content in gases from a stratified downdraft gasifier: Part II –Model

sensitivity, in: Gasification: Chemistry, Processes and Applications (Ed: Michael D. Baker), Nova Science Publishers, Inc. 2011.

[3] Neeft JPA, Knoef HAM, Zielke U, Sjőstrőm, Hasler P, Simell PA, Dorrington MA, Abatzoglou N, Deutch S, Greil C, Buffinga GJ, Brage C, Suomalainen M. Guideline for sampling and analysis of tar and particles in biomass producer gases (Version 3.3). Energy project EEN5-1999-00507 (Tar protocol); 2002.

[4] Brage C, Sjőstrőm K. An outline of R&D work supporting the tar guideline. Report related to the project Tar Guideline: (project no EEN5-00507); 2002.

[5] British Standards Institution. Methods for the measurement of fluid flow in pipes: Part1. Orifice Plates, Nozzles and Venturi Tubes, [B.S. 1042 : Part 1 : 1964]; 1965.

[6] Reed TB, Agua D. Handbook of biomass downdraft gasifier engine systems. Biomass Energy Foundation Press, Washington; 1988.

[7] Bui T, Loof R, Bhattacharya SC. Multi-stage reactor for thermal gasification of wood. Energy 1994;19:397-404.

[8] Walawender WP, Chern SM, Fan LT. Influence of operating parameters on the performance of a wood-fed downdraft gasifier. In: Energy from Biomass and Wastes X, ed. LK Donald; 1987. p. 607-27.

[9] Barriga A, Essenhigh RH. A mathematical model of a combustion Pot: Comparison of theory and experiment, Paper 80- WA/HT-32, ASME Meeting, Chicago; 1980.

[10] Barriom M, Fossum M, Hustad, JE. A small-scale stratified downdraft gasifier coupled to a gas engine for combined heat and power production. In: Progress in Thermochemical Biomass Conversion' ed. AV Bridgwater; 2001. p. 217-54.

[11] García-Bacaicoa P, Bilbao R, Arauzo J, Salvador ML. Scale-up of downdraft moving bed gasifier (25-300 kg/h) – design, experimental aspects and results. Bioresource Technology 1994:48:229-35.

[12] Salazar CM. The influence of particle size and shape on the mechanisms of decomposition of wood during pyrolysis. PhD Dissertation. The University of Melbourne, Australia; 1987.

In: Gasification: Chemistry, Processes and Applications
Editor: Michael D. Baker
ISBN: 978-1-61209-681-0

Chapter 8

GASIFICATION OF BIOMASS FOR THE PRODUCTION OF SYNGAS WITH REDUCED TAR FORMATION

***Chunbao (Charles) Xu*[*a,b] and Scott Hurley*[a]**

[a]Department of Chemical Engineering, Lakehead University, Thunder Bay, Ontario, Canada

[b]Department of Chemical and Biochemical Engineering, The University of Western Ontario, London, Ontario, Canada

ABSTRACT

Among all the biomass conversion technologies, gasification has attracted increasing research interest in recent decades, as biomass gasification combined with gas turbine, internal combustion engines and gas burners has great potential for large-scale green power generation (through IGCC, integrated gasification combine cycle) at higher efficiencies than the conventional pulverized coal-fired power generation. Gasification of biomass can be achieved by partial oxidation of the feedstock using air, O_2 and steam or CO_2, producing a low to medium-BTU syngas gas (4-15 MJ/Nm^3), consisting primarily of CO, H_2, CH_4, CO_2 and trace higher hydrocarbons (C_2+C_3). However, a major challenge for biomass gasification is the high yields of tars that are composed mainly of aromatics and complex poly-aromatic hydrocarbons (PAHs). In an air-blown fluidized bed gasifier for instance, typical tar contents in the producer gas are between 0.5 to 100 g/m^3. For many applications, with the exception of direct and immediate syngas combustion for heat or electricity production, these tar levels must be reduced, often to below 50 mg/m^3. Tars can have significant negative effects on gasification with respect to efficiency and operation. Specifically, the production of tars instead of combustible gases reduces the gasification efficiency, and condensation and deposition of tars at temperatures below 350°C can lead to fouling and potential blockage of downstream equipment and piping. Although the biomass-derived syngas with tar can be effectively cleaned up with the wet scrubbing technology, hot gas clean-up technology, preferably employing using catalysts, to remove these contaminants at the "hot" state is advantageous with respect to energy efficiencies as it eliminates the need to cool the

* Corresponding author. Fax: 1(807)343-8928; Email: cxu@lakeheadu.ca

product gas and re-heat again for the syngas applications (internal combustion, gas engines, fuel cells and methanol synthesis, etc.). The hydrocarbons and tars formed during the gasification process could be removed/suppressed at a hot state through steam/dry reforming or cracking/hydro-cracking reactions using catalysts. This chapter overviews various biomass gasification technologies and tar formation during gasification, tar reduction technologies, and catalysts for biomass gasification and tar reduction.

1. INTRODUCTION

Due to the concerns over climate changes and depleted fossil fuels reserves, it is of strategic significance to explore alternatives to fossil resources for both energy and chemical production. Among all the potential alternatives to fossil resources, biomass is promising as it represents an immense renewable and hence sustainable energy source (the global annual biomass production (estimated at 2000 billion tonnes) is about 6-8 times that the world energy consumption [1,2]. Practically, useful biomass resources include wood and forestry residues (sawdust, limbs, bark, tree tops), energy crops and agricultural residues (wheat/rice straws and corn waste). Currently, the use of biomass accounts for approximately 12% of the world's total primary energy consumption [2, 3]. Although biomass resources are renewable, carbon-neutral, and massive in amount, they are bulky and difficult to transport, handle, and store. Therefore, appropriate biomass conversion technologies are required to upgrade them into gas (hydrogen, syngas), liquid (ethanol, bio-diesel, bio-oils) and other solid fuels (pellets, bio-chars, bio-carbon) through a variety of bio-chemical and thermo-chemical conversion technologies. Among all the conversion technologies, biomass gasification has attracted increasing interest in the recent decades. Gasification of biomass can be achieved by partial oxidation of the feedstock using air, O_2 and steam or CO_2, producing a low to medium-BTU synthetic gas or syngas (4-15 MJ/Nm^3). Syngas contains primarily CO, H_2, CH_4, CO_2 and trace contents of higher hydrocarbons (C_2+C_3) [3]. Syngas can be cleaned and used directly for power generation or catalytically modified to produce various chemical products such as methanol, dimethyl ether, Fischer-Tropsch oils, etc [4]. Biomass gasification, combined with gas turbine, internal combustion engines and gas burners has great potential for large-scale green power generation (through IGCC, integrated gasification combine cycle, etc.) at much higher efficiencies than the conventional pulverized coal-fired power generation.

In biomass gasification, some contaminants such as tars, NH_3, H_2S and SO_2, etc are also generated and present in the product gas. The major obstacle for commercialization of biomass gasification technology may be associated with the high yields of tars [4-6]. In an air-blown fluidized bed gasifier, typical tar contents in the producer gas have been reported between 0.5 to 100 g/m^3 [4, 7, 8]. The formation of tars instead of combustible gases results in a decrease in gasification efficiency, and condensation and deposition of tars at temperatures below 350°C can lead to fouling and potential blockage of downstream equipment and piping [8]. The biomass-derived tars consist of a wide range of condensable hydrocarbon and oxygen containing hydrocarbons compounds, which are mostly aromatics and complex poly-aromatic hydrocarbons (PAHs) [9]. In order to achieve better efficiencies of syngas applications, the contaminants must be removed/cleaned before the syngas can be used for internal combustion, gas engines, and in particular for fuel cells or methanol synthesis. For many applications, with the exception of direct and immediate syngas combustion for heat or electricity production, the tar levels must be reduced often to below 50 mg/m^3 [4]. The biomass-derived syngas with tar can be effectively cleaned up via the wet scrubbing technology, and by thermal/catalytic reforming/cracking reactions as shown below [9]:

Steam reforming: $$C_nH_m + nH_2O \leftrightarrow nCO + (n + \frac{m}{2})H_2 \quad (1)$$

Dry reforming: $$C_nH_m + nCO_2 \leftrightarrow 2nCO + (\frac{m}{2})H_2 \quad (2)$$

Cracking: $$C_nH_m \leftrightarrow nC + (\frac{m}{2})H_2 \quad (3)$$

Hydro-cracking: $$C_nH_m + (\frac{4n-m}{2})H_2 \leftrightarrow nCH_4 \quad (4)$$

Although the above tar reducing reactions can be carried out without a catalyst, a high reaction temperature >900°C is needed due to the endothermic nature of these reactions and a high activation energy >250-350 kJ/mol. The use of catalysts allows for tar removal at lower temperatures, generally as low as 600-800°C [5], which increases the thermal efficiency of the process, and prevents combustion or thermal degradation of the syngas products at high temperatures. Dolomite and group VIII metals such as Ni, Fe, Co, etc. have been most commonly used as the catalysts for the tar removing reactions [10, 11]. For instance, the use of dolomite and NiMo/Al_2O_3 could significantly reduce the activation energy of the steam reforming reaction of toluene at 650-850°C to 123 and 56 kJ/mol, respectively [12]. Depending on the stage of catalyst addition, tar cracking catalysts can be the primary catalysts, added to the gasification system (gasifier) as reactive bed materials [14-16] or directly to the biomass feedstock, and the secondary catalysts, added to secondary guard beds and tar cracking reactors downstream of the gasifier [5].

This chapter overviews the various aspects of biomass gasification and tar reduction technologies, focusing on the formation and reduction of tars using catalysts.

2. Biomass Gasification Technologies

Biomass gasification is a thermochemical process in which biomass is heated under partially oxidizing atmosphere (air, O_2, steam, supercritical water or CO_2) to form syngas products (mainly CO, H_2, and CO_2) and small amounts of CH_4, C_2-C_3 hydrocarbons and water, as well as trace contaminants such as tars, NH_3, H_2S and SO_2, etc [4-6, 17, 18]. The overall reaction of biomass gasification may be described by the following general reaction:

$$Biomass + O_2 \text{ or } H_2O \rightarrow CO + CO_2 + H_2O + H_2 + CH_4 + other\ HCs \rightarrow tar + char + ash \quad (5)$$

$$\rightarrow HCN + NH_3 + HCl + H_2S + other\ sulfur\ gases(SO_2, COS, etc.)$$

Biomass gasification is a staged process involving pyrolysis, occurring at temperatures > 300°C during which the biomass is degraded into volatile components and char, and cracking/partial oxidation/reforming of the volatile components into gas products (CO, H_2, CO_2). This char can also be gasified by partial oxidation/reforming reactions [19]. Some main reactions associated with biomass gasification are outlined in Table 1. It is clearly seen that the reactions that contribute to the CO and H_2 formation are mainly highly endothermic, thus they are thermodynamically favorable at high temperatures.

Table 1. Major reactions in gasification of biomass [2, 20]

Reaction	ΔH_{298} (kJ mol^{-1})
Volatiles $\rightarrow CH_4 + C$	Exothermic (mildly)
$C + 0.5O_2 \rightarrow CO$	-111
$CO + 0.5O_2 \rightarrow CO_2$	-254
$H_2 + 0.5O_2 \rightarrow H_2O$	-242
$C + H_2O \rightarrow CO + H_2$	+131
$C + CO_2 \rightarrow 2CO$	+172
$C + 2H_2 \rightarrow CH_4$	-75
$CO + 3H_2 \rightarrow CH_4 + H_2O$	-206
$CO + H_2O \rightarrow CO_2 + H_2$	-41
$CO_2 + 4H_2O \rightarrow CH_4 + 2H_2O$	-165

Re-condensation of the thermally cracked volatile components or intermediates would yield tars, which are aromatic-rich mixture of heavy hydrocarbons with molecular weights greater than benzene. Differing from coal gasification, biomass gasification tends to produce greater tar yields, and the resulting tars are heavier and more stable aromatic species [21]. In an air-blown fluidized bed gasifier, for example, typical tar contents in producer gas have been reported between 0.5 to 100 g/m^3 [4, 7, 8]. Tar condenses at reduced temperatures, thereby fouling the downstream pipes/lines and equipment, so tar formation is regarded as one of the main barriers to commercialization of biomass gasification technologies [22-25]. However, tar formation, the compositions and heating values of the resulted gas are significantly affected by the operating conditions of the gasification process such as temperature, pressure, residence time, gasifying agent, type of feedstock, feed rate of biomass, flow rate of gasification medium, and catalysts, etc. [18, 26-28].

2.1 Air/Oxygen Gasification of Biomass

The common gasifying agents are air, oxygen, steam, and CO_2 or their combinations. Compared with stream-blown gasification (highly endothermic), air-blown gasification is exothermic and hence can achieve an authothermal (energy self-sustaining) process. As such, it has been widely practiced on a research or industrial demonstration scale. Nevertheless, air-blown gasification produces a gas product with a lower calorific value (4-6 MJ/Nm^3) than steam gasification (13-20 MJ/Nm^3) [3]. This is mainly due to the presence of inert N_2 from air and a high concentration of CO_2 in the product gas from air-blown gasification [18]. Calorific values of gas products from different gasification processes in comparison to that of natural gas are shown in Table 2.

Table 2. Heating values of gas products from different gasification processes in comparison to that of natural gas [2]

Gasification process	Calorific value of the gas product (MJ/Nm^3)
Air	4-6
Oxygen	10-15
Steam	13-20
Natural Gas	36

Gasification in an environment of pure oxygen is advantageous as it produces a syngas with a much higher calorific value (10-15 MJ/Nm^3) than that produced by air due to the absence of N_2 in the gas products as shown in Table 2. In addition, the ease in separation of high purity CO_2 from the product gas stream by absorption/adsorption or membrane technology makes this technology more attractive with respect to CO_2 emission control. However, the cost for separation of oxygen from air, and the safety issue associated with pure oxygen utilization are the main barriers for commercialization of the oxygen-blown gasification technology [21]. For air and oxygen gasification, an operating parameter, the equivalence ratio (ER), is defined as the ratio of O_2 supplied to the reactor compared to the O_2 required for complete combustion of the feedstock. For air/oxygen-blown gasification, ER values of 0.25 to 0.35 are very common [29]. In air-blown gasification, the products are strongly influenced by varying ER. At a higher ER, the tar yield was found to be reduced significantly, likely due to the greater availability of oxygen to oxidize the volatiles from the biomass feedstock [30, 31]. Increasing ER, however, could decrease the product gas quality by decreasing the concentrations of combustible gases (H_2, CO) while simultaneously increasing CO_2 concentration in the gas product.

Air-blown fluidized bed gasification may be the most cost effective technology for biomass gasification [32]. Fluidized bed reactors, such as circulating fluidized bed (CFB) reactors, provide excellent mixing and therefore better temperature control than fixed bed reactors [33, 34]. Another major advantage of employing fluidized bed gasifiers is the some catalytically active bed materials can be employed directly within the fluidized bed to simultaneously reduce tar formation and increase the yield of combustible gases [5, 6, 21, 33, 35].

2.2 Steam and CO_2 Gasification of Biomass

Steam gasification produces a gaseous product that is free of N_2 but enriched with H_2 of a concentration of >50 vol.% [30, 36]. As an important operational parameter used for steam gasification, the steam-to-biomass ratio (SB) is defined as the ratio of H_2O feeding rate (kg h$^-$

^{1}) to the biomass feeding rate ($kg_{daf}\ h^{-1}$). Herguido et al [36] observed that increasing SB from 0.5 to 2.5 led to a significant decrease in CO and tar yields, with slight increases in H_2 and CH_4 yields. The decreases in tar and CO formation may be accounted for by the steam reforming reactions of tars and CO as shown below. These reactions could be catalyzed by nickel catalysts [37, 38]:

$$C_nH_x + nH_2O \rightarrow (n + x/2)H_2 + nCO \quad (6)$$

$$CO + H_2O \rightarrow H_2 + CO_2 \quad (7)$$

Pure steam gasification is highly endothermic, and the supply of sufficient heat to sustain the gasification reactions may be problematic. Therefore, steam gasification is often carried out in the presence of a small amount of air/oxygen in order to provide the necessary energy via partial combustion of the feedstock. In operation, an H_2O/O_2 ratio of approximately 3.0 was recommended for efficient air/steam gasification at approximately 850°C [39]. It was observed that under similar operating conditions, tar yield was greater in pure steam gasification than in air/steam gasification, and was the lowest under air/oxygen-blown gasification [30]. Increasing the O_2 content in the gasifying agent, however, would generally reduce the contents of the combustible gases in the gas product [37, 40, 41].

Carbon dioxide can also be used as a gasifying agent. Scott et al. [42] investigated the rate of CO_2 gasification of various chars from waste materials. Devi et al. [30] considered CO_2 a promising gasifying agent due to its potential in the tar reduction by dry reforming, as shown in the following reaction (8) [30]. The reaction of CO_2 with carbonaceous material at atmospheric pressure was however found to be relatively slower than the reactions of steam or oxygen [42-44]. As such, pure CO_2 gasification processes are limited due to the low reaction rate. Gasification of biomass was investigated using a steam/CO_2 mixed atmosphere with a co-precipitated Ni/Al catalyst [38]. When compared to pure steam gasification [37,38], catalytic gasification with steam/CO_2 led to decreases in both CH_4 and tar yields accompanied by increases in H_2 and CO yields, implying the dry reforming of CH_4 and tar to the combustible gas products. Iwaki et al. [45] attempted CO_2 gasification of wastepaper using novel molten carbonate catalysts, where it was found that CO_2 gasification in the presence of the molten carbonate catalysts was more effective than steam gasification. For CO_2 gasification, it shall be noted that the reverse Bouduourd reaction is extremely difficult even at very high temperatures [45-49].

$$C_nH_x + nCO_2 \rightarrow (x/2)H_2 + 2nCO \quad (8)$$

2.3 Supercritical Water Gasification of Biomass

Depending on the quality and moisture contents of the biomass feedstock, an energy-intensive dewatering/drying process may be needed. Therefore biomass with high moisture content will require a higher energy input, and it is much advantageous if these types of low grade wet waste biomass feedstock can be gasified in their wet forms. This challenge can be addressed by the research and development on supercritical water gasification (SCWG) of biomass. Supercritical water (SCW) is water or vapour at above water's critical points, i.e.,

374°C and 22.1 MPa. SCW can be an ideal medium for gasification of biomass or lignocellulosic waste materials (wastewater sludge) as water at above its critical points decreases its polarity and dielectric constant value (ε). Hence SCW can dissolve many non-polar organics derived from biomass. Moreover, hot compressed water has an increased value of its ion product (K_w = $[H^+][OH^-]$) [50], which may efficiently catalyze the biomass hydrolysis reactions [51]. A wide range of biomass such as glucose and lignocellulosic materials were successfully gasified in SCW with or without catalyst (carbon or nickel, Pt) to produce hydrogen-rich gas products [52, 53]. Taking glucose as an example, up to 12 mol of pure hydrogen can be generated by SCWG of one mole of feedstock, as shown in the following stoichiometric reaction:

$$C_6H_{12}O_6\ (s) + 6H_2O\ (l) \rightarrow 6\ CO_2\ (g) + 12H_2\ (g) \qquad \Delta H° = 625.0\ kJ/mol \qquad (9)$$

Compared to the conventional steam gasification or reforming processes, SCWG can obtain higher gasification-efficiency, a larger hydrogen yield and less tar and coke formation [52]. In addition, SCWG can utilize wet biomass directly, eliminating the energy- and capital-intensive drying process. Thus, SCWG is particularly suitable for gasifying biomass feedstock with a high moisture content such as wastewater sludge [54-56]. The SCWG has attracted significant interest since the 1980's especially in Europe and Japan, but so far there is no process demonstrated on a large/commercial scale. The major challenges for SCWG of biomass are related to the difficulty in feeding slurry feedstock into a continuous flow-type reactor system operated at high-pressure, and the problems of reactor plugging caused by deposition of solid biomass particles, ashes, tars and chars. These challenges might be addressed by properly designing of the reactor system and developing highly active catalysts so as to minimize the formation of tars and chars in the process.

3. Tar Reduction Technologies

For tar reduction and control, there are two strategies, i.e., reducing tar formation inside a gasifier (primary treatment), and cleaning the gas products outside the gasifier (secondary treatment). The technical approaches and the advantages/disadvantages for these two strategies are summarized in Table 3. More details of the related tar controlling technologies will be discussed in the following sections.

3.1 Tar Formation from Various Biomass Gasification Processes

Gasification operating parameters (temperature, pressure, residence time, gasifying agent and feedstock) have significant effects on tar formation, the compositions and heating values of the resulting gas products. For example, gasification of different biomass feedstocks would produce different tar yields under similar conditions. This is likely due to the varying compositions of cellulose, hemicelluloses and lignin in the feedstocks. High cellulose and hemicellulose concentrations might lead to greater tar yields, while high-lignin containing feedstocks would produce a lower tar yield [28]. In a gasification study, it was found that waste paper produced a much higher tar yield that sawdust and wood [4]. As discussed previously, tar formation is strongly dependent on the gasifying agents as well as temperature.

Table 3. Tar controlling strategies

Tar controlling strategies	Approaches	Advantages/ disadvantages	Ref.
Primary treatment -reducing tar formation inside a gasifier	(a) optimization of operating parameters for the gasification process; (b) addition of active bed materials (tar reduction catalysts).	(a) simple and relatively less expensive treatment; (b) lower tar reduction efficiencies, which makes the gas products difficult to meet the strict requirements for some downstream applications.	[4-6, 30, 33, 63-66]
Secondary treatment - gas cleaning outside the gasifier	(a) mechanical removal using cyclones, scrubbers and filters; (b) hot-gas catalytic cracking.	(a) higher tar reduction efficiencies, up to >95%; (b) higher capital and operational costs as additional equipment / infrastructure is needed; (c) lower overall energy efficiencies as cooling or heating of the syngas stream may be needed.	[4, 5, 21, 61, 67-70]

In an air-blown gasification study it was found that increasing gasification temperature led to a higher concentration of CO_2 due to the oxygenation of the combustible gases, accompanied with a decrease in H_2 concentration in the gas product [57]. Compared with air/oxygen-blown gasification, steam gasification of wood chips in a bubbling fluidized bed increased the tar yield from 4-62 g/kg to 60-95 g/kg [58]. The steam gasification however produced higher yields of H_2 and CO, leading to an increase in the heating value of the gas product (LHV of 10-14 MJ/Nm^3, compared with only 4-8 MJ/Nm^3 for air-blown gasification) [58]. Typical gas composition and tar yields from gasification of woody biomass in a bubbling fluidized bed with different gasifying agents are comparatively shown in Table 4.

Table 4. Typical gas composition and tar yields from gasification of pine wood chips in a bubbling fluidized bed with different gasifying agents (summarized from Gil et al. [58])

Gasifying agent	T(°C)	Gas Composition (vol%, db[a])							Tar yields
		H_2	CO	CO_2	CH_4	C_2H_n	N_2	H_2O^b	(g/kg)
Air	780-830	5-16	10-22	9-19	2-6	0-3	42-62	11-34	4-62
Steam	750-780	38-56	17-32	13-17	7-12	2	0	52-60	60-95
Steam+ O_2	785-830	14-32	43-52	14 -36	6-8	3-4	0	38-61	2-46

[a] On a dry basis; [b] Moisture content in the producer gas.

Tar formation is also strongly dependent on the type of gasifier. Basic types of biomass gasifier include counter-flow fixed-bed gasifier, concurrent flow fixed-bed gasifier, bubbling fluidized bed gasifier, and circulating fluidized bed (CFB) gasifier, as schematically illustrated in Figure 1.

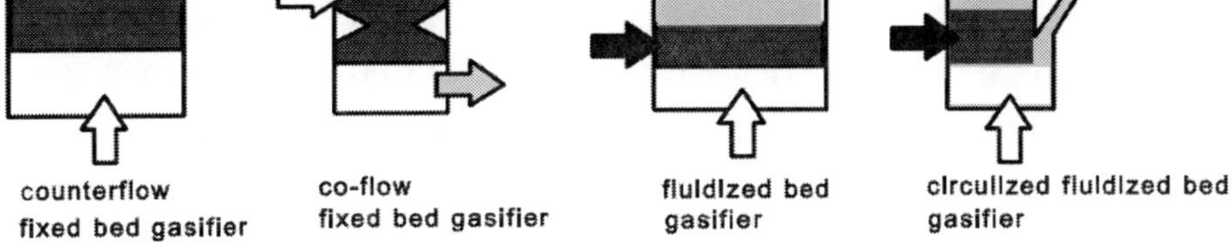

Figure 1. Basic types of biomass gasifier (modified from [59]).

The quality of the gas products (tar, particulate contents and LHV values) from various air-blown biomass gasifiers are summarized in Table 5. The co-flow fixed-bed gasifiers generally produce a much lower tar yield compared with other types of gasifiers (e.g., counter-flow fixed bed and fluidized bed gasifiers). This is owing to thermal/catalytic cracking of tar vapor as it flows through the high temperature zone within the co-flow fixed bed gasifier (where the char and ash minerals may catalyze the cracking reactions of tar).

Table 5. Comparison of quality of the gas products from various air-blown biomass gasifiers [60, 61]

Gasifier	Co-flow fixed bed	Counter-flow fixed bed	CFB
Tars (g/Nm^3)	0.01-6	10-150	2-30
Particles (g/Nm^3)	0.1-8	0.1-3	8-100
LHV (MJ/Nm^3)	4-6	4-5	4-6

3.2 Primary Methods for Tar Reduction

Tar in the syngas products must be cleaned before use of syngas for many applications such as internal combustion systems, direct-fired industrial gas turbines and methanol synthesis [8]. The allowable limits for tar in the product gas depend on the end-use applications. For example, a tar content of less than 50-100, 5, and <0.1 mg/Nm^3 is recommended for internal combustion systems, direct-fired industrial gas turbines, and methanol synthesis [61,62].

Tar reduction technologies can be divided into two types: primary methods – treatment inside a gasifier, and secondary methods – gas cleaning outside the gasifier, as summarized previously in Table 3. The primary methods may be more advantageous over thesecondary methods with respect to the process simplicity and cost. In primary treatment, the gasification process is optimized to minimize tar formation. Primary treatment methods include different approaches: (a) optimization of the gasification operating parameters (temperature, pressure, residence time, gasifying agent, ER and feedstock), and (b) addition of active bed materials to the gasifier (catalysts for tar reduction). For instance, the tar formation was decreased by increasing the operating temperature to > 600°C, and by increasing ER [4]. Use of catalysts within the biomass gasification process has proved to be very effective for in-situ reduction of tar and catalytic gasification of biomass. Dolomite-based and nickel-based catalysts have been widely used [30,33]. Mogtaderi [63] reported that the maximum hydrogen production could be obtained at a low temperature of 600°C in steam gasification of biomass with a nickel-based catalyst. The in-furnace use of catalysts allows for tar removal at a lower temperature (generally as low as 600-800°C) than the thermal tar cracking technologies, which increases the overall thermal efficiency of the process and prevents thermal degradation of the combustible gases at high temperatures [5, 6]. Olivine and dolomite are particularly useful as reactive bed materials for fluidized bed gasifiers [14]. While dolomite (calcined prior to use) tends to show a greater catalytic activity than olivine, the use of dolomite in a fluidized bed gasifier seems to be problematic as it causes severe operating problems of attrition loss in the process [14, 64-66].

3.3 Secondary Methods for Tar Reduction

Primary measures as discussed above are advantageous in terms of the process simplicity and treatment costs, but employing them alone is difficult to meet the strict requirements on tar content for the gas products for some downstream applications. Secondary methods prove to be more effective for tar removal. Common secondary measures for tar reduction include (i) mechanical removal using cyclones, scrubbers and filters, and (ii) hot gas catalytic cracking of tars.

Mechanical tar removal is achieved using scrubbers, filters, cyclones, and electrostatic precipitators [61]. Some typical mechanical methods for tar removal are compared with hot gas catalytic tar cracking in Table 6. Sand bag filter and venturi scrubbers are highly efficient. They are able to remove up to 95% of tar in the syngas at ambient or a low temperature [4, 61], and reduce the tar content in the gas stream to below 20-40 mg/Nm^3 [21]. Other mechanical methods such as wet scrubbers and electrostatic precipitators can also be used for removing tars and particulates from the gas stream [4,67]. A common problem with the mechanical methods is that they are cold-gas cleanup methods requiring cooling of the producer gas to <150°C. This thus reduces the overall energy efficiency of the whole process [5]. In this regard, hot-gas cleanup in a secondary catalytic reactor is more advantageous. As shown in Table 6, a catalytic tar cracker operating at a high temperature can obtain high tar removal efficiency (95%) [68-70]. Ma et al [69] used porous alumina discs impregnated with nickel metal as catalytic filters, and achieved 99% reduction of tars at >800°C. As discussed in the following section, the most commonly used catalysts for tar cracking/reforming are nickel-based materials [4, 5, 70], olivine [4, 5] and dolomite [4, 5].

Table 6. Comparison of tar/particulate reduction efficiencies between mechanical methods and the catalytic tar cracking method [61]

Method	Temp. (°C)	Tar Reduction (%)	Particle reduction (%)
Sand bed filter	10-20	60-95	70-95
Wash tower	50-60	10-25	70-90
Venturi scrubber		50-90	
Fabric filter	~200	0-25	
Wet ESP	40-50	0-60	>99
Catalytic tar cracker	900	>95	

4. Catalysts for Catalytic Biomass Gasification and Tar Reduction

Sutton [6] suggested the following criteria for developing a useful catalyst for catalytic cracking/reforming of biomass derived tars:

- It is effective for the reduction of tars;
- It is effective for methane reforming if syngas ($CO + H_2$) is the desired product;

- It should yield a suitable syngas composition to meet the desired end-use requirement;
- It should be resistant to fouling, deactivation and mechanical attrition;
- It should be easily regenerated;
- It should not be prohibitively expensive.

The most commonly used primary and secondary catalysts for tar reduction include nickel-based materials, dolomite, limestone, olivine and iron-based materials [4-6], whose characteristics are summarized in Table 7. In addition, some clay minerals such as kaolinite, montmorillonite, and illite were also tested for tar cracking, but their activity was found to be lower than those of dolomite or nickel based catalysts [6, 24, 71].

Table 7. Comparison of the some common tar cracking catalysts

Catalysts	Characteristics	Ref.
Commercial Ni-based	Highly Effective tar conversion; Very expensive; Readily poisoned and deactivated by sulphur and coke.	[4-6, 24, 30, 68, 70, 74-76]
Calcined dolomites and limestones	Effective tar conversion; Low mechanical strength, hence a short catalyst life.	[6, 15, 27, 70, 77, 78]
Olivine & iron based minerals	Relatively lower tar conversion; High fracture resistance, hence a long catalyst life; High calcination temperature required.	[22, 24, 25, 79-85]

4.1 Nickel-based catalysts

Nickel is the most widely used industrial catalyst for many applications [72]. The use of commercially available nickel catalyst for biomass gasification at temperature greater than 740˚C effectively increased the syngas production at the expense of methane, higher hydrocarbon gases and tar [6]. Zhang et al. investigated the use of nickel catalysts for tar reduction, and found that the use of Ni-based catalysts in a secondary catalytic reactor destructed 99% of tar, while simultaneously increased the hydrogen yield by 6-11% [5]. Nickel catalysts are commonly supported on alumina, silicate, ZrO_2, TiO_2, dolomite and MgO, etc. [27, 71], and can be used as primary catalysts within a gasifier. Interestingly, the presence of oxygen-containing functional groups (-OH, -COOH, $-CH_3$, C=O) in the feedstock appears to be a pre-requisite for the activity of the supported nickel catalyst in biomass gasification. Srivastava et al. [73] tested steam gasification of coal containing tehse functional groups and chars essentially free of these functional groups with a nickel catalyst. They observed that the gasification of coal produced a significantly higher gas yield than the char gasification. Supported nickel (Ni/MgO) catalysts were also active in supercritical water gasification of biomass. The catalyst of 10 wt% Ni/MgO proved to be very active for the gasification of lignin in supercritical water at about 600°C [74].

Nickel-based catalysts can degrade through mechanical erosion and attrition, particularly in a fluidized bed gasifier. These catalysts can also be deactivated by coke deposition [24], and poisoned by H_2S [30]. Catalyst poisoning and deactivation remains a major problem for the nickel-based catalysts [4, 5, 70], although the problem could be mitigated by increasing the reaction temperature [72, 75], using non-alumina catalyst support (such as MgO and CeO_2) or promoters (such as WO_3) [70, 76].

4.2 Dolomite and Limestone

Dolomite is a magnesium ore, i.e., $CaMg(CO_3)_2$. It has been widely used as a cheap and active catalyst for biomass gasification [6, 27, 70]. Dolomite can be used as either a bed material (i.e., a primary catalyst) within a fluidized bed reactor or a secondary catalyst in a downstream reactor. Corella et al. [15] compared these two addition methods and noted that there was no significant difference between the two with regard to product gas heating value and tar removal efficiency. Both methods were reported to reduce tar yield significantly to a similar extent [15]. Sutton et al [6] concluded that dolomite could catalyze tar reforming, resulting in increased gas yields, and can be more active when it is calcined and used downstream at temperatures exceeding 800°C. However, dolomite can be readily deactivated by carbon deposition and attrition.

Dolomite can also be used as a support material for metallic catalysts, notably nickel [27, 70]. Sato and Fujimoto [70] developed a nickel-based catalyst supported on dolomite material, doped with a sulphur-resistant WO_3 promoter. The catalsyst achieved > 90% tar reduction efficiency during biomass gasification at 800-850°C in the presence of H_2S.

Similar to dolomite, limestone ($CaCO_3$) can be used as an effective primary catalyst in fluidized bed gasification of bioamss [77, 78]. Calcined limestone appeared to be an excellent catalyst for both combustible gas production and tar reduction, and the use of calcined limestone has the potential for simultaneous hydrogen production and CO_2 capture in biomass gasification. When calcined, however, the material is fragile and may not be able to withstand the fluidization for long periods of time.

4.3 Olivine and Iron-based Catalysts

Olivine, a naturally occurring mineral containing magnesium, iron and silicon, is a promising reactive bed material for tar cracking in a fluidized bed gasifier [79]. A study by Rapagnà et al. [22] suggested that the catalytic activity of olivine was comparable to that of calcined dolomite, leading to 90% reduction in tar formation when compared to a fluidized bed of sand material. Although dolomite showed a slightly better catalytic efficiency than olivine, olivine had significantly better resistance to physical attrition within a fluidized bed reactor [22]. Calcined olivine has been tested for steam reforming of tar (using naphthalene as a model compound) [79], and for air-blown fluidized bed gasification of biomass [80]. However, natural olivine alone without calcination showed little or no catalytic activity for dry reforming of methane and tar [81].

Iron-based catalysts are less expensive, although they exhibit a lower catalytic activity than nickel and other metal (Ru, Pt, etc.) catalysts and can be deactivated in the absence of hydrogen [24]. According to Nordgreen et al. [25], metallic iron obtained from hematite exhibited very high catalytic activity at temperatures exceeding 900°C. Iron oxide in the form of hematite (Fe_2O_3) demonstrated to be a useful catalyst for hydro-cracking of coal tar, pyrolysis of biomass and steam gasification of biomass [82-84]. Uddin et al. [85] investigated the use of iron oxides as secondary catalysts for reducing the tar from gasification of a woody biomass. The iron oxide catalyst showed high activity for tar cracking, increasing H_2 and CO_2 concentrations in the syngas by approximately 5 times. Natural limonite ores containing mainly hematite and goethite (FeOOH), with a small amount of nickel have been used successfully as a catalyst for direct coal liquefaction [86] and coal tar decomposition [87]. In a study by Li et al [87], an Indonesian limonite first reduced at temperatures above 650°C was used for the decomposition of coal tars. The reduced limonite catalyst showed high activity for tar conversion [87]. Recently, the authors' group studied catalytic gasification of a woody biomass in an air-blown fluidized-bed reactor using a Canadian limonite iron ore as the bed material in comparison to some conventional bed materials (olivine, limestone and dolomite). The Canadian limonite iron ore, limestone and dolomite were found to be more active than olivine for tar reduction in the fluidized bed gasification at a low equivalence ratio (ER $\leq$0.3), as displayed in **Figure 2**. However, olivine was the most attrition-resistant and thermally stable bed material among the four bed materials tested for biomass gasification in the fluidized bed reactor, leading to the best temperature control in the gasification operation.

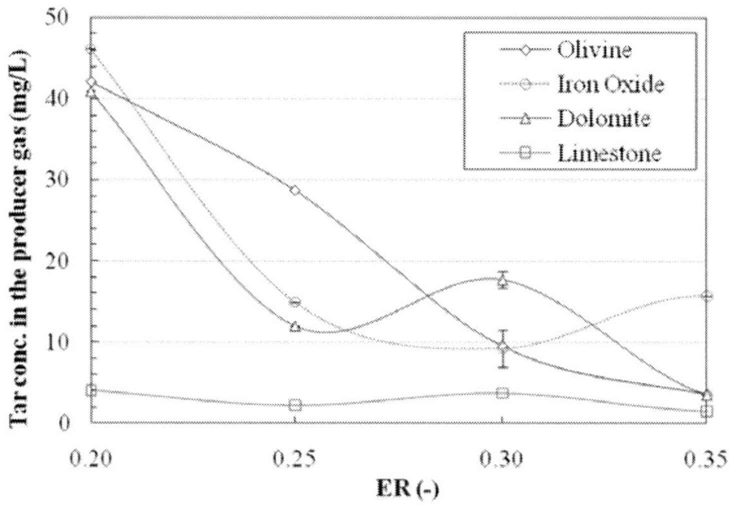

Figure 2. Tar concentrations in the producer gas from air-blown gasification of pine sawdust at various equivalence ratios and with different bed materials.

Conclusion

1) Gasification represents a promising pathway for conversion of biomass into energy and chemicals. Tar formation during biomass gasification, however, remains the major obstacle to commercialization of biomass gasification technologies.

2) Gasification operating parameters (temperature, pressure, residence time, gasifying agent, the type of feedstock and gasifier, etc.) have significant effects on tar formation, the compositions and heating values of the resulting gas products. A higher temperature and higher air-to-fuel ratio in air-blown gasification tend to produce less tar. Steam gasification of biomass would produce much higher tar yields than air/oxygen-blown gasification. Co-flow fixed-bed gasifiers generally produce a much lower tar yield than other types of gasifiers.
3) Primary measures for tar reduction (by optimizing the operating parameter and addition of primary catalysts in the gasifier) are more advantageous with regard to the process simplicity and treatment costs. Secondary methods, by mechanical removal using cyclones, scrubbers and filters, and by hot gas catalytic cracking of tars, prove to be more effective for tar removal.
4) The most commonly used primary and secondary catalysts for tar reduction include nickel-based materials, dolomite, limestone, olivine and iron-based materials. Poisoning and deactivation by sulphur and coke deposition is the major problem for the use of nickel-based catalysts. The major challenge of using dolomite, limestone and some other iron-based minerals in fluidized bed gasifiers is the low mechanical strength of these materials, being fragile when calcined and unable to withstand fluidization for long periods of time. In contrast, calcined olivine had superb resistance to physical attrition.

ACKNOWLEDGMENT

This research was financially supported by the Ontario Ministry of Energy and Ontario Centres of Excellence through the Atikokan Bioenergy Research Centre Program and partially by NSERC through the Discovery Grant awarded to one of the authors (C. Xu). The authors would also like to acknowledge researchers at CanmetEnergy (Dr. Jinsheng Wang, Dr. Preto Fernando, Mr. Guy Tourigny and Mr. Dave Wambolt) and Lakehead University Green Energy Laboratory (Ms. Yuanyuan Shao, Dr Hanning Li and Mr. Chadi Badour) for their contribution to this work.

REFERENCES

[1] Kucuk M, Demirbas A. Biomass conversion processes. Energy Conversion Management 1997; 38: 151-165.

[2] McKendry P. Energy production from biomass (Part 1): overview of biomass. Bioresource Technology 2002; 83: 37-46.

[3] X.T. Li, J.R. Grace, C.J. Lim, A.P. Wakinson, H. Chen, J. Kim. Biomass gasification in a circulating fluidized bed. Biomass and Bioenergy 26 (2004) 171-193.

[4]] J. Han, H. Kim. The Reduction and control technology of tar during biomass gasification/pyrolysis: An overview. Renewable and Sustainable Energy Reviews 12 (2008) 397-416.

[5] R. Zhang, R. Brown, A. Suby, K. Cummer. Catalytic destruction of tar in biomass derived producer gas. Energy Conversion and Management 45 (2004) 995-1014.

[6] D. Sutton, B. Kelleher, J. Ross. Review of literature on catalysts for biomass gasification. Fuel Processing Technology 73 (2001) 155-173.

[7] M. Asadullah, T. Miyazawa and K. Kunimori, Catalyst development for the gasification of biomass in the dual-bed gasifier, Applied Catalysis A: General 255 (2003) 169–180.

[8] D. Lopamudra, K.J. Ptasinski and F.J.J.G. Janssen. A review of the primary measures for tar elimination in biomass gasification processes, Biomass and Bioenergy 24 (2003) 125–140.

[9] L. Devi, K.J. Ptasinski, F. J. Janssen .Pretreated olivine as tar removal catalyst for biomass gasifiers: investigation using naphthalene as model biomass tar. Fuel Processing Technology 2005; 86: 6: 707-730.

[10] Coll, J. Salvado, X. Farriol, D. Montane. Steam reforming model compounds of biomass gasification tars: conversion at different operating conditions and tendancy towards coke formation. Fuel Processing Technology 2001; 74: 19-31.

[11] P. A. Simell; J.O. Hepola; A. Krause. Effects of gasification gas components on tar and ammonia decomposition over hot gas cleanup catalysts. Fuel 1997; 76: 1117-1127.

[12] G. Taralas, M.G. Kontominas, Kinetic modelling of VOC catalytic steam pyrolysis for tar abatement phenomena in gasification/pyrolysis technologies. Fuel, 83(2004): 1235–1245.

[13] G. Hu, S. Xu, S. Li, C. Xiao, S. Liu. Steam gasification of apricot stones with olivine and dolomite as downstream catalysts. Fuel Processing Technology 87 (2006) 375-382.

[14] J. Pecho, T.J. Schildhauer, M. Sturzenegger, S. Biollaz, A. Wokaun. Reactive bed materials for improved biomass gasification in a circulating fluidized bed. Chemical Engineering Science 63 (2008) 2465-2476.

[15] J. Corella, M. Aznar, J. Gil, M. Caballero. Biomass gasification in fluidized bed: Where to locate the dolomite to improve gasification? Energy Fuels 13 (1999) 1122-1127.

[16] Orío, J. Corella, I. Narvàez. Performance of differne dolomites on hot raw gas cleaning from biomass gasification with air. Industrial Engineering Chemistry Research 36 (1997) 3800-3808.

[17] Faaij, A., Van Ree, R., Waldheim, L., Olsson, E., Oudhuis, A., Van Wijk, A., Daey-Ouens, C., Turkenburg, W. Gasification of biomass wastes and residues for electricity production. Biomass and Bioenergy 12 (1997) 387-407.

[18] Mahishi MR, Goswami DY. Thermodynamic optimization of biomass gasifier for hydrogen production. Int J Hydrogen Energy (2007), doi: 10.1016/j.ihydene.2007.05.018.

[19] Henrich, E., Bürkle, S., Meza-Renken, Z., Rumpel, S. Combustion and gasification of pyrolysis chars from waste and biomass. Journal of Analytical and Applied Pyrolysis 49 (1999) 221-241.

[20] J. Vélez, F. Chejne, C. Valdés, E. Emery, C. Londoño. Co-gasification of Columbian coal and biomass in fluidized bed: An experimental study. Fuel (2008), doi:10.1016/j.fuel.2008.10.1018.

[21] Bridgewater, A. The technical and economic feasibility of biomass gasification for power generation. Fuel 14 (1995) 651-655.

[22] S. Rapagnà, N. Jand, A. Kiennemann, P. U. Foscolo. Steam-gasification of biomass in a fluidised-bed of olivine particles. Biomass and Bioenergy 19 (2000) 187-197.

[23] Kirubakaran, V., Sivaramakrishnan, V., Nalini, R., Sekar, T., Premalatha, M., Subramanian, P. A review on gasification of biomass. Renewable and Sustainable Energy Reviews (2007), doi:10.1016/j.rser.2007.07.001.

[24] Abu El-Rub, Z., Bramer, E., Brem, G. Review of Catalysts for Tar Elimination in Biomass Gasification Processes Ind. Eng. Chem. Res. (2004) 43, 6911-6919.

[25] Nordgreen, T., Liliedahl, T., Sjöström, K. Metallic iron as a tar breakdown catalyst related to atmospheric, fluidised bed gasification of biomass. Fuel 85 (2006) 689–694.

[26] Lv, P., Yuan, Z., Wu, C., Ma, L., Chen, Y., Tsubaki, N. Bio-syngas procuction from biomass catalytic gasification. Energy Conversion and Management 48 (2007) 1132-1139.

[27] Srinakruang, J., Sato, K., Vitidsant, T., Fujimoto, K. A highly efficient catalyst for tar gasification with steam. Catalysis Communications 6 (2005) 437–440.

[28] T. Hanaoka, S. Inoue, S. Uno, T. Ogi and T. Minowa, Effect of woody biomass components on air–steam gasification. Biomass and Bioenergy 28 (2005) 69–76.

[29] Narvàez, A Orío, M. Aznar, J. Corella. Biomass gasification with air in an atmospheric bubbling fluidized bed. Effect of six operational variables on the quality of produced raw gas. Industrial and Engineering Chemistry Research 35 (1996) 2110-2120.

[30] Devi, L., Ptasinski, K., Janssen, F. A review of the primary measures for tar elimination in biomass gasification processes. Biomass and Bioenergy 24 (2003) 125 – 140.

[31] C. Kinoshita, Y. Wang, J. Zhou. Tar formation under different biomass gasification conditions. Journal of Analytical and Applied Pyrolysis 29 (1994) 169-181.

[32] M. Campoy, A. Gómez-Barea, F. Vidal, P. Ollero. Air-steam gasification in a fluidized bed: Process optimization by enriched air. Fuel Processing Technology (2009), doi: 10.1016/j.fuproc.2008.12.007.

[33] J. Pencho, T. Schildhauer, M. Sturzenegger, S. Biollaz, A. Wokaun. Reactive bed materials for improved biomass gasification in a circulating fluidized bed reactor. Chemical Engineering Science 63 (2008) 2456-2476.

[34] P. Garcia-Ibanez, A. Cabanillas and J.M. Sanchez. Gasification of leached orujillo (olive oil waste) in a pilot plant circulating fluidised bed reactor. Preliminary results. Biomass and Bioenergy 27 (2004) 183–194.

[35] C. Courson, L. Udron, A. Kienneman. Grafted NiO on natural olivine for dry reforming of methane. Science and Technology of Advanced Materials 3 (2002) 271-282.

[36] J. Herguido, J. Corella, J. Gonzàlez-Saiz. Steam gasification of lignocellulosic residues in a fluidized bed at a small pilot scale: Effect of the type of feedstock. Industrial and Engineering Chemistry Research 31 (1992) 1274-1282.

[37] L. Garcia, M. Salvador, J. Aruzo, R. Bilbao. CO2 as a gasifying agent for gas production from pine sawdust at low temperatures using a Ni/Al coprecipitated catalyst. Fuel Processing Technology 69 (2001) 157-174.

[38] L. García, M. Salvador, J. Arauzo, R. Bilbao. Catalytic steam gasification of pine sawdust. Effect of catalyst weight/biomass flow rate and steam/biomass ratios on gas production and composition. Energy and Fuels 13 (1999) 851-859.

[39] J. Gil, M. Aznar, M. Caballero, E. Francés, J. Corella. Biomass gasification fluidized bed at pilot scale with steam-oxygen mixtures. Product distribution for very different operating conditions. Energy and Fuels 11 (1997) 1109-1118.

[40] P.M. Lv, Z.H. Xiong, J. Chang, C.Z. Wu and Y. Chen. An experimental study on biomass air–steam gasification in a fluidized bed. Bioresource Technology 95 (2004) 95–101.

[41] Z.A. Zainal, A. Rifau, G.A. Quadir and K.N. Seetharamu. Experimental investigation of a downdraft biomass gasifier. Biomass and Bioenergy 23 (2002) 283–289.

[42] S.A. Scott, J.F. Davidson, J.S. Dennis, P.S. Fennell, A.N. Hayhurst, The rate of gasification by CO_2 of chars from waste. Proceedings of the Combustion Institute 30 (2005) 2151-2159.

[43] N.M. Laurendau, Prog. Energy Combustion Science. 4 (1978) 221-270.

[44] S. Kajitani, N. Suzuki, M. Ashizawa, S. Hara, CO_2 gasification rate analysis of coal char in entrained flow coal gasifier. Fuel 85 (2006) 163-169.

[45] H. Iwaki, S. Ye, H. Katagiri, K. Kitagawa. Wastepaper gasification with CO2 or steam using catalysts of molten carbonates. Applied Catalysis 270 (2004) 237-243.

[46] S. Carabineiro, D. McKee, I. Silva. Uncatalysed and catalysed CO2 reaction using metal catalysts and binary vanadium mixtures supported on activated carbon. Carbon 39 (2001) 451-463.

[47] H. Liu, C. Luo, S. Kato, S. Uemiya, M. Kaneko, T. Kojima. CO_2/char gasification at elevated temperatures Part I: Experimental results. Fuel Processing Technology 87 (2006) 775-781.

[48] H. Liu, C. Luo, S. Kato, S. Uemiya, M. Kaneko, T. Kojima. CO_2/char gasification at elevated temperatures Part II: Clarification of mechanism through modelling and char characterization. Fuel Processing Technology 87 (2006) 769-774.

[49] B.H. Hameed, M.I. El-Khaiary. Equilibrium, kinetics and mechanism of malachite green adsorption on activated carbon prepared from bamboo by K_2CO_3 activation and subsequent gasification with CO2. Journal of Hazardous Materials 157 (2008) 344-351.

[50] Marshall WL, Franck EU. Ion product of water substance, 0-1000 °C, 1-10,000 bars new international formulation and its background. J. Phys. Chem. Ref. Data. 1981; 10 (2): 295-304.

[51] Bobleter O. Hydrothermal degradation of polymers derived from plants. Prog. Polym. Sci. 1994; 19: 797-841.

[52] Xu X., Antal M.J., 1998. Environmental Progress 17, 215-220.

[53] Williams R.T., Onwudili J., 2006. Energy and Fuels 20, 1259-1265.

[54] G.J. Dileo, P.E. Savage. Catalysis during methanol gasification in supercritical water, Journal of Supercritical Fluids 39 (2006) 228–232.

[55] K. Arai, R.L. Smith Jr., T.M. Aida, Decentralized chemical processes with supercritical fluid technology for sustainable society, Journal of Supercritical Fluids 47 (2009) 628–636.

[56] Y. Matsumura. Evaluation of supercritical water gasification and biomethanation for wet biomass utilization in Japan. Energy Conversion and Management 43 (2002) 1301-1310.

[57] P.L. Spath, D.C. Dayton. Preliminary Screening—Technical and Economic Assessment of Synthesis Gas to Fuels and Chemicals with Emphasis on the Potential for Biomass-

Derived Syngas; National Renewable Energy Laboratory: Golden, CO., 2003. http://www.nrel.gov/docs/fy04osti/34929.pdf.
[58] J. Gil, J. Corella, M. P. Aznar, M. A. Caballero. Biomass Gasification in Atmospheric and Bubbling Fluidized Bed: Effect of the Type of Gasifying Agent on the Product Distribution. Biomass and Bioenergy 1999; 17: 389-403.
[59] U. Zuberbuhler, M. Specht, A. Bandi. Gasification of biomass – An overview on available technologies. http://www.zsw-bw.de/topics/fuel/REGbiomassconversion/renew_zu_en.pdf.
[60] R.G. Graham, R. Bain. Biomass gasification: hot-gas cleanup. Biomass Gasification Working Group Report, International Energy Agency (IEA), 1993.
[61] Ph. Hasler, R. Buehler, Th. Nussbaumer. Evaluation of gas cleaning technologies for small scale biomass gasifiers. In Biomass for Energy and Industry, 10th European Conference and Technology Exhibition, June 8-11, 1998, Wurzburg (Germany).
[62] T.A. Milne, R.J. Evans. Biomass gasification "tars": their nature, formation and conversion. NREL, Golden, CO, USA, Report no. NREL/TP-570-25357, 1998.
[63] Moghtaderi, B. Effects of controlling parameters on production of hydrogen by catalytic steam gasification of biomass at low temperatures. Fuel 86 (2007) 2422-2430.
[64] S. Paasen, Rabou. Tar removal with wet ESP: parametric study. The Second World Conference and Technology Exhibition on Biomass for Energy, Industry and Climate Protection, (2004) 205–210
[65] J. Corella, M. Aznar, J. Gil, M. Caballero. Biomass gasification in fluidized bed: Where to locate the dolomite to improve gasification? Energy Fuels 13 (1999) 1122-1127.
[66] Orío, J. Corella, I. Narvàez. Performance of differne dolomites on hot raw gas cleaning from biomass gasification with air. Industrial Engineering Chemistry Research 36 (1997) 3800-3808.
[67] J.C. Jansen, K. Jönsson and M. HagmanLa. Biological detoxification of tar-water. Water Sci Technol 46 (2002), pp. 59–65.
[68] Aznar, M., Caballero, M., Gil, J., Martın, J., and Corella, J. Commercial Steam Reforming Catalysts To Improve Biomass Gasification with Steam-Oxygen Mixtures. 2. Catalytic Tar Removal. Ind. Eng. Chem. Res. (1998) 37, 2668-2680.
[69] Ma, L., Verelst, H., Baron, G. Integrated high temperature gas cleaning: Tar removal in biomass gasification with a catalytic filter. Catalysis Today 105 (2005) 729–734.
[70] K. Sato, K. Fujimoto. Development of new nickel based catalyst for tar reforming with superior resistance to sulphur poisoning and coking in biomass gasification. Catalysis Communications 8 (2007) 1697-1701.
[71] Kimura T, Miyazawa T, Nishikawa J, Kado S, Okumura K, Miyao T, Naito S, Kunimori K, Tomishige K. Development of Ni catalysts for tar removal by steam gasification of biomass. Applied Catalysis B: Environmental 68 (2006) 160–170.
[72] P. Simell, E. Kurkela, P. Stahlberg, J. Hepola. Catalytic hot gas cleaning of gasification gas. Catalysis Today 27 (1996) 55-62.
[73] R. Srivastava, J. Sinha, S. K. Srivastava, S. Tripathi. Nickel catalysed steam gasification of chars obtained from coal–alkali reaction at 600 °C. Fuel 82 (2003) 93-96.

[74] T. Furusawa, T. Sato, H. Sugito,Y. Miura,Y. Ishiyama, M. Sato, N. Itoh, N. Suzuki. Hydrogen production from the gasification of lignin with nickel catalysts in supercritical water. International Journal of Hydrogen Energy 32 (2007) 699 – 704.
[75] W. Wang, N. Padban, Z. Ye, A. Andersson, I. Bjerle. Kinetics of ammonia decomposition in hot gas cleaning. Industrial and Engineering Chemistry Research 38 (1999) 4175-4182.
[76] M. Asadullah, S. Ito, K. Kunimori, M. Yamada, K. Tomishige, Biomass gasification to hydrogen and syngas at low temperature: Novel catalytic system using fluidized-bed reactor. J. Catal., 208 (2002): 255–259.
[77] M.J. Llorente, R. Cuadrado, J.M. Laplaza, J.E. Garcia. Combustion in bubbling fluidized bed with bed material of limestone to reduce the biomass ash agglomeration and sintering. Fuel 85 (2006) 2081-2092.
[78] T. Weimer, R. Berger, C. Hawethorne, J.C. Abanades. Lime enhanced gasification of solid fuels: Examination of a process for simultaneous hydrogen production and CO_2 capture. Fuel 87 (2008) 1678-1686.
[79] L. Devi, M. Craje, P. Thune, K. Ptanski, F. Janssen. Olivine as tar removal catalyst for biomass gasifiers: Catalyst characterization. Applied Catalysis 294 (2005) 68-79.
[80] J. Corella, J. Toledo, R. Padilla. Olivine or dolomite as in-bed additive in biomass gasification with air in a fluidized bed: Which is better? Energy & Fuels 18 (2004) 713-720.
[81] C. Courson, L. Udron, A. Kienneman. Grafted NiO on natural olivine for dry reforming of methane. Science and Technology of Advanced Materials 3 (2002) 271-282
[82] V. I. Sharypov, B. N. Kuznetsov, N. G. Beregovtsova, O. L. Reshetnikov, S. V. Baryshnikov. Modification of iron ore catalysts for lignite hydrogenation and hydrocracking of coal-derived liquids. Fuel 75 (1996) 39-42.
[83] V.I. Sharypov, N.G. Beregovtsova, B.N. Kuznetsov, S.V. Baryshnikov, V.L. Cebolla, J.V. Weber, S. Collura, G. Finqueneisel, T. Zimny. Co-pyrolysis of wood biomass and synthetic polymers mixtures: Part IV: Catalytic pyrolysis of pine wood and polyolefinic polymers mixtures in hydrogen atmosphere. Journal of Analytical and Applied Pyrolysis 76 (2006) 265-270
[84] Khelfa, V. Sharypov, G. Finqueneisel, J.V. Weber. Catalytic pyrolysis and gasification of Miscanthus Giganteus: Haematite (Fe_2O_3) a versatile catalyst. Journal of Analytical and Applied Pyrolysis 84 (2009) 84-88.
[85] M. A. Uddin, H. Tsuda, S. Wu, E. Sasaoka. Catalytic decomposition of biomass tars with iron oxide catalysts. Fuel 87 (2008) 451–459.
[86] T. Kaneko, S. Sugita, M. Tamura, K. Shimasaki, E. Makino, L. Sikalalhi. Highly active limonite catalysts for direct coal liquefaction. Fuel 81 (2002) 1541-1549.
[87] L. Li, K. Morishita, T. Takarada. Light fuel gas production from nascent coal volatiles using a natural limonite ore. Fuel 2007; 86: 1570-1576.

In: Gasification: Chemistry, Processes and Applications
Editor: Michael D. Baker
ISBN: 978-1-61209-681-0

Chapter 9

GASIFICATION OF BIOMASS AND WASTE PLASTICS IN A TWO-STAGE GASIFIER USING ACTIVATED CARBON

Joo-Sik Kim and Tae-Young Mun
Faculty of Environmental Engineering, University of Seoul, Dongdaemun-Gu, Seoul, Republic of Korea, Korea

INTRODUCTION

Among alternative energy resources, biomass, such as wood wastes and sewage sludge, and waste plastics are very attractive due to its abundance, accessibility, and ease of use. The conversion of lignocellulosic biomass or waste plastics to a fuel or chemicals usually can take place with thermo-chemical conversion technologies. Among the thermo-chemical conversion processes, gasification is a complex combination of pyrolysis and oxidation reactions in the condensed and vapor phases. Under typical gasification conditions, oxygen levels are restricted to less than 30 % of that required for complete combustion [1]. Gasification converts biomass or waste plastics to a uniform gaseous mixture that mainly comprises H_2, CO, and CH_4. The gaseous mixture is called producer gas. Producer gas can be used to run internal combustion engines to generate electricity, can be used as substitute for fuel oils in direct heat applications and can be used to produce chemicals which are useful as fuel, as well as chemical feedstock for industries [2].

Gasification itself is very old technology, which flourished during the 2nd world war. Today, there is renewed interest in the gasification, due to the increased petroleum prices. The main advantage of gasification lies in that it is decentralized conversion system which can be operated even in a small scale.

In Korea, an average of 5 million tons/yr of wood waste is being generated of late, much of which has been treated simply through incineration and recycling, such as into wood pellets for boiler fuel. Therefore, the need for more efficient treatment of wood wastes has recently gained strong national recognition.

Meanwhile, plastics are widely used in everyday life, as well as in high-tech industries, as an indispensable ingredient. Consequently, the amount of waste plastics has also increased. Therefore, there is a growing need for waste plastics management. The traditional ways for the disposal of waste plastics have been either to bury it in landfills, or to burn it in incinerators. Landfill and incineration, however, are associated with serious environmental problems. Therefore, there is an increasing need to utilize waste plastics environmental friendly and economically.

In this regard, the present work gives detailed experimental results of gasification of forest wood waste (FWW), construction wood waste (CWW), dried sewage sludge (DSS), and post-consumed mixed waste plastics (MPW) in a newly developed two-stage gasifier. The aim of the research is to directly reduce the tar yield in the gasifier and to produce a high caloric producer gas. In accordance with the aim, this article reports the development of a producer gas composition according to experimental conditions, such as the reactor temperatures and the equivalence ratio, the role of activated carbon in tar removal, and the chemical and thermal characteristics of the tar.

THEORY OF BIOMASS GASIFICATION

During biomass gasification, drying, pyrolysis, combustion, and gasification take place in a gasifier. These reactions in a moving-bed gasifier take place in a distinct region. In a fluidized bed gasifier, in contrast, they occur almost simultaneously. In drying, moisture in a biomass evaporates as the biomass is heated. As temperature increases, the dried biomass generally produces char, tar, and light gases such as H_2, CO, CH_4, CO_2, H_2O, and light hydrocarbons through pyrolysis. The condensable and non-condensable vapors and char that are produced during pyrolysis undergo combustion and gasification reactions. The combustion reactions [Equations (1) and (2)] are exothermic; and along with the partial-combustion reaction [Equation (3)],

$$C + O_2 \leftrightarrow CO_2 - 408.4 \text{ KJ/mol} \quad (1)$$

$$H_2 + 1/2O_2 \leftrightarrow H_2O - 242 \text{ KJ/mol} \quad (2)$$

$$C + 1/2O_2 \leftrightarrow CO - 122.9 \text{ KJ/mol} \quad (3)$$

The oxidation reactions effectively provide the thermal energy that is needed for the other endothermic reactions in the gasifier.

The gasification reactions are those that take place between the different gaseous and solid reactants (char) and water (steam). They contribute to the formation of the producer gas mixture. The following are the major gasification reactions [3-7]:

$$C + CO_2 \leftrightarrow 2CO + 160.5 \text{ KJ/mol} \quad (4)$$

$$C + H_2O \leftrightarrow CO + H_2 + 118 \text{ KJ/mol} \quad (5)$$

$$C + 2H_2O \leftrightarrow CO_2 + 2H_2 + 76 \text{ KJ/mol} \quad (6)$$

$$2C + 2H_2O \leftrightarrow CH_4 + CO_2 - 11 \text{ KJ/mol} \quad (7)$$

$$C + 2H_2 \leftrightarrow CH_4 - 88\ KJ/mol \tag{8}$$

$$CO + H_2O \leftrightarrow H_2 + CO_2 - 42.2\ KJ/mol \tag{9}$$

$$CH_4 + H_2O \leftrightarrow CO + 3H_2 + 206\ KJ/mol \tag{10}$$

$$CH_4 + 2H_2O \leftrightarrow CO_2 + 4H_2 + 165\ KJ/mol \tag{11}$$

The reactions from Equations (4) to (8) are char or carbon-gas phase reactions such as the Boudouard reaction [Equation (4)], water-gas reactions [Equations (5) and (6)], hydromethanation [Equation (7)], and methane formation reaction [Equation (8)]. On the contrary, the reactions from Equations (9) to (11) are gas-gas phase reactions. The reaction shown in Equation (9) is a water-gas shift reaction, and the reactions in Equations (10) and (11) are steam reforming reactions for hydrogen gas production using a natural gas (CH_4). In all these reactions, the value of the reaction enthalpy refers to the temperature of 298.15 K.

In this article, the LHV (Lower Heating Value) of the producer gas was calculated, and its definition is shown in Equation (12).

LHV (MJ/Nm^3) in producer gas: (12)

$$= \{121.60 \times H_2 + 10.10 \times CO + 50.57 \times CH_4 + 47.93 \times C_2H_6 + 47.49 \times C_2H_4 + 46.76 \times C_3H_8 + 46.11 \times C_3H_6 + 45.30 \times C_4H_8 + 40.76 \times C_6H_6 + 41.14 \times C_7H_8\} \times \{producer\ gas\ density\ (kg/Nm^3)\}$$

The unit of the composition of the producer gas is wt. %.

Among the gasification operation variables, the gasifying agent is important to the caloric value of a producer gas. The main gasifying agent is usually air, but oxygen and steam are also used as gasifying media. According to the gasifying agent, three types of producer gases with different calorific values (CVs) can be produced. Low-CV (4-6 MJ/Nm^3) gas, which is obtained from air gasification, is used directly in combustion or as an engine fuel; medium-CV (10-15 MJ/Nm^3) gas is obtained with oxygen; and high-CV (13-20 MJ/Nm^3) gases, which are obtained with steam, can be utilized as feedstock for subsequent conversion into basic chemicals, principally hydrocarbons and methanol [8, 9]. Although the caloric value of producer gas is higher than that with air when oxygen and steam are used as gasifying agents, their use is not economical due to the need for additional equipment such as steam and oxygen generation systems.

TAR PROBLEM

During biomass gasification, some obstacles must be overcome. Tar formation is one of the major problems that must be dealt with because it blocks and fouls the process equipment such as the pipes and valves in the gas engine and the turbine [10, 11]. Figure 1 shows tar that is condensed on the pipe line of the gasification process.

Figure 1. Tar condensed on a gasification apparatus.

Tar can be defined in various ways and can be classified on the basis of its aromatic structure as follows [12, 13].

Tar definitions:

1) All organic contaminants with a molecular weight larger than that of benzene.
2) The sum of compounds with a boiling point that is typically higher than 150 °C.
3) The mixture of chemical compounds that condense on metal surfaces at room temperature.

Tar classifications:

1) Primary tars are fragments that originate directly from cellulose, hemicelluloses, and lignin structures.
2) Secondary tars consist mainly of phenolics and olefins and are products of the conversion of primary tars.
3) Alkyl tertiary tars are methyl derivates of aromatics.
4) Tertiary tars are polyaromatic hydrocarbons (PAHs).

Tar removal technologies can be divided mainly into two methods: primary methods, which are treatments inside the gasifier, and secondary methods, which are measures downstream of the gasifier. The primary methods include optimal operation parameters, the use of bed additives or catalysts, and gasifier modifications. The secondary methods are tar-cracking, either thermally or catalytically, and mechanical methods such as the use of a cyclone, a ceramic filter, an electrostatic precipitator, and a scrubber. The secondary methods have been well studied, but until now, they seem ineffective and too expensive. In contrast, the primary methods have an advantage in that they can eliminate or strongly reduce the need for downstream cleanup.

Two-stage Gasifier

A schematic diagram of the two-stage gasifier applied in the experiments and a picture of the plant are shown in Figure 2 and 3, respectively.

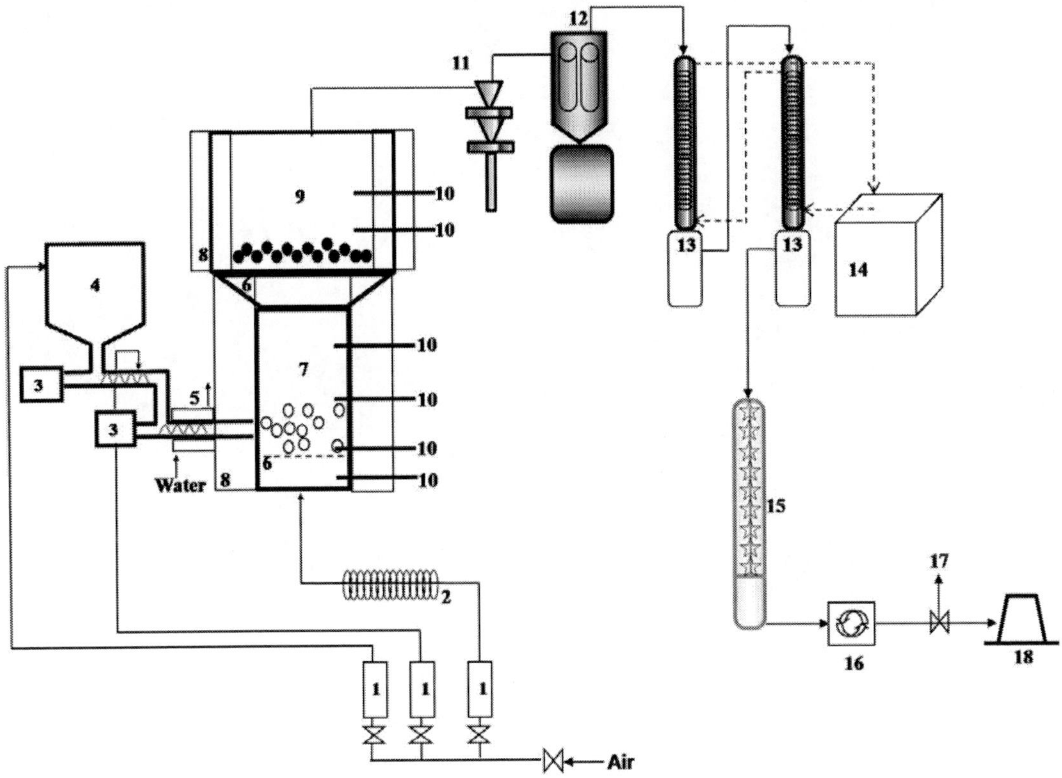

Figure 2. Diagram of the two-stage gasification plant.

(1.Flow meter, 2.Pre-heater, 3.Screw feeder, 4.Silo, 5.Water jacket, 6.Tuyere cap distributor, 7.Fluidized bed, 8.Electricity furnace, 9.Activated carbon zone, 10.Thermocouple, 11.Two-stage cyclone, 12.Hot filter, 13.Condensers, 14.Chiller, 15.Electrostatic precipitator, 16.Gas meter, 17.Gas sampler, 18.Flare stack)

The plant has two reactors in a series, which are heated indirectly using electricity furnaces, and which are made of STS-316. The bottom reactor has a height of 390 mm and an inner diameter of 110 mm. The bottom reactor has four thermocouples and among them, the two that are just above the distributor are used to check the fluidization stability in the bottom reactor. The bed material is used 2500 g silica sand (99 wt.% SiO_2, 150-300 μm). The bed height is 180 mm and the freeboard height in bottom reactor is 210 mm. The upper-reactor is 340 mm tall, its inner diameter is 160 mm, and it has two thermocouples. Additives, such as activated carbon, can be applied in the upper-reactor to reduce the tar yield. The two reactors are separated using a perforated or tuyere distributor that is made of 29 bubble caps. Both reactors are originally designed as fluidized beds, but the upper reactor can be operated as a fixed bed reactor if the filling material used has a high density.

Figure 3. Picture of the two-stage gasification plant.

A feed material can be entered into the sand bed in the bottom-reactor through two screw feeders. As a fluidizing medium, air is heated at 750 °C using a pre-heater to reduce the heat loss in the bottom reactor, after which it is put inside the bottom reactor. The producer gas produced by the two-stage gasifier is immediately passed through a cyclone and a hot filter made of Al_2O_3 to collect its particles. The cyclone and the hot filter are designed to capture particles bigger than 100 μm and 10 μm, respectively. After particle separation, the condensate liquid is captured in a series of stainless and glass condensers that are cooled up to 10 °C, using water as the cooling medium. Finally, to capture the aerosols and untreated particles in the producer gas, an electrostatic precipitator is used in the experiments. The remaining producer gas is either burned in a flare stack or sampled using Teflon gas bags at 3-min. intervals, to analyze their composition.

Experimental Results and Discussion

In the gasification experiments, four feed materials, namely dried sewage sludge (DSS), forest wood waste (FWW), construction wood waste (CWW), mixed waste plastic (MWP), were used. Firstly, DSS was digested and dried sewage sludge, which was sampled from waste treatment facility in Seoul. Secondly, a fraction of forest wood waste (FWW) from a manufacturing facility that uses wood residues from forest to make wood pellets was applied.

The CWW was collected in a construction site, which grinded, sieved, and then used to conduct gasification. Finally, a fraction of pellet-type mixed waste plastics that were obtained from a recycling center in South Korea. According to manufacturer, the mixed waste plastics consisted of polyethylene, polypropylene, PS, PVC and other miner plastics, such as PMMA and PET.

Experiments were conducted to investigate the effects of the main reaction conditions, such as the reactor temperature and the equivalence ratio on the characteristics of the producer gas. The effects of the activated carbon on the development of the producer gas and on tar reduction were also estimated. In addition, experiments with spent activated carbons were performed to investigate the reusability of used activated carbon. In MWP gasification, experiments varying the amount of activated carbon were conducted to find the effect of the amount of activated carbon on tar reduction and the development of producer gas. In order to find the influence of one operation parameter, other operation parameters were kept almost unchanged during gasification.

In the experiments, equivalence ratio was determined by the following equation:

$$ER = \frac{\textit{Actually fed air weight during gasification operation}}{\textit{Stoichiometrical air weight to fed feed materials}} \tag{13}$$

5.1. Mass balance

Mass balance of each of four experiments carried out with DSS, FWW, CWW and MWP is shown in Fig. 4. The four experiments indicated here were performed under similar reaction conditions (See Tables 1-4: Run7 in DSS gasification, Run7 in FWW gasification, Run7 in CWW gasification and Run2 in MWP gasification). As product fractions, producer gas, condensate liquid, char and tar were obtained.

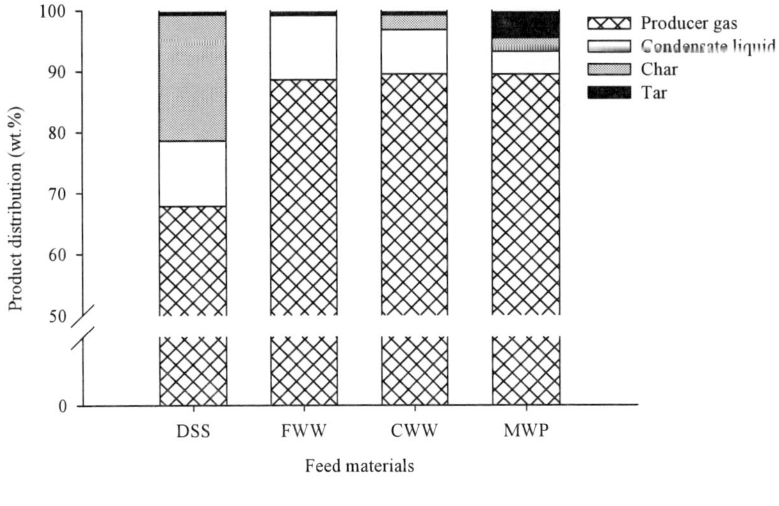

Figure 4. Mass balance obtained from the gasification of DSS, FWW, CWW, and MWP.

In all the experiments, the largest portion of the products was producer gas up to 90 wt.%. However, the second largest portion of four mass balances was different. In the case of the DSS, char was the second largest portion because of high ash content in the DSS. In the cases of the FWW and CWW, the second largest portion was condensate liquid. Tar amount indicated here is the sum of tar adsorbed the apparatus, tar captured by the electrostatic precipitator and tar in condensate liquid, and it is defined in this article as total tar. What is noticeable is the high amount total tar of the gasification of the MWP. Detailed investigation of total tar yield will be given in the next section.

5.2. Producer gas composition and LHV

The reaction conditions, producer gas composition, tar contents in the producer gases and LHVs of the producer gases are shown in Tables 1 to 4. In each experiment, the main components of the producer gases are CO_2, H_2, CO, CH_4, C_2H_4, and N_2. In the case of the MWP gasification, the LHV of the producer gas is high up to 13 MJ/Nm^3. In the biomass gasification, the LHV of the producer gas goes to 11 MJ/Nm^3. The producer gases obtained with activated carbon in the experiments contain especially high concentration of hydrogen, compared to the results without activated carbon under similar reaction conditions. The high concentrations of hydrogen and methane seem to result from the cracking of tar with the help of activated carbon.

A comparison between biomass gasification and plastic gasification in the two-stage gasifier can be seen in the Tables. It is observed that a high concentration of hydrocarbons including methane in the gasification of the MWP. This is mainly due to the fact that plastics have a higher amount of volatiles, compared with biomass. The volatiles produce high amount of hydrocarbons through decomposition during gasification resulting in the higher production of hydrocarbons in producer gas. In the DSS gasification, a high concentration of hydrogen could be observed. This can be mainly due to the fact that the DSS has a high moisture concentration, which leads to the enhanced hydrogen formation by the stimulating water-gas reaction.

The produce gases had very low tar contents with values below 8 mg/Nm^3, when activated carbon was applied. This low tar value, however, does not directly mean the high efficiency of activated carbon in tar removal, because tar removal can be achieved not only by activated carbon, but also by downstream measurements, such as condensers and the electrostatic precipitator. The net effect of activated carbon on the tar removal is shown in the next paragraph.

Table 1. Reaction conditions and experimental results of the DSS gasification

Parameter	Run1	Run2	Run3	Run4	Run5	Run6	Run7	Run8	Run9	Run10	Run11	Run12
Bottom reactor temperature (°C)	**751**	**809**	**821**	**853**	756	759	755	864	865	877	871	863
Upper reactor temperature (°C)	**643**	645	645	646	**698**	**750**	**793**	684	677	682	681	685

Parameter	Run1	Run2	Run3	Run4	Run5	Run6	Run7	Run8	Run9	Run10	Run11	Run12
Equivalence Ratio	0.27	0.30	0.31	0.30	0.27	0.26	0.26	**0.17**	**0.20**	**0.28**	**0.42**	0.23
Active Carbon (g)	500	500	500	500	500	500	500	500	**500**	500	500	**0**
Experimental results (vol.%)												
N_2	59.2	55.7	55.3	53.2	59.1	55.2	58.0	39.4	48.6	50.3	68.2	54.3
CO_2	14.2	12.9	12.4	10.5	14.1	14.4	14.2	11.0	9.3	12.3	11.3	9.1
H_2	8.2	11.1	11.9	15.6	9.6	14.4	15.2	24.8	20.3	16.2	8.7	10.8
CO	6.6	6.3	7.0	6.3	6.5	6.6	6.2	7.6	6.5	6.3	4.9	6.7
CH_4	6.4	8.0	8.2	9.1	6.2	6.3	5.1	12.4	10.5	10.2	4.7	10.0
C_2H_6	0.6	0.5	0.6	0.4	0.4	0.2	0.1	0.43	0.4	0.3	0.1	0.3
C_2H_4	2.9	3.6	3.1	3.4	2.7	1.9	1.0	3.11	3.2	3.1	1.4	6.0
$C_3+C_4+C_5$	1.3	1.0	0.9	0.5	0.9	0.4	0.1	0.3	0.4	0.4	0.2	1.2
Benzene	1.5	0.7	0.6	0.8	0.5	0.4	0.2	0.9	0.8	0.9	0.4	1.1
Above C_6	0.2	0.3	0.2	0.2	0.1	0.1	0.0	0.1	0.1	0.1	0.1	0.3
Tar (mg/Nm^3)	8	8	6	7	6	4	1	6	7	5	2	13
LHV (MJ/Nm^3)	8.0	9.2	8.9	9.6	7.3	6.8	5.2	11.6	10.3	9.5	4.9	11.9

Table 2. Reaction conditions and experimental results of the FWW gasification

Parameter	Run1	Run2	Run3	Run4	Run5	Run6	Run7	Run8	Run9	Run10	Run11
Bottom reactor temperature (°C)	**763**	**820**	**873**	823	818	823	829	829	828	816	819
Upper reactor temperature (°C)	799	**794**	805	**675**	**732**	798	796	796	796	796	792

Table 2. (Continued)

Parameter	Run1	Run2	Run3	Run4	Run5	Run6	Run7	Run8	Run9	Run10	Run11
Equivalence Ratio	0.37	**0.34**	0.34	0.33	0.31	**0.20**	**0.25**	**0.46**	**0.68**	0.32	0.33
Active Carbon (g)	450	**450**	450	450	450	450	450	450	450	**0**	**spent**
Experimental results (vol.%)											
N_2	24.8	28.0	28.9	28.1	26.6	18.4	22.1	31.7	38.1	28.1	25.4
CO_2	24.7	22.6	23.0	23.2	21.1	19.0	20.4	28.9	31.6	22.2	20.9
H_2	12.2	11.9	11.4	9.7	11.3	13.2	12.4	9.8	8.0	6.4	10.7
CO	31.6	30.5	30.6	31.1	32.0	38.5	35.1	25.0	19.6	33.9	33.9
CH_4	5.6	5.9	5.00	6.2	7.1	8.1	7.4	4.1	2.6	6.4	6.9
C_2H_6	0.1	0.1	0.04	0.1	0.2	0.2	0.1	0.02	0.01	0.1	0.1
C_2H_4	08	0.8	0.7	1.1	1.3	1.9	1.7	0.3	0.17	2.3	1.6
C_3+C_4+C_5	0.1	0.1	0.1	0.1	0.2	0.2	0.2	0.02	0.02	0.3	0.2
Benzene	0.2	0.3	0.3	0.3	0.3	0.4	0.4	0.1	0.1	0.4	0.4
Above C_6	0.03	0.03	0.02	0.1	0.1	0.1	0.1	0.01	0.01	0.1	0.04
Tar (mg/Nm^3)	1	1	1	2	2	2	2	0.4	0.3	2	2
LHV (MJ/Nm^3)	7.5	7.6	7.1	8.0	8.6	10.3	9.5	5.6	4.2	8.8	8.8

Table 3. Reaction conditions and experimental results of the CWW gasification

Parameter	Run1	Run2	Run3	Run4	Run5	Run6	Run7	Run8	Run9	Run10
Bottom-reactor temperature (°C)	777	775	786	816	825	818	821	819	819	812
Upper-reactor temperature (°C)	**645**	**722**	**799**	703	700	700	802	800	798	794
Equivalence ratio	0.28	0.28	0.28	**0.25**	**0.34**	**0.37**	0.24	0.26	0.26	0.26
Activated carbon (g)	450	450	450	450	450	450	**450**	**1st Spent**	**2nd spent**	**0**
Experimental results (vol.%)										
N_2	39.9	30.4	38.5	37.3	49.0	50.4	38.3	46.0	44.6	46.2
CO_2	8.5	9.2	9.0	9.1	9.4	9.2	8.2	7.6	7.9	7.3
H_2	15.3	20.9	22.6	17.6	13.4	13.2	18.3	12.7	12.2	11.3
CO	25.8	28.4	22.1	24.8	19.6	19.2	24.9	23.1	24.1	24.0
CH_4	8.0	9.8	7.0	8.8	6.9	6.4	8.3	7.5	7.8	7.8
C_2H_6	0.3	0.3	0.1	0.3	0.2	0.2	0.1	0.2	0.2	0.1
C_2H_4	1.5	0.8	0.55	1.4	1.0	1.0	1.3	2.1	2.4	2.4
$C_3+C_4+C_5$	0.2	0.04	0.03	0.2	0.05	0.05	0.1	0.2	0.2	0.2
Benzene	0.3	0.2	0.2	0.5	0.4	0.3	0.4	0.5	0.6	0.6
Above C_6	0.1	0.03	0.01	0.1	0.1	0.05	0.1	0.1	0.1	0.1
Tar (mg/Nm^3)	3	1	0.5	3	2	2	2	3	4	3
LHV (MJ/Nm^3)	8.9	9.5	7.7	9.5	7.3	7.0	9.0	8.8	9.3	9.1

Table 4. Reaction conditions and experimental results of the MWP gasification

Parameter	Run1	Run2	Run3	Run4	Run5	Run6	Run7	Run8	Run9
Bottom-reactor temperature (˚C)	798	824	825	827	796	805	795	804	819
Upper-reactor temperature (˚C)	**806**	800	803	804	**739**	**860**	797	802	798
Equivalence ratio	**0.21**	**0.26**	**0.31**	**0.62**	0.21	0.20	0.21	0.20	0.21
Activated carbon (g)	**430**	430	430	430	430	430	**500**	**640**	**0**
Experimental results (vol.%)									
N_2	51.0	56.9	64.0	71.5	53.7	49.8	48.5	46.8	51.9
CO_2	4.0	5.0	5.2	6.7	4.5	3.7	4.3	4.2	5.5
H_2	14.2	13.3	12.7	12.6	13.5	18.2	17.5	21.3	11.0
CO	6.6	5.6	5.4	5.0	5.5	7.2	6.9	6.0	5.0
CH_4	15.7	12.6	8.7	3.6	14.1	15.0	15.1	15.5	17.1
C_2H_6	0.5	0.4	0.2	0.02	0.8	0.2	0.5	0.5	0.5
C_2H_4	6.1	4.4	2.6	0.3	5.6	4.0	5.3	4.1	6.7
$C_3+C_4+C_5$	0.5	0.3	0.2	0.02	0.6	0.5	0.4	0.3	0.7
Benzene	1.5	1.3	1.0	0.2	1.5	1.5	1.5	1.3	1.6
Above C_6	0.1	0.1	0.1	0.00	0.1	0.03	0.1	0.1	0.1
Tar (mg/Nm^3)	3	2	2	0.2	6	2	3	2	4
LHV (MJ/Nm^3)	13.4	10.9	7.8	3.5	12.9	12.3	13.1	12.6	13.8

5.3. Effect of the temperature on the producer gas composition

As mentioned earlier, the gasifier has two reactors in series. Each of the two reactors has a separate temperature control system. During gasification, the temperature of the bottom reactor lay between 700 to 900 °C. In the experiments, it was revealed that the influence of the bottom-reactor temperature on the producer gas composition was insignificant. In contrast, the effect of the upper-reactor temperature was great. The development of producer gas composition with respect to the upper-reactor temperature is shown in Figure 5. Thermodynamically, when the reaction temperature increased at the atmospheric pressure, the CO and H_2 formations increased and the CO_2 and CH_4 formations decreased [14, 15]. The experiments in the two-stage gasifier along with activated carbon, however, result in a different gas development, due to the reactions between gas components and activated carbon or tar cracking on activated carbon. As can be seen in Figure 5, the H_2 concentration increased up to 23 vol.% at 799 °C with increasing temperature. The higher concentration of H_2 at elevated temperatures can be explained on one hand by the enhanced cracking of tar at these temperatures. In particular, the tar cracking seems to have been stimulated by the presence of activated carbon, which might have adsorbed the tar that was produced in the bottom reactor. The prolonged residence of the tar in the upper reactor might have given it more chances to be further thermally cracked. On the other hand, the formation of H_2 from the water-gas reactions in Equations (5) and (6), which are endothermic reactions, seems to have been more active at the elevated temperatures, due to Le Chatelier's principle. The temperature influence on the other components change was relatively small.

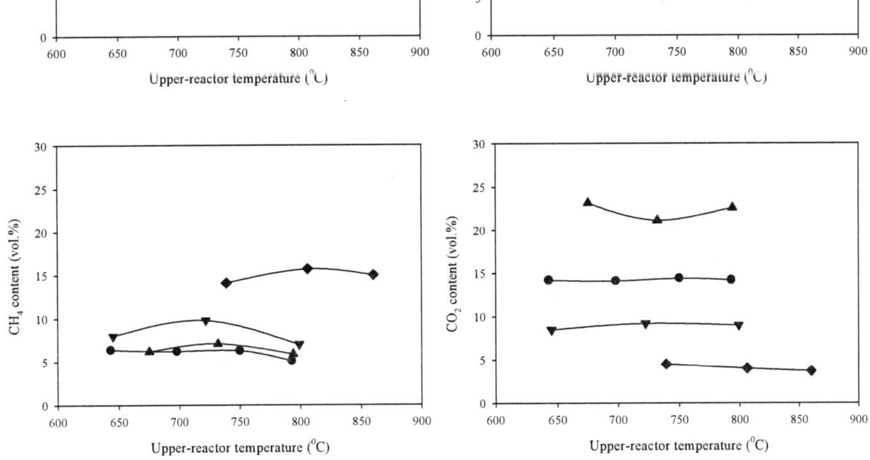

Figure 5. Producer gas development with respect to the upper-reactor temperature. (● DSS, ▲ FWW, ▼ CWW, ◆ MWP).

5.4. Effect of the equivalence ratio (ER)

The effect of the ER variation on the composition of the producer gas is shown in Figure 6. The ER effect on the producer gas composition is very clear.

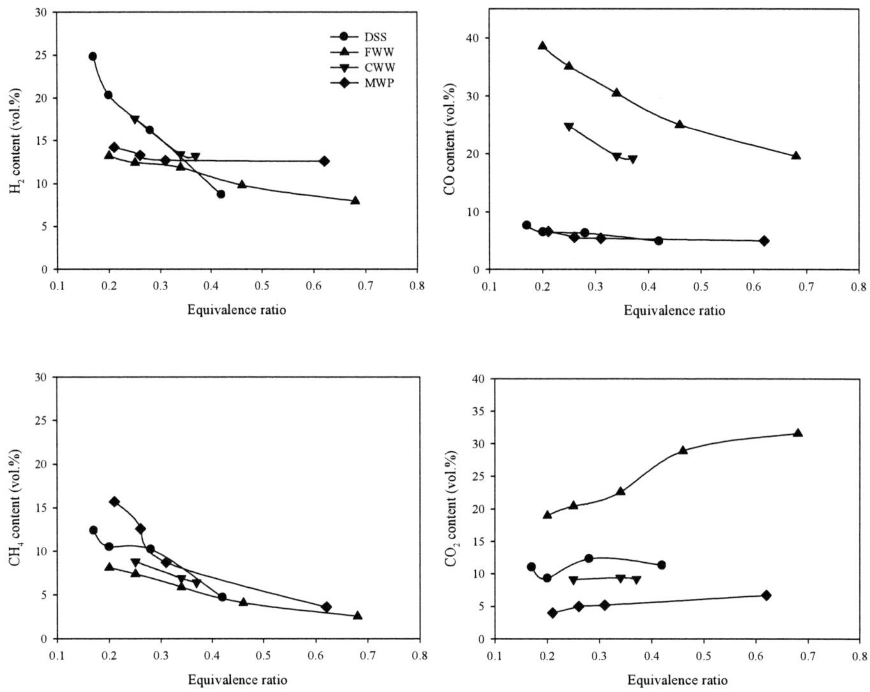

Figure 6. Producer gas composition at various ERs. (● DSS, ▲ FWW, ▼ CWW, ◆ MWP).

The increase in the ER means that the feeding air or oxygen increased. At a higher ER, more active oxidization reactions will occur, normally to increase the CO_2 and H_2O concentrations at the expense of CO and H_2 [16]. Figure 6 shows an increase in the concentration of CO_2 and a decrease in the H_2, CO, CH_4, and other hydrocarbons at a higher ER. A considerable increase in CO_2 at a higher ER, which may be attributed to its strengthened oxidation reactions with the hydrocarbons, including CH_4. In addition, the activated carbon in the upper reactor might have been consumed in the oxidation to produce CO_2. The significant decrease in CO and H_2 at a higher ER might have resulted from their strengthened oxidation. The maximum concentration of H_2 (25 vol.%) was achieved at the minimum ER (0.17) in the gasification of the DSS. In the Figure 6, it can be seen that at the same ER, plastic gasification yields higher concentration of CH_4 and less concentration of CO and CO_2, compared to the biomass gasification.

A higher ER value causes a decrease in LHV, due to the excessive formation of the products of complete combustion, such as CO_2 and H_2O. In an experiment of A. van der Drift et al, producer gases that were produced through air gasification of 10 residual biomass fuels mostly had an LHV of 4-6 MJ/Nm^3 [17]. The LHVs of the producer gases obtained in the experiments, however, were above 10 MJ/Nm^3. In particular, it is desirable that the high caloric value of the producer gas is mainly not due to tar components that have usually high caloric values, but due to the high concentrations of H_2 and CH_4. As can be seen in Tables 1-4, at a higher ER, the producer gas has a very low a tar concentration.

5.5. Role of activated carbon, reusability and the effect of the amount of activated carbon.

This section deals with the effect of activated carbon on the producer gas composition and the reusability of spent activated carbon. In addition, the effect of the amount of activated carbon is also discussed. Fig. 7 represents the gas development according to the presence and amount of activated carbon.

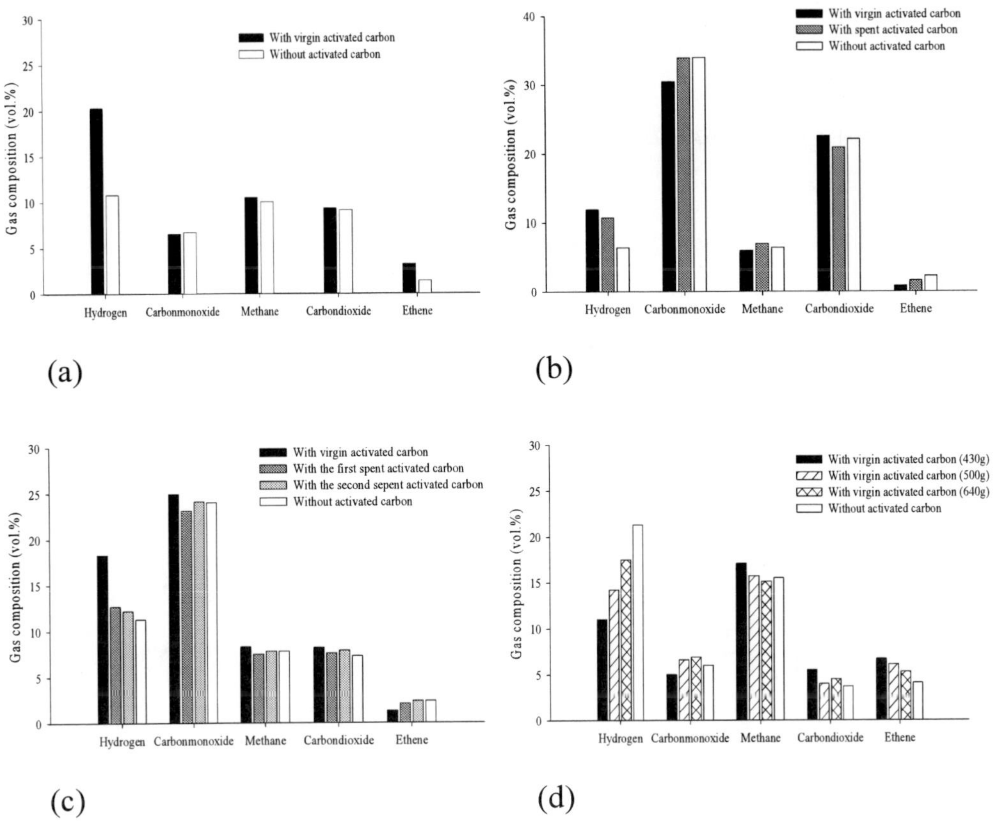

Figure 7. Composition of the producer gases obtained by the gasification of DSS (a), FWW (b), CWW (c), MWP (d).

It can be seen in Figure 7 that the H_2 production with the activated carbon was more significantly enhanced than in the experiment without activated carbon. The increase in H_2 with activated carbon was achieved not only through the longer adsorption of tar on the activated carbon surface, which could have led to the enhanced tar cracking reactions, thereby producing hydrogen, but also through the stimulation of the hydrogen formation reactions as followings:

Thermal cracking: $pC_xH_y \rightarrow qC_nH_m + rH_2$ (14)
Carbon formation: $C_nH_m \rightarrow nC + m/2H_2$ (15)
Steam reforming: $C_xH_y + nH_2O \rightarrow (n + y/2)H_2 + xCO$ (16)
C_xH_y means tar and C_nH_m is hydrocarbons with smaller carbon number than C_xH_y.

In Figure 7, the H_2 production with the spent activated carbon is still higher than that without activated carbon. In the MWP gasification, the H_2 production increases, as the amount of activated carbon increases. In the experiments, a decrease in tar yield in producer gas is also observable, as the amount of activated carbon increased. This can be an example of more active tar cracking when the amount of activated carbon increased. In spite of difference of H_2 content in producer gases, the LHVs of the producer gases obtained varying the amount of activated carbon have no significant difference, because the production of hydrocarbons including methane increases with the decrease of the amount of activated carbon.

5.6. Total tar generation and tar analysis

The tar that was produced in the experiments existed in the producer gas, in the condensate liquids collected by the scrubbers and the electrostatic precipitator, and on the surface of the equipment, such as the pipe lines. Until now, only the tar in the producer gases has been mentioned (in Section 5.2). Only the amount of tar in the producer gases, however, cannot demonstrate the efficiency of activated carbon and the two-stage gasifier in tar removal, because the tar removal in the experiments took place via adsorption on the activated carbon and via mechanical methods, such as via the condensers and electrostatic precipitator. This section discusses the total tar yield during the gasification, the major tar components, and the thermal characteristics of the tar. At first, the total tar amount was determined from the sum of the tar that was collected in the apparatus, such as condensers and tubes after quenching and the tar captured by the electrostatic precipitator. The amount of tar in the producer gases was so small that it was excluded. Figure 8 shows the effect of activated carbon and its amount on the total tar removal.

As shown in Figure 8, much of the tar in each experiment except DSS gasification is collected by the electrostatic precipitator. In the FWW gasification, the total amount of tar that is produced with activate carbon is six times less (approximately 4 g) than the amount of tar that is produced without activated carbon (24 g). The total amount of tar that is produced with spent activated carbon is about 8 g, which is three times less than the amount of tar that is produced without activated carbon. In the CWW gasification, the total tar produced with the virgin activated carbon (Run 7) was three times less (about 9 g) than that without activated carbon (about 29 g in Run10). The total tar yields produced with the spent activated carbons of Run8 (the first time) and Run9 (the second time) were about 20 and 21g, respectively. The efficiency of the spent activated carbons in tar removal decreases to some extent, but seems to be nevertheless still excellent. The slightly reduced tar removal efficiency of the spent activated carbons may be due to the lower accessibility of the surface, as for example, due to the blocking of the pores of the activated carbon by the char particles that are produced during the gasification. Additionally, in MWP gasification, when the amount of activated carbon increases (Run 9 (0 g), Run 1 (430 g), Run 7 (500 g) and Run 8

(640 g)), the total amount of tar decreases steeply from 50.99 g (Run9) to 19.95 g (Run8), due to the strengthened tar adsorption followed by tar cracking.

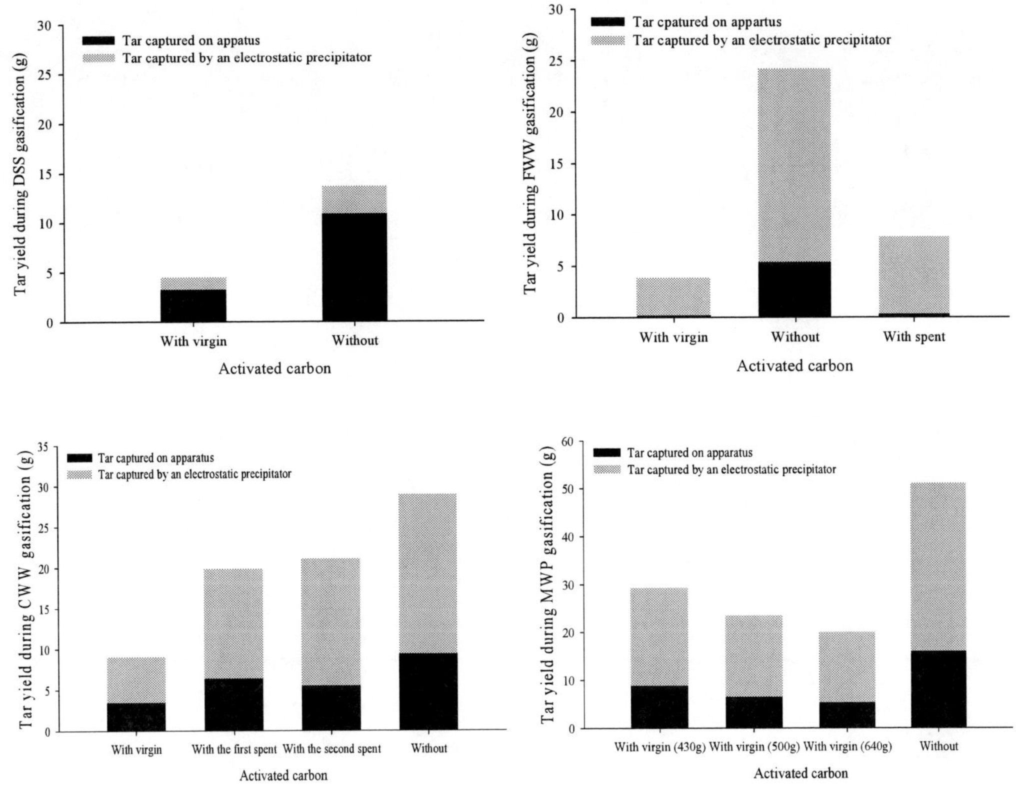

Figure 8. Effects of activated carbon and its amount on tar removal.

To confirm this hypothesis, FE-SEM (Field Emission Scanning Electron Microscopy) and BET surface area analysis were conducted with the virgin activated carbon and the spent activated carbon. Figure 9 represents the structures of the virgin and the spent activated carbons.

The left figure (the virgin activated carbon) of FWW gasification has a relatively smooth surface. The middle figure is the activated carbon after one experiment, and the right figure is the activated carbon that was obtained after the experiment with the spent activated carbon. The activated carbons in the middle and right figures adhered to some char particles and covered the surface of the original activated carbon. The specific surface area of the virgin activated carbon was 950 m^2/g, but that of the activated carbon after one experiment decreased to 589 m^2/g. The specific surface area, however, did not further decrease with the activated carbon that was obtained after the experiment with the spent activated carbon (681 m^2/g). Table 5 shows the elemental analysis of the activated carbons.

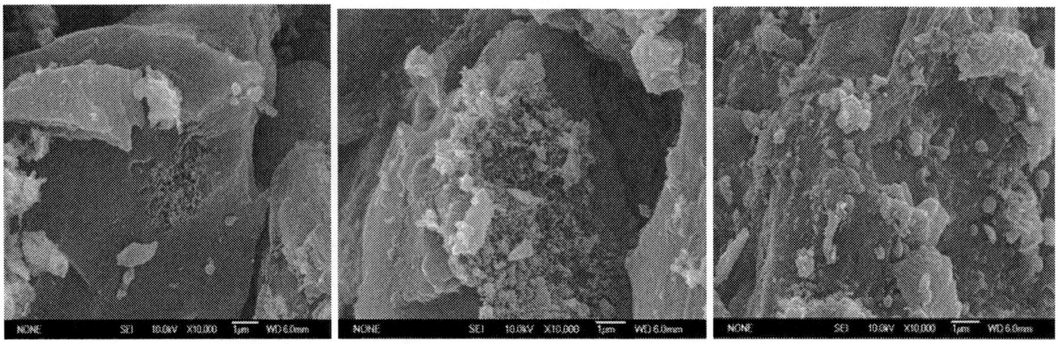

Figure 9. SEM graphs of the virgin activated carbon and the spent activated carbon after FWW gasification.

Table 5. Elemental analysis of the activated carbons

Components (wt.%)	Virgin	FWW gasification		CWW gasification		
		After 1st Run	After 2nd Run	After 1st Run	After 2nd Run	After 3rd Run
	Activated carbon					
C	63.8	82.7	70.1	87.7	90.3	88.2
H	1.6	0.4	0.4	0.6	0.4	0.4
N	0.05	0.03	0.01	0.7	0.4	0.5

The carbon contents of the activated carbons after the 1st and 2nd Runs are higher than those of the virgin activated carbon. This can be an evidence of adhesion of carbonaceous particles on the activated carbon.

Tar that was captured by the electrostatic precipitator in experiment was analyzed using a GC-MS, with acetone as the solvent. Although tar may have many high-molecular components that cannot be identified with GCs, the analysis will give an insight on the characteristics of tar. Table 6 shows the major components of tar.

The tar consisted mainly of polycyclic aromatic hydrocarbons, such as 2-ring (naphthalene), 3-ring (acenaphthylene, fluorene, and phenanthrene) and 4-ring (pyrene, fluoranthene, and triphenylene). These PAHs belong to tertiary tars that could have been mainly formed during the biomass gasification at above 800 °C.

TGA experiments were conducted to investigate the thermal characteristics of the produced tars. The results are presented in Figure 10.

The TGA curves present the temperature range at which mass loss took place. A significant mass loss of tar is shown to have occurred within the range of 150-400 °C, which implies that a secondary cracking unit after the two-stage gasifier and before the quenching unit would help reduce the tar content of the producer gas, if the temperature of the cracking unit is kept above 400 °C.

Table 6. Major components of the tars captured by the electrostatic precipitator

Components	Formula	DSS	FWW	CWW	MWP
		Aromatics (1-ring)			
Benzene	C_6H_6	-	-	0.19	15.15
Toluene	C_7H_8	-	-	0.34	4.58
Styrene	C_8H_8	-	-	1.04	8.78
Indene	C_9H_8	-	2.39	4.06	3.70
		Light PAHs (2-ring)			
Azulene	$C_{10}H_8$	9.5	-	-	0.18
Naphthalene	$C_{10}H_8$	-	11.52	17.31	34.62
1-methyl-naphthalene	$C_{11}H_{10}$	-	2.35	1.48	-
2-methyl-naphthalene	$C_{11}H_{10}$	4.49	-	2.70	2.46
Acenaphthylene	$C_{12}H_8$	-	8.14	-	-
Biphenyl	$C_{12}H_{10}$	3.37	2.19	2.52	4.57
Biphenylene	$C_{12}H_{10}$	5.28	-	-	7.08
Acenaphthalene	$C_{12}H_{10}$	-	-	-	0.56
2-Ethenly-naphthalene	$C_{12}H_{12}$	1.67	1.18	-	1.50
		Heavy PAHs (3 and 4-ring)			
Acenaphthylene	$C_{12}H_8$	-	-	10.55	-
Fluorene	$C_{13}H_{10}$	2.58	7.06	7.25	1.16
Phenanthrene	$C_{14}H_{10}$	5.27	15.06	14.97	-
Anthracene	$C_{14}H_{10}$	0.62	-	1.43	5.57
4H-cyclophenta[def] phenanthrene	$C_{15}H_{10}$	-	3.45	3.50	-
2-Phenyl-naphthaelene	$C_{16}H_{12}$	0.84	1.73	-	-
Fluoranthene	$C_{16}H_{10}$	2.42	6.92	5.36	-
Pyrene	$C_{16}H_{10}$	0.91	10.11	8.26	-
1,2-Benzofluorene	$C_{17}H_{12}$	2.50	-	-	-
Benz[a]-anthracene	$C_{18}H_{12}$	-	-	3.50	-
Triphenylene	$C_{18}H_{12}$	-	3.63	-	-
		Oxygen containing compounds			
Phenol	C_6H_6O	0.37	-	0.70	-
Benzofuran	C_8H_6O	-	-	0.79	0.29
Dibenzofuran	$C_{12}H_8O$	1.09	2.20	3.22	0.09
		Nitrogen containing compounds			
Phthalonitrile	$C_6H_4(CN)_2$	1.09	-	-	-
Benzonitrile	C_7H_5N	1.23	0.15	0.07	0.16
Indole	C_8H_7N	0.66	-	-	-
Benzeneacetonitrile	C_8H_7N	0.75	-	-	-
Methyl-benzonitrile	C_8H_8N	1.34	-	-	-
Quinoline	C_9H_7N	0.82	0.17	-	-
Isoquinoline	C_9H_7N	5.20	0.04	0.11	-
2-Naphthalene carbonitrile	$C_{11}H_7N$	7.68	-	-	-
Carbazole	$C_{12}H_9N$	0.59	-	-	-
Benzo[f]quinoline	$C_{13}H_9N$	1.74	-	-	-
		Sulfur containing compounds			
Benzo-thiophene	C_8H_6S	1.21	-	-	-
Dibenzo-thiophene	$C_{12}H_8S$	0.84	-	-	-

Unit: area% of GC-MS

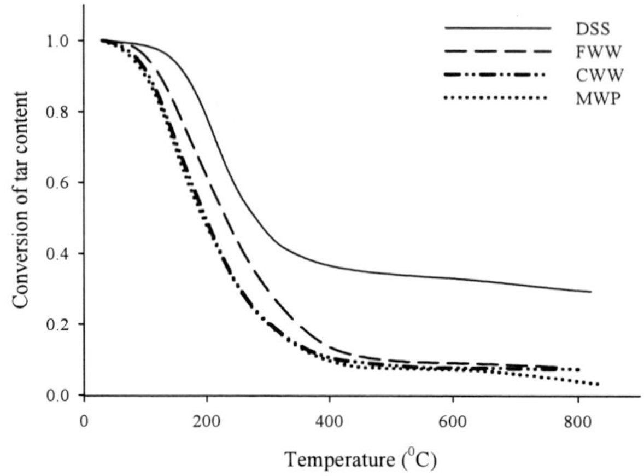

Figure 10. TGA curves of the tar that was captured by the electrostatic precipitator at 20 °C/min.

CONCLUSION

In this chapter, characteristics of the gasification of biomass and waste plastics in a newly developed two-stage gasifier were investigated. The aim of the work was to produce a clean and high caloric producer gas. In the experiments, the effects of the reaction conditions, such as the reactor temperatures, equivalence ratio, and the effect of activated carbon and its amount in reducing the tar content of producer gases were investigated. In addition, characteristics of the tar captured during gasification were investigated.

The producer gases that were obtained in all the experiments were mainly made up of CO, CO_2, H_2, CH_4, and C_2H_4. The tar contents of all the producer gases were below 10 mg/Nm^3, when activated carbon was applied. The maximum LHVs of producer gases from plastics and biomass were about 13 MJ/Nm^3 and 11 MJ/Nm^3, respectively. The bottom-reactor temperature had almost no influence on the producer gas composition. The effect of the upper-reactor temperature, however, was significant. As the upper-reactor temperature increased, the H_2 concentration strongly increased. The composition of the producer gas was strongly influenced by the equivalence ratio. An increase in the equivalence ratio resulted in a higher CO_2 production and a decrease in the concentrations of H_2, CH_4, and hydrocarbons. In the gasification of forest wood wastes, a producer gas with a maximum LHV was obtained at the lowest equivalence ratio (ER = 0.2). The producer gas that was obtained at the ER of 0.2, however, contained only 2 mg/Nm^3 of tar concentration.

Activated carbon influenced the producer gas composition. When activated carbon was applied, the hydrogen production was strongly enhanced. It was also revealed that the increase of hydrogen production depended on the amount of activated carbon, due to enhanced tar cracking. Furthermore, activated carbon also showed excellent performance in tar removal. The total tar that was produced with activated carbon was much less than that produced without activated carbon. Also, the spent activated carbon still showed a good performance in tar removal. The tar that was produced in an experiment consisted mainly of polycyclic aromatic hydrocarbons that belonged to tertiary tars.

In summary, activated carbon and the two-stage gasifier resulted in a producer gas with high caloric value and low contaminant content. The producer gas that was obtained in the experiments seems to be appropriate for use as a fuel gas in gas engines for electricity generation.

REFERENCES

[1] Milne T. A.; Evans R. J. Biomass gasification "tars": their nature, formation and conversion. NREL, Golden, CO, USA, Report no. NREL/TP-570-25357, 1998.

[2] Dogru, M. Midilli, A. Howard, C. R. Gasification of sewage sludge using a throated downdraft gasifier and uncertainty analysis. Fuel processing Technology 2002, 75, p.55-82.

[3] Basu P. Combustion and gasification in fluidized bed. New York: Taylor and Francis Group; 2009. p. 59–101.

[4] Flecher DF, Haynes BS, Christo FC, Joseph SDACFD. Based combustion model of an entrained flow biomass gasifier. Appl. Math Model 2000, 24, p.165–182.

[5] Gao N, Li A, Quan C. A novel reforming method for hydrogen production from biomass steam gasification. Bioresource Technology 2009,100, p.4271–4277.

[6] Rezalyan J, Cheremisinoff NP. Gasification technologies. New York: Taylor and Francis; 2005. p.1–33.

[7] Watanabe H, Otaka M. Numerical simulation of coal gasification in entrained flow coal gasifier. Fuel 2006, 85, p.1935–1943.

[8] Gil J, Corella J, Aznar MP, Caballero MA. Biomass gasification in atmospheric and bubbling fluidized bed: effect of the type of gasifying agent on the product distribution. Biomass and Bioenergy 1999, 17, p.389–403.

[9] Mckentry P. Energy production from biomass (part 3): gasification technologies. Bioresource Technology 2002, 83, p.55–63.

[10] Devi L, Ptasinski KJ, Janssen FJJG. A review of the primary measures for tar elimination in biomass gasification processes. Biomass and Bioenergy 2003, 24, p.125–140.

[11] Manya JJ, Sanchez JL, Abrego J, Gonzalo A, Arauzo J. Influence of gas residence time and air ratio on the air gasification of dried sewage sludge in a bubbling fluidized bed. Fuel 2006, 85, p.2027–2033.

[12] Dayton D. A review of the literature on catalytic biomass tar destruction. Milestone completion report, NREL/TP-510–32815 (project No. DE-AC36–99-GO10337). Golden, Colorado (USA): National Energy Laboratory; 2002.

[13] Kiel JHA, van Paasen SVB, Neeft JPA, Devi L, Ptasinski KJ, Janssen FJJG, et al. Primary measures to reduce tar formation in a fluidised-bed biomass gasifier. Report ECN-C–04–014, ECN. Petten, The Netherlands, March 2004.

[14] Mathieu P, Dubuisson R. Performance analysis of a biomass gasifier. Energy Convers. Manage. 2002, 43, p.1291–1299.

[15] Petersen I, Werther J. Experimental investigation and modeling of gasification of sewage sludge in the circulating fluidized bed. Chem. Eng. Process 2005,44,p.717–736.

[16] Lia XT et al. Biomass gasification in a circulating fluidized bed. Biomass and Bioenergy 2004, 26, p.171–193.
[17] Van der Drift A, Van Doorn J, Vermeulen JW. Ten residual biomass fuels for circulating fluidized-bed gasification. Biomass and Bioenergy 2001, 20, p.45–56.

In: Gasification: Chemistry, Processes and Applications
Editor: Michael D. Baker
ISBN: 978-1-61209-681-0

Chapter 10

COMPONENTS AND OPERATION OF THE FIXED BED DOWNDRAFT SYSTEM JOHANSSON BIOMASS GASIFIER

Ntshengedzeni Sampson Mamphweli* and Edson Leroy Meyer
University of Fort Hare, Institute of Technology, Private, Alice, South Africa

INTRODUCTION

The chapter presents a detailed description of the System Johansson Biomass Gasifier and its operation; it also deals with the factors influencing the efficiency of fixed bed biomass gasifier systems since the latter gasifier is a fixed bed type. Basically, the fuel is fed into the reactor where the gasification process takes place resulting in the formation of syngas. Syngas is channeled through a system of pipes to the downstream processes that consists mainly of the purification unit where it is cleaned of impurities such as fine carbon particles. The gas is then used to drive the engine that drives the generator and generates electricity. The gasifier is mainly used for electricity generation. A 150kVA pilot project aimed at improving the socio-economic status of a rural community has been implemented. The project aims to provide low-cost electricity to stimulate the establishment of small businesses in the area. Figure 1 shows the components of the gasifier.

* e-mail:smamphweli@ufh.ac.za

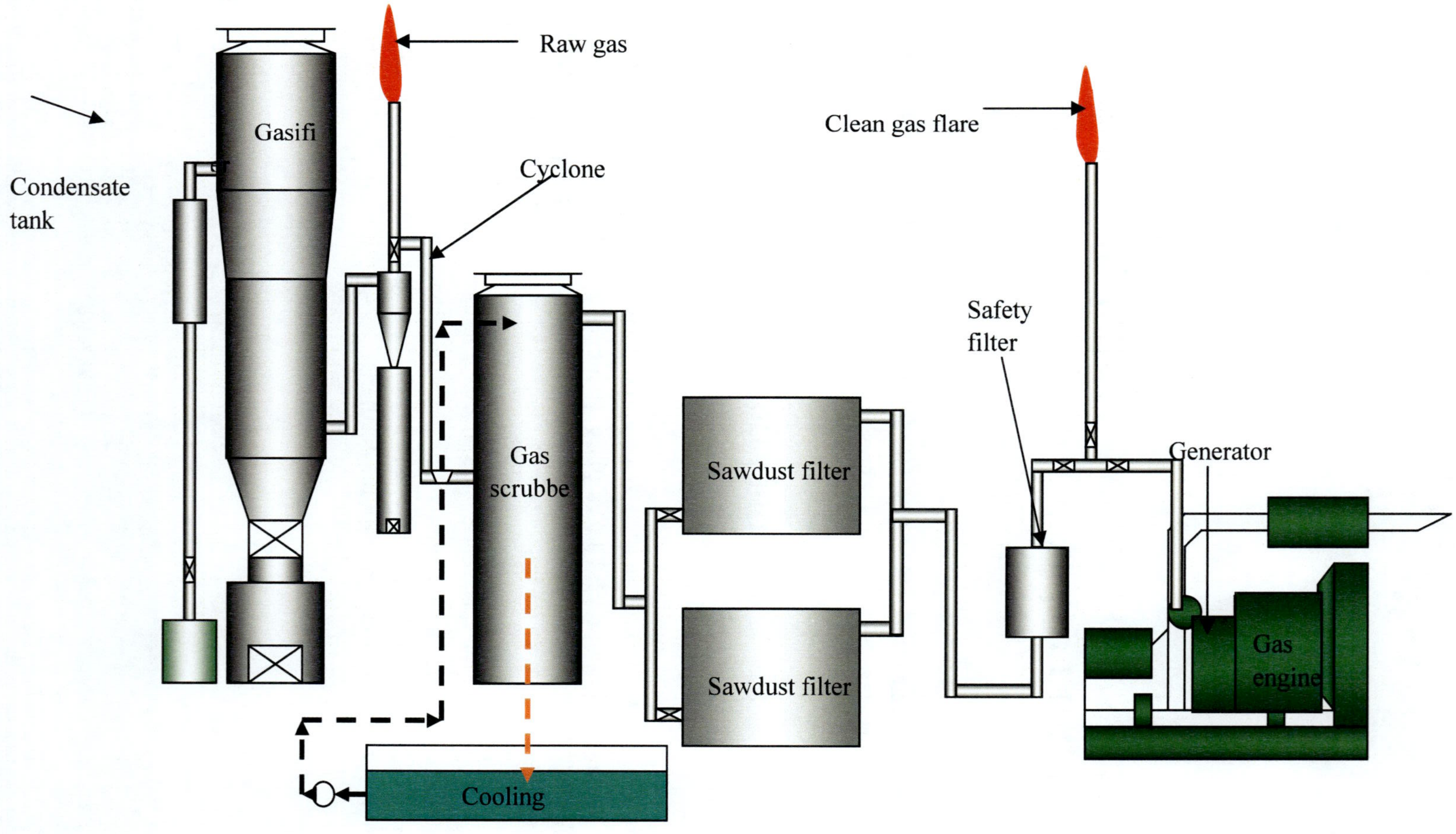

Figure 1. Schematic diagram of the gasifier.

Reactor/gasifier

The Johansson Biomass Gasifier reactor is of a downdraft type developed through modification of the Imbert downdraft gasifier [Johansson, 2002]. The reactor/gasifier is available in five different standard design sizes from $120Nm^3/h$ to $850Nm^3/h$ gas production, and five different special non-standard sizes from 1100 to $2400Nm^3/h$ gas production [Johansson, 2002]. The one installed at Melani village is a $300Nm^3/h$ gas production unit. The fuel is fed into the reactor through the top loading zone. To start the gasifier, the ignition of the reactor is done by inserting two or three sparklers, locked in a sparkler holder with handle, through the igniter sleeve. In the reactor, biomass is heated by combustion. Four chemical processes can be distinguished namely drying, pyrolysis/carbonization, oxidation and reduction reactions.

The gasifier/reactor is divided into four zones according to these chemical reactions. Combustion occurs in the oxidation zone. Introduced air in the oxidation zone contains (besides oxygen and water vapours) inert gases such as nitrogen and argon. These inert gases are considered to be non-reactive with fuel constituents. The oxidation takes place at the temperature of 700-2000°C.

Heterogeneous reaction takes place between oxygen in the air and solid carbonized fuel (Charcoal), producing carbon monoxide. Hydrogen in the fuel reacts with oxygen in the air blast, producing steam. Combustion is described by the following chemical formulae:

$$C + O_2 \rightleftharpoons CO_2 + 401.9kJ/mol \qquad (1)$$

$$H2+½O_2 \rightleftharpoons H_2O+241.1 kJ/mol \qquad (2)$$

The gas forming reactions that take place in the reduction zone of the gasifier are as follows:

Boudouard reaction

$$CO_2 + C \rightleftharpoons 2CO-164.9kJ/mol \qquad (3)$$

Water-gas reaction

$$C + H_2O \rightleftharpoons CO+H_2-122.6kJ/mol \qquad (4)$$

Water shift reaction

$$O_2+C+ 3H_2 \rightleftharpoons CO + H_2O-42.3kJ/mol \qquad (5)$$

Methane production reaction

$$C + 2H_2 \rightleftharpoons CH_4 + 75kJ/mol \qquad (6)$$

$$O_2+C+3H_2 \rightleftharpoons CH_4+ H_2O+205.9kJ/mol \qquad (7)$$

Chemical reactions represented by equations 2.3 and 2.4 are the main reactions taking place in the reduction zone and they are endothermic, this results in temperature decreasing during these reactions [Quaak *et al*, 1999, Reed and Das, 1988 and Stassen, 1995]. The hot gases and charcoal coming from the oxidation zone provide the energy required for the reduction chemical reactions to take place. As these reactions proceed the temperature inside the reactor sinks progressively until it becomes as low as 700°C. This implies that the extent of reduction reactions is dependent on the amount of energy entering the reduction zone and consequently also on the heat losses from the reactor [Barrio *et al*, 2007].

Carbonization is the thermal decomposition of biomass fuels in the presence of 26-30% of oxygen at temperature ranging from 450°C to 600°C. This process results in the release of charcoal, organic vapours and gasses [van de Beld, 2004 and Oregon Department of Energy, 2004]. The ratio of products is influenced by the chemical composition of biomass fuels and the operating conditions of the gasifier. Yields of primary carbonization products depend on temperature. For instance charcoal yield decreases and the gas yield increases with temperature [van de Beld, 2004]. This is because the charcoal is converted to gas, but at extremely high temperature the gas formed gets combusted resulting in low gasifier efficiency as explained in this chapter.

The heating value of gas produced during the gasification process is low (4-6 MJ/Nm^3), or about 10% to 15% of the heating value of natural gas [Stassen, 1995].

Automatic variable speed ash grate activator

The gasifiers are fitted with an automatic variable speed ash grate activator and with ash removal and refueling systems for non-interrupted, continuous working. They are also supplied with electronic fuel level indicators and with an early refueling warning system. An electronic flashing red light ash removal warning system is also provided [Johansson, 2002]. This allows for a smooth uninterrupted 24 hours operation.

The cyclone

The raw gas is passed through the cyclone, which removes the coarse carbon particles from the raw gas. When operating at full gasifier/engine power, the conventional fixed cyclone removes about 80% of the carbon particles and soot, or about 4g/m^3 of gas, leaving the remaining about 20% fine carbon and soot particles or about 1g/m^3 of gas and this is carried through to the gas scrubber/cooler. If the power output is reduced, the cyclone starts to lose efficiency [Johansson, 2002]. This is basically because of the reduced centrifugal forces as explained later in this section.

The SJBG cyclone is a conventional cyclone. Typically, a particulate-laden gas enters tangentially near the top of the cyclone. The gas flow is forced into a downward spiral simply because of the cyclone's shape and the tangential entry. Another type of cyclone (a vane axial cyclone) employs an axial inlet with fixed turning vanes to achieve a spiraling flow. Centrifugal forces and inertia cause the particles to move outward, collide with the outer wall, and then fall downward to the bottom of the device. Near the bottom of the cyclone, the gas reverses its downward spiral and moves upward in a smaller inner spiral. The cleaned gas exits from the top through a vortex finder tube, and the particles exit from the bottom of the cyclone through a pipe sealed by a spring-loaded flapper valve or rotary valve [Cooper and Alley, 1986 and Gradon *et al*, 1998]. The clean gas exiting the cyclone is termed the overflow while the retained particles collected at the bottom are termed the underflow.

The collection efficiency of cyclones vary with particle size and cyclone design. The efficiency of particle collection is generally good for particles that are larger than 5 microns. Other cyclones have collection efficiency greater than 98% for particles larger than 5 microns and others do achieve efficiencies of 90% for particles larger than 15-20 microns [Cooper and Alley, 1986 and Gradon *et al*, 1998]. High efficiency requires higher inflow pressure. Three categories of cyclones are available and these are the high efficiency, conventional and high throughput. The high efficiency attains a higher efficiency followed by the conventional and high throughput respectively.

The cyclone performance is rated in terms of particle cut diameter (d_p) or cut size. The cut size, D_p50 for instance is the particle size which is captured 50% [Reed and Das, 1988]. The impact of particle size on collection efficiency is quantified in the following section.

Cyclone collection Efficiency model

A model can be used to determine the effects of both cyclone design and operation on collection efficiency. In this model, gas spins through a number of revolutions N_e in the outer vortex. The value of N_e can be approximated by [Cooper and Alley, 1986]:

$$N_e = \frac{1}{H}\left(L_b + \frac{L_c}{2}\right) \qquad (8)$$

where:

N_e = number of effective turns
H = height of inlet duct (m)
L_b = length of cyclone body (m)
L_c = length (vertical) of cyclone cone (m).

Figure 2 shows the various dimensions of the cyclone for better understanding of the equations.

To be collected, particles must strike the wall within the amount of time that the gas travels in the outer vortex. The gas residence time in the outer vortex is

$$\Delta t = \frac{\pi D N_e}{V_i} \qquad (9)$$

where:
Δt = time spent by gas during spiraling descent (sec)
D = cyclone body diameter (m)
V_i = gas inlet velocity (m/s) = Q/WH
Q = volumetric inflow (m^3/s)

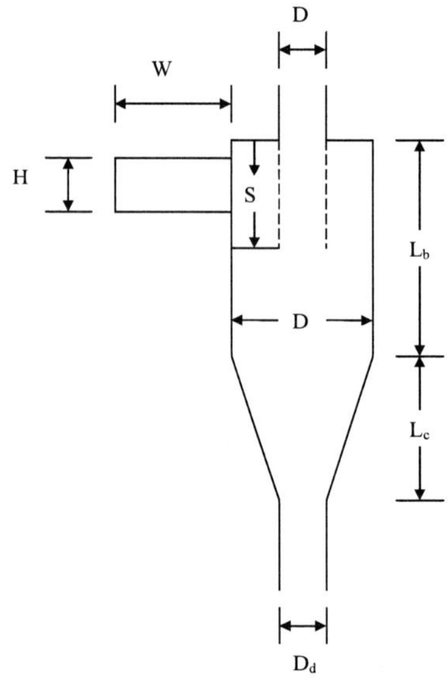

Figure 2. Various dimensions of the cyclone.

The maximum radial distance travelled by any particle is the width of the inlet duct (W). The centrifugal force quickly accelerates the particle to its terminal velocity in the outward (radial) direction, with the opposing drag force equaling the centrifugal force.

The terminal velocity V_t of the particle in a radial direction that will just allow a particle initially at distance (W) away from the wall to be collected in time Δt is

$$V_t = \frac{W}{\Delta t} \tag{10}$$

where

W = width of inlet (m).
V_t = particle terminal velocity in the radial direction (m/s).

In addition V_t is a function of particle size. Assuming Stokes' regime flow (drag force = $3\pi\mu d_p V_t$) and spherical particles subjected to a centrifugal force F.

$$F=mv^2/r \tag{11}$$

where:

m = mass of particle in excess of mass of gas displaced
v = V_i of inlet flow, and
r =D/2= radius

We obtain:

$$V_t = \frac{(\rho_p - \rho_g) d_p^2 V_i^2}{9 \mu D} \qquad (12)$$

where

d_p = diameter of the particle (m)
ρ_p = density of the particle (kg/m^3)
ρ_g= gas density (kg/m^3)
μ = gas viscosity (kg/m·s).

The density and viscosity values of air and syngas are similar, since most of syngas constituents are similar to air (CO, N_2, and to a lesser extent CO_2). Only the hydrogen portion is substantially great in syngas.

Substitution of equation 2.9 into the 2.10 eliminates Δt. Then, equating 10 and 12 and rearranging to solve for particle diameter, we obtain [Cooper and Alley, 1986]:

$$d_p = \left[\frac{9 \mu W}{\pi N_e V_i (\rho_p - \rho_g)} \right]^{1/2} \qquad (13)$$

It is worth noting that in this expression, d_p is the size of the smallest particle that will be collected if it starts at the inside edge of the inlet duct. Thus, in theory, all particles of size d_p or larger should be collected with 100% efficiency [Cooper and Alley, 1986].

There are three factors that determine the centrifugal force; these are radius of spiral motion, particle mass, and particle velocity. The relationship between these factors and the centrifugal force can be described by equation 11. In equation 11, increasing the velocity of the particle (V) and/or decreasing the radius of the vortex flow (r) can increase the efficiency of separation [Holley *et al*, 2006]. However care should be taken when decreasing the size of the vortex finder as this might results in low gas flow rates impacting negatively on the efficiency of the engine as it will have to apply more power in sucking the gas.

The gas scrubber/cooler

After the coarse dust has been removed in the cyclone, about 0.8g/m^3 gas is removed in the gas scrubber/cooler, where the gas is cooled to between 20°C to 25°C with water sprayed over a suitable low resistance, large surface area scrubbing media (charcoal). The water is normally recycled through an ambient cooling pond for a long period of time [Johansson, 2002b].

Particles larger than 1μm settle by gravity and inertia in accordance with Stoke's law and can be captured by impaction, gravitational or centrifugal means. For particles smaller than 0.1μm, motion is dominated by molecular collisions in accordance with Brownian motion principles. These particles may be collected by diffusion onto a liquid surface. Particles with diameters between 0.1 and 1μm are the most difficult to capture either by diffusion or inertial mechanisms. They are too large to diffuse well but too small to settle, however they can be made to agglomerate into larger particles that are easier to collect. Agglomeration is also assisted by the presence of droplets that act as nuclei [Reed and Das, 1988].

Particles tend to move towards a surface on which condensation is taking place. This phenomenon is referred to as Stefan motion. Particles tend to migrate away from a hot surface towards a cold surface, this phenomenon is called thermophoresis. Wetted particles tend to stick together when they collide, thereby assisting agglomeration [Reed and Das, 1988]. At the gas scrubber, the gas enters at high temperature (approximately 500°C); the gas is cooled to room temperature (approximately 25°C) before it exits the scrubber. The condensable part of the gas (water vapour) reaches dew point temperature and condenses against the charcoal and walls of the scrubber. Therefore particles will move towards these two surfaces where condensation will be taking place. The particles entering with hot gas will also move towards the cold surface in the wet scrubber enhancing their collection efficiency.

Wet scrubbing requires that the water remain in the liquid phase, which requires that the syngas be cooled to below 100°C. This loss of heat may be undesirable in some systems. Most biomass gasification systems that currently use wet scrubbers do so primarily as a means to remove tars rather than particulates from the gas stream. Removing the particulates separately can prevent condensation of the sticky tars on the particulate surface, which further prevents fouling and plugging of filter surfaces. Figure 3 shows the packed bed gas scrubber used in the System Johansson Biomass Gasifier.

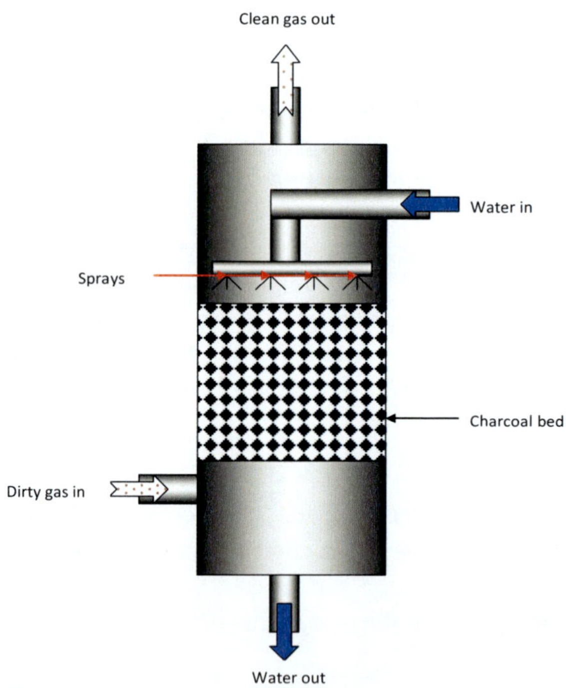

Figure 3. The packed bed scrubber.

Long life particle interference sawdust filters

After the gas cooler/scrubber, the cooled gas is passed through the long life filters, filled with particle interference sieved course sawdust filter media, where the remaining very fine non-wettable lampblack carbon dust of about 0.2g or $1g/m^3$ gas is absorbed [Johansson, 2002]. This filters the fine carbon particles through adsorption.

Engine safety filters

Before the engine, the clean gas is finally passed through a standard 5 micron double cartridge Donaldson air filter, modified as safety gas filter with much reduced flow capacity. It is provided with a special gas-tight purpose-made seal between the dust bowl and the filter body [Johansson, 2002].

Gas engine

Figure 4 shows the gas engine that is powered by syngas to drive a 150kVA synchronous generator to produce electricity.

Figure 4. The gas engine coupled to a 150kVA synchronous generator.

The classical control principle of the large power synchronous generators with excitation winding is well known [Burth *et al*, 1999, Fard *et al*, 2005, Karrari, and Menhaj, 2000, IEEE Std 1110-2002, 2003, Karrari, and Malik, 2004, Karayaka, 2003, Melgoza *et al*, 2001,

Shamsollahi and Malik , 1996, Stefopoulos *et al*, 2005, Venayagamoorthy, and Harley, 2000, and Wright, 1931], considering the frequency and voltage control by means of the active (P) and reactive power (Q) adjustment, respectively. The two control loops are usually operating separately from each other. Consequently, it may be considered as a scalar control (SC) procedure, which disregards some fundamental phenomena, i.e. the coupling effect inside the synchronous generator [Kelemen and Imecs, 1990, Kelemen and Imecs, 1993, and Kelemen and Imecs, 1989].

FACTORS INFLUENCING THE EFFICIENCY OF FIXED BED BIOMASS GASIFIER SYSTEMS

Definition of gasifier efficiency

Reactor efficiency

The average energy conversion efficiency of wood gasifiers is about 60-70% and is defined as:

$$\eta = \frac{H_g}{H_s} \tag{14}$$

where:

η is the gasification efficiency (%)
H_g is the higher heating value of syngas (MJ/m^3),
H_s is the average Higher Heating value of fuel (MJ/kg) [Rajvanshi, 1986].

If the gas is used for engine applications the gasification efficiency is defined as follows:

$$\eta_m = \frac{H_g \times Q_g}{H_s \times M_s} \times 100\% \tag{15}$$

where:

η_m = gasification efficiency (%) (Mechanical)
H_g = higher heating value of the syngas (kJ/m^3),
Q_g = volume flow of syngas (m^3/s)
H_s = lower heating value of gasifier fuel (MJ/kg)
M_s = gasifier solid fuel consumption (kg/s)

If the gas is used for direct burning, the gasification efficiency is sometimes defined as [Food and Agricultural Organization, 1986]:

$$\eta_{th} = \frac{(H_g \times Q_g) + (Q_g \times \rho_g \times C_p \times \Delta T}{H_s \times M_s} \times 100\% \quad (16)$$

where:

η_{th} = gasification efficiency (%) (thermal)
ρ_g = density of the gas (kg/m³)
C_p = specific heat of the gas (kJ/kg°C)
ΔT = temperature difference between the gas at the burner inlet and the fuel entering the gasifier (°C).

Depending on the type and design of the gasifier as well as on the characteristics of the fuel mechanical efficiency may vary between 60 and 75% whereas the thermal efficiency can be as high as 93% [Food and Agricultural Organization, 1986].

Co-generative efficiency

The electrical efficiency of the gasifier power plant depends on the efficiency of engine/ generators. This can be defined as:

$$\eta_{ep} = \frac{EP_{net}}{HHV_{bm}} \quad (17)$$

where

η_{ep}= Electrical power efficiency
EP_{net}= Net electrical power
HHV_{bm}= High heating value of biomass materials utilized.

The co-generative efficiency of the biomass power plant can be defined as:

$$\eta_{cg} = \frac{P_{thermal} + EP_{net}}{HHV_{bm}} \quad (18)$$

where:

η_{cg}= Co-generative efficiency/overall power plant efficiency
$P_{thermal}$= Thermal power [Fermeglia *et al*, 2008].

The co-generative efficiency of the power plant depends on the efficiency of the gasifier and that of the engines/generators.

Impact of gasifier/reactor design on efficiency

There are a number of gasifier designs; each gasifier design has its advantages and disadvantages. Attention will be given to fixed bed gasifiers, which are updraft, downdraft and crossdraft gasifiers.

An updraft gasifier has clearly defined zones for partial combustion/oxidation, reduction and carbonization. Figure 5 shows a diagram of the updraft gasifier. The feedstock is introduced at the top and air is introduced at the bottom and acts as countercurrent to fuel flow. The gas is drawn at the top of the gasifier. The updraft gasifier achieves the highest thermal efficiency as the hot gas passes through the fuel bed and leaves the gasifier at low temperature of around 200–300°C. The sensible heat given off by the gas is used to preheat and dry the fuel; as a result wet biomass up to 50% moisture content can still be gasified without any pre-drying. The disadvantages of this type of gasifier are excessive tar in the raw syngas and poor loading capacity. The tar can interfere negatively with the operation of internal combustion engines. This makes this gasifier not a likely candidate for power applications [Quaak *et al*, 1999, Rajvanshi, 1986 and Mckendry, 2002]. Large quantities of liquid effluents are produced in updraft gasifiers, the liquid effluent can be highly toxic and their disposal can pose environmental and health hazards. Additional study is needed on treatment options for these liquid fuels [Stassen, 1995]. The tar and condensates contains a certain proportion of energy that leaves the gasifier without converted to useful form, there is need to quantify the amount of energy leaving this type of gasifier with tar and condensates to establish its impact on the energy and mass balance of the gasifier.

The higher efficiency that the updraft gasifier type can achieve is due to low exit gas temperature. Considering the fact that some of the energy is lost during tar removal this gasifier efficiency could be overestimated when compared to the other gasifier types. The internal energy usage in the drying of fuel results in higher condensates quantities; the condensates contain chemical energy which is lost during their removal, this energy lost also needs to be quantified.

Updraft gasifiers can be scaled up while the maximum size of downdraft gasifier is probably limited to 1MWe [Mckendry, 2002] because of the uneven distribution of the gasification agent around the large gasifier throat, leading to uneven heat distribution and excessive tar production. However, for electricity generation applications, the very high level of tars in the product gas must be greatly reduced prior to the internal combustion engine to avoid problems of deposition and ultimate blockage. In the Wellman configuration studied by Brammer and Bridgewater, this is achieved by two stage cracking of the tar-laden gas, first thermally in an air-fired secondary oxidation reactor and finally in a catalytic cracker [Brammer and Bridgewater, 2002].

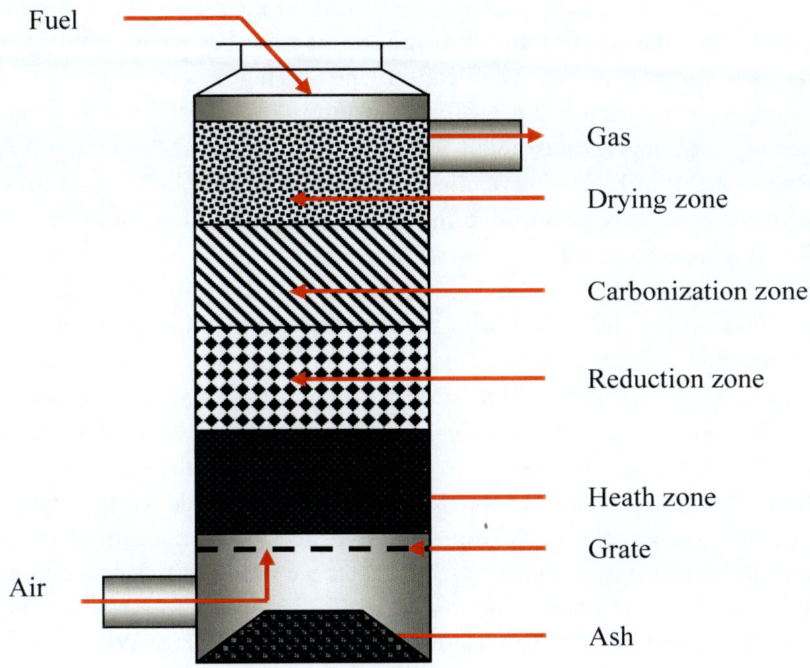

Figure 5. Updraft gasifier.

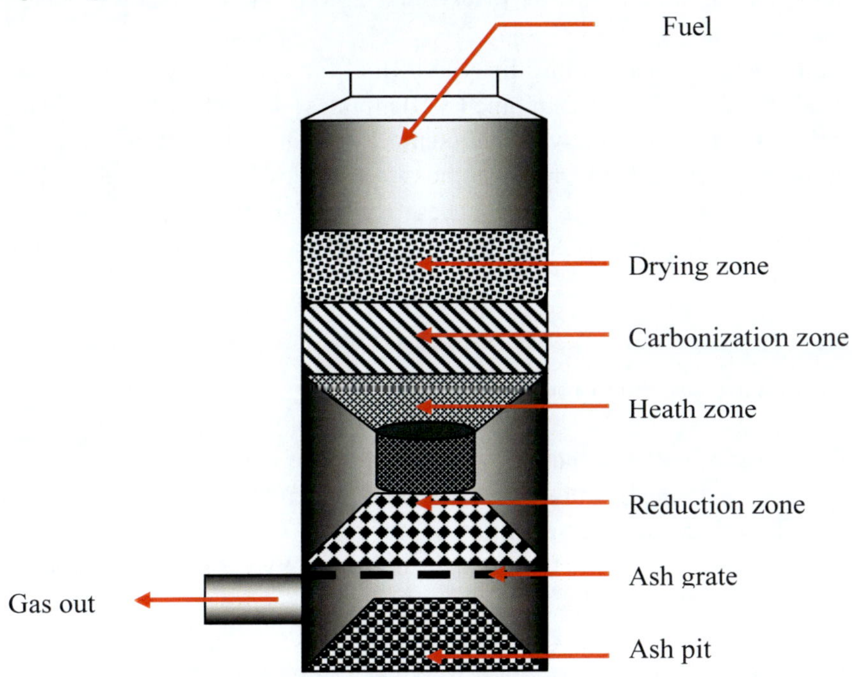

Figure 6. Downdraft gasifier.

The problem of high tar content in raw gas is minimized in downdraft gasifiers compared to updraft gasifiers. In these gasifiers, air is introduced through orifices into downward flowing packed bed or solid fuels at the oxidation zone and gas is drawn off at the bottom. The gases leave the gasifier after passing through the hot zone, enabling the partial cracking of the tars formed during gasification and giving a gas with low tar content. The disadvantage of the downdraft gasifier is a lower overall efficiency because gases leave the gasifier unit at temperatures about 900–1000°C [Food and Agricultural Organization, 1986 and Rajvanshi, 1986]. The thermal energy in the exit gas need to be extracted and utilized in downstream processes to improve the efficiency of gasification systems using this type of gasifier. Figure 6 shows the downdraft gasifier.

The downdraft gasifier also experiences difficulties in handling higher moisture contents in fuel; typically requires fuel with maximum 20% moisture content. Higher ash contents pose problems in small downdraft gasifiers. The time (20-30 minutes) needed to ignite and bring the plant to working temperature with good gas quality (4-6MJ/kg) is shorter than in updraft gas producers. This gasifier is preferred to updraft gasifier for internal combustion engines [Beeneckers, 1999] because of low levels of tar in the final gas.

In a crossdraft gasifier the feedstock moves downwards while the air is introduced from the side, the gas is withdrawn from the opposite side of the unit at the same level. A hot combustion/gasification zone forms around the entrance of the air; with the pyrolysis and drying zones being formed at the top section of the gasifier. Ash is removed at the bottom and the temperature of the gas leaving the unit is about 800–900°C; as a consequence this gives low overall energy efficiency for the process like in downdraft gasifier and a gas with high tar content like in updraft gasifier [Mckendry, 2002]. Figure 7 shows a diagram of the crossdraft gasifier.

These design characteristics limit the type of fuel for operation to low ash fuels such as wood, charcoal and coke. The startup time (5-10 minutes) is much faster than that of updraft and downdraft gasifiers. The relatively high temperature in crossdraft gasifiers has an effect on gas composition such as high carbon monoxide, and low hydrogen and methane content when dry fuels such as charcoal is used. These types of gasifiers operate well on dry fuel [Tripod, 2006], which is normally hard to find and mostly processed from wet fuel through drying.

Impact of equivalence ratio on gasifier efficiency

Equivalence Ratio (ER) is a measure of the amount of external oxygen (or air) supplied to the gasifier. ER is obtained by dividing the actual oxygen (or air) to biomass molar ratio to the stoichiometric oxygen (or air) to biomass molar ratio. Oxygen is generally supplied as a gasifying and fluidizing medium. Using air in place of oxygen though economical has the negative effect of diluting the syngas due to the presence of nitrogen. Madhukar *et al* [2007] conducted simulations to investigate the impact of ER on equilibrium composition for operating conditions of Temperature (T) = 827°C and moles of steam supplied per mole of biomass (β) =0. They established that higher ER results in reduced CO and H_2 yield while that of CO_2 increases, this is due to the oxidation of H_2 and CO to H_2O and CO_2. At low values of ER, small amounts of solid carbon ($C_{(s)}$) and CH_4 are formed in the gasifier, both of which get oxidized as more air is supplied.

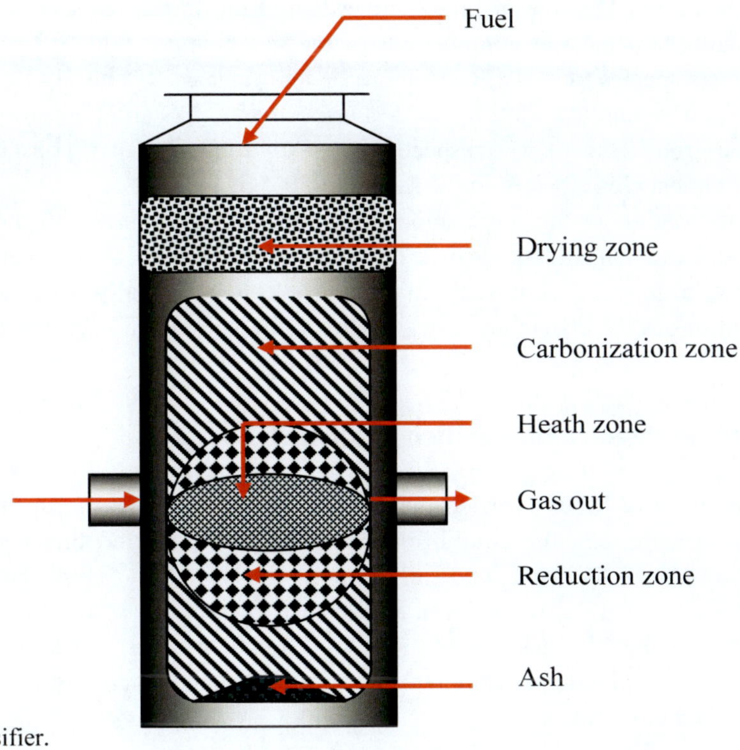

Figure 7. Crossdraft gasifier.

Higher gasification efficiencies are achieved at lower ER for fuel with higher moisture content, and higher char burnout is achieved earlier in the case of higher moisture content compared to the case of feedstock having low moisture content [Sharma, 2008]. It is not clear how this happens because a considerable amount of energy is lost in driving off the moisture in the feedstock and therefore this energy is not available for oxidation and reduction reactions. The origin of the energy consuming the char in case of fuel with high moisture content is not accounted for in the reference. There is need for further investigations of this aspect.

The heating value of the gas decreases with increasing equivalence ratio [Sharma, 2008 and van den Eden and Silva Lora, 2004] after reaching its peak at about 0.26. For fuel with higher moisture content, the heating value of the gas is higher at lower equivalence ratios, and decreases slightly at higher equivalence ratios, the heating value of the gas decreases slightly [Sharma, 2008]. Higher equivalence ratios lead to complete combustion of the feedstock and combustion of the combustible gas species in the reactor resulting in low heating value of the exit gas.

Gasification efficiency decreases at higher equivalence ratios (above 26%). This is due to the increase in temperature with an increase in equivalence ratio leading to complete combustion of the char and gas formed. Below 20% the gas heating value is reduced drastically as there is not enough air for partial combustion, therefore the temperature decreases resulting in less heat available for reduction reactions to take place. Reduction reactions are the ones that result in the major part of syngas as indicated earlier in this chapter.

Mathieu and Dubuisson [2002] investigated the impact of ER on the gas composition; they established that the composition of the gas changes with ER. The variations of the various gas species versus ER are more or less linear. N_2 and H_2O increase with ER from 35% up to 53% and from 5% up to 15% respectively. CO and H_2 decrease from 28% down to 15% and from 21% to 7% respectively. CO_2 remains fairly constant (10%) whilst CH_4 remains almost close to zero in the range 20-50% of ER.

The variation in the gas composition with ER is in line with the findings by Sharma [2008] and van den Eden and Silva Lora [2004] that the efficiency increases with ER until it reaches its peak at 26% and starts to decrease. This is basically because of the changes in gas compositions as the efficiency is dependent on the volume and composition of the gas.

Impact of pressure on gasifier efficiency

Altafini *et al* [2003] carried out simulations to establish the impact of pressure on gas compositions through the equilibrium model. They reported that the increase in pressure results in reduced hydrogen and carbon monoxide volumes. They also established that very low pressures (10.13kPa) results in an increase in the yield of H_2, however the increase was found to be neglible (less than 0.2%). Low pressure does not provide substantial improvements and high pressure reduces H_2 yield. Atmospheric pressure is the best condition for fixed bed gasifiers.

Impact of fuel properties on gasifier efficiency

A study was undertaken to establish the impact of fuel properties on gasification. Wyodak coal and cellulose (which accounts for half the weight of biomass) were used to conduct the study. A pyrolysis experiment was conducted. Nearly complete de-volatilization of cellulose was found to occur below 500°C. In contrast, only about 40% of coal was de-volatilized and only after heating to close to 900°C. The slower weight loss with coal reflects its inherently lower thermochemical reactivity and the much higher fraction of weight remaining even after heating to 900°C reflects the much lower content of volatile components in coal compared to cellulose. The remaining char was gasified [Williams and Larson, 1996].

Char gasification is one of the major processes involved in converting solid fuels into combustible gases. Because of the higher reactivity of biomass chars, these gasify much more rapidly and at lower temperatures than coal chars. Thus, lower temperatures can be used in biomass gasifiers compared to coal gasifiers to achieve the same level of char conversion to gas [Williams and Larson, 1996].

Higher ash or moisture content results in lower conversion efficiency [Faaij *et al*, 1997]. There is an increase in conversion efficiency with a decrease in moisture content. This is because a high quantity of energy is consumed during the drying process and the energy is no longer available for reduction reactions [Jayah *et al*, 2003] mentioned earlier in this chapter.

Effect of air temperature on efficiency

Air pre-heating has a positive impact on gasification efficiency, the high air inlet temperature (T_a) results in higher gasification efficiency as the sensible heat brought into the reactants induces an increase in reaction temperature (T_r). Mathieu and Dubuisson [2002] conducted an experiment to investigate the impact of air pre-heating on gasification efficiency. The reaction temperature was found to increase from 775°C up to 1025°C when air inlet temperature (T_a) increases from 25°C up to 800°C. Gasification efficiency increases with air inlet temperature significantly from 76.6% up to 79.5% when the feeding air is preheated from 25°C up to 300°C. Beyond 300°C, gasification efficiency still increases but only slightly from 79.5% to 80.1% when T_a goes up from 300°C to 825°C. The composition of the gas varies and the corresponding heating value is 5169 kJ/kg fuel when the air is at 25°C and increases up to 5402 kJ/kg with the air at 800°C [Mathieu and Dubuisson, 2002].

All these evolutions can be explained by the following conflicting trends:

- The CH_4 production from C and H_2 being exothermic is decreased when T_r and hence T_a increase;
- The consumption of CH_4 in endothermic reactions with H_2O and CO_2 is increased when T_r and hence Ta increase;
- The shift reaction $CO + H_2O \rightleftharpoons CO_2 + H_2$ being exothermic, water and CO consumption decrease when the temperature increases;
- The Boudouard reaction is endothermic, CO production increases at the expense of carbon and CO_2 when the temperature increases [Mathieu and Dubuisson, 2002].

High temperature air increases gasification yields leading to high gasification efficiencies. Zubtsov *et al* [2001] conducted an experiment with air pre-heated to 1000°C when gasifying Skyline coal, the resulting syngas was found to have a heating value of about 1400Kcal/m^3; without the preheated air, only 850Kcal/m^3 could be achieved. When ambient temperature air was used, the resulting low combustion temperature would prevent the reactions from reaching completion resulting in low heating value gases and low conversion efficiency [Zubtsov *et al*, 2001]; the same applies to gasification of biomass materials as indicated by Mathieu and Dubuisson [2002].

The gasification temperature not only affects the product yield but also governs the process energy input. High gasification temperature produces a gas mixture rich in H_2 and CO with small amounts of CH_4 and higher hydrocarbons. At low temperatures, solid carbon ($C_{(s)}$) and CH_4 are present in the syngas. Solid carbon is carried away and is deposited on the downstream processes. It is necessary to ensure that the syngas is free of any solid carbon. As temperature increases, both carbon and methane are reformed. At about 727°C both are reduced to very small amounts and in the process get converted into CO and H_2. This explains the increase in hydrogen moles from 627°C to 757°C. At about 757°C, the H_2 yield reaches a maximum value of about 1.33 mol. At still higher temperatures, the H_2 yield starts reducing. This is attributed to the water–gas shift (WGS) reaction (equation 2.4). According to Le-Chatelier's principle, high temperature favors reactants in an exothermic reaction thus explaining the increase in CO and reduction in H_2 (and CO_2 yield) at higher temperature.

Hence, gasification temperature of about 757°C gives the highest equilibrium hydrogen yield with negligible solid carbon in the product gas [Madhukar *et al,* 2007].

Conclusion

The fixed bed downdraft biomass gasifiers are preferred in electricity generation because of their production of gas that has low tar content compared to their updraft and crossdraft counterparts. However the main problem with fixed bed downdraft gasifier systems is their low conversion efficiency due to their low charcoal burnout rate. The other challenge facing these gasifier types is the fact that they are limited in size; they cannot be scaled up for large scale project applications. The system Johansson Biomass gasifier is a downdraft gasifier system that can be applied in small scale for electricity generation purposes because of its production of tar-free gas. It can also be applied at medium scale in modular form.

References

Altafini C. R. Paulo R. Wander and Ronaldo Barreto, M. Prediction of the working parameters of a wood waste gasifier through an equilibrium model, Energy Conversion and Management 44 (17), (2003), pp. 2763-2777.

Barrio, M. Fossum, M. Hustad J. E. (2007) A small-scale stratified downdraft gasifier coupled to a gas engine for combined heat and power production. Norwegian University of Science and Technology, Department of Thermal Energy and Hydro Power, 7491 Trondheim, Norway Sintef Energy Research, Department of Thermal Energy, 7465 Trondheim, Norway.

http://www.tev.ntnu.no/MariaBarrio/publikasjoner/article%20gasifier.pdf, Last accessed February 2007.

Beeneckers A. A. C. M. Biomass gasification in moving beds, a review of European technologies, Renewable energy 16 (1999), pp. 1180-1186.

Brammer J. G. and Bridgwater A. V. The influence of feedstock drying on the performance and economics of a biomass gasifier engine CHP system, Biomass and Bioenergy 22 (2002) (4), pp. 271-281.

Burth, M. Verghese G. C. and Velez-Reyes, M. Subset selection for improved parameter estimation in on-line identification of a synchronous generator, *IEEE Trans. Power Syst.* 14 (1) (1999), pp. 218–225.

Carbo Consult, 2008, http://www.carboconsult.com/fuel-compartment.asp, Last accessed October 2008.

Cooper, C. D. and Alley, F. C. (1986) Air Pollution Control: A Design Approach. Prospect Heights, Ill.: Waveland Press.

Faaij, A, van Ree, R. Waldheim, L. Olsson E. Oudhuis A. F. van Wijk, AD. Daey-ouwensll, C. and Turkenburg, W. Biomass and Bioenergy 12 (6) (1997), pp. 387-417.

Fard, R. D. Karrari, M. and Malik, O. P. Synchronous generator model identification for control application using Volterra series, *IEEE Trans. Energy Convers.* 20 (4) (2005), pp. 852–858

Food and Agricultural Organization. (1986) Wood gas as engine fuels. FAO forestry papers. http://www.fao.org/documents/show_cdr.asp?url_file=DOCREP/TO512eOb.htm. Last accessed May 2006.

Food and Agricultural Organization, FAO Forestry paper-72. (1986) Wood gas engine fuel, http://www.fao.org/documents/show_cdr.asp?url_file=/docrep/t0512e/T0512e09.htm, last visited, 29/05/2006.

Fermeglia, M. De Simon, G. Donald G. Energy production from biomass gasification by molten carbon fuel cells process simulation, proceedings of the International conference on Hydrogen Age, Palermo, Italy, 18-20 October 2005.

Gradon, L. Luckner, H. Podgorski, A. Wertejuk, Z. Separation of neutral and charged aerosol particles in cyclones with external electric field, Journal of Aerosol Science 29 (1998), pp S927-S928.

IEEE Std 1110-2002, IEEE Guide for Synchronous Generator Modeling Practices and Applications in Power System Stability Analysis, 2003, pp. 1–72.

Jayah, T. H. Lu, A. Fuller, R. J and Stewart, D. F. Computer simulation of a downdraft wood gasifier for tea drying, Biomass and Bioenergy 25 (4) (2003) 459-469.

Johansson, K. G. (2002) System Johansson gas producers: Brief description of wood gas producer system for tar free continuous gas generation. Midrand, South Africa. Unpublished document.

Johansson, K. G. (2002) (b) System Johansson gas producers: General description and operator's instructions. Midrand, South Africa. Unpublished manual.

Karrari, M. and Menhaj, M. B. Application of different neural networks for identification of power systems, *UKACC Conference on Control 2000* University of Cambridge, UK (2000).

Karrari, M. and Malik, O.P. Identification of physical parameters of a synchronous generator from on-line measurements, *IEEE Trans. Energy Convers.* 19 (2) (2004), pp. 407–415.

Karrari, M. and Malik, O. P. Identification of Heffron–Philips model parameters for synchronous generators using online measurements, *IEE Proc. Gener. Transm. Distrib.* 151 (3) (2004), pp. 313–320.

Karayaka, H. B. Keyhani, A. Heydt, G. T Agraval B. L. and Selin, D. A. Synchronous generator model identification and parameter estimation from operating data, *IEEE Trans. Energy Convers.* 18 (1) (2003), pp. 121–126

Kelemen Á, Imecs M: (1993) Vector Control of AC Drives, Vol. 2: Vector Control of Synchronous Machine Drives, Ecriture Publisher, Budapest.

Kelemen, Á. Imecs M. Vector control system for frequency and voltage of synchronous generators, Proceedings of International Conference on Electrical Machines, ICEM '90, (1990), Cambridge, Massachusetts, USA.

Kelemen, Á. Imecs M. Procedure and Driving Device for Unitary Automatic Vector Control of the Active and Reactive Power of Synchronous Generators by Means of Frequency and Voltage Regulation (in Romanian), Patent of Invention Nr. 104278/30.10.1989, Romania.

Kerekezi, S. and Ranja, T. (1997) Renewable energy technologies in Africa, Zed Books ltd. London, UK.

Madhukar, R. Mahishi, Y. and Goswami, D. Thermodynamic optimization of biomass gasifier for hydrogen production, International Journal of Hydrogen Energy 32 (16) (2007), pp. 3831-3840.

Mathieu, P. Dubuisson, R. Performance analysis of a biomass gasifier. Energy Conversion and Management 43 (2002), pp. 1291–1299.

Mckendry, P. Energy production from biomass (part3): gasification technologies, Bioresource Technology 83 (2002) (1), pp. 55-63.

Quaak, P. Knoef, H. and Stassen, H (1999) Energy from Biomass: A review of Combustion and Gasification Technologies, The World Bank, Washington DC, USA.

Rajvanshi A. K. (1986) Biomass Gasification., Published as chapter 4 in book: Goswami D.Y., Ed (1986), Alternative Energy in Agriculture Vol. II., CRC Press, pp 83-102. http://nariphaltan.virtualave.net/gasbook.pdf, last visited 11/11/ 2008.

Reed, T. B. and Das, A., (1988) Handbook of Biomass Downdraft Gasifier Engine Systems, SERI/SP-271-3022, UC Category: 245, Solar Energy Research Institute, 1617 Cole Boulevard, Golden, Colorado 80401-3393.

Stefopoulos, G. K. Georgilakis, P. S. Hatziargyriou N. D. and Sakis A. P. Meliopoulos, A genetic algorithm solution to the governor-turbine dynamic model identification in multi-machine power systems, *IEEE Conference on Control Application* Toronto, Canada (2005), pp. 245–250.

Stassen, H. E. (1995) Small-Scale Biomass Gasifiers for Heat and Power: A Global Review, World Bank Technical Paper No. 296, Energy series, World Bank, Washington DC, USA.

Sharma, A. K., Equilibrium and kinetic modeling of char reduction reactions in a downdraft biomass gasifier: A comparison, Solar Energy 82 (2008), pp 918–928.

Sharma, A. Kr. Equilibrium modeling of global reduction reactions for a downdraft (biomass) gasifier. Energy Conversion and Management 49 (4) (2008), pp. 832-842.

Tripod, Biomass Gasification: Technology and utilization, http://members.tripod.com/cturare/his.htm, last visited, 29/05/2006.

van de Beld Ir. L, (2004) Flash pyrolysis. Biomass Technology Group. http://www.btgworld.com/technologies/pyrolysis.html, Last accessed on December 2006.

van den Enden, P. J. Silva Lora, E. B. Design approach for a biomass fed fluidized bed gasifier using the simulation software CSFB. Biomass and Bioenergy 26 (2004), pp. 281 – 287.

Williams, R.H. and Larson E. D. Biomass gasifier gas turbine power generating technology (1996) Biomass and Bioenergy 10 (2-3) (1996), pp. 149-166.

Wright, S. H. Determination of synchronous machine constants by test, *AIEE Trans.* 50 (1931), pp. 1331–1351.

Zubtsov, V.M., Pian, C.C.P., and Yoshikawa, K. Potential applications of high-temperature air/steam blown gasification and pyrolysis systems. Energy 30 (2005), pp. 2229-2242.

In: Gasification: Chemistry, Processes and Applications
Editor: Michael D. Baker
ISBN: 978-1-61209-681-0

Chapter 11

STUDY OF THE CATALYTIC EFFECT OF MAGNETITE IN THE GASIFICATION OF CHARS FROM THE LIGNITE/RUBBER MIXTURES

Pavel Straka*
Institute of Rock Structure and Mechanics, v.v.i, Academy of Sciences of the Czech Republic, V Holešovičkách, Praha, Czech Republic

INTRODUCTION

The utilization of waste rubber is still a topical issue. It is proven by a number of studies devoted to its treatment, especially thermal [1,2], to the process conditions [3], the products [4,5] as well as a description of the behavior of this material in pyrolysis [6,7] and gasification [8], and kinetics of reaction of the rubber char with CO_2 [9]. A realistic solution could be the co-gasification of waste rubber with lignite in the Integrated Gasification Combined Cycle. Recent experiments on the laboratory but predominantly commercial scale [3] have demonstrated that waste rubber can without any technological difficulties thus be gasified at a content of 10% in a gasified mixture, in the short term at 20%. Higher content however causes an increase in the unburned carbon, whose content in ash grows significantly with the rising share of rubber in the mixture, as demonstrated in Fig. 1. The reason is apparently the significant difference between the reactivities of the lignite and rubber chars in the gasification zone of the generator, where the reactivity of the rubber char is low in comparison with the reactivity of the lignite char. The removal of the negative phenomenon of the unburned carbon would contribute to the more effective treatment of waste rubber, avert the losses of energy inserted into the process, improve the quality of the ash for its further use and facilitate the maintenance of the surface of the disposal site of the generator ash. From the environmental perspective, the possibility to treat greater volumes of waste rubber (e.g. waste-tire) than before is significant, whereas from the economic perspective it is important to avert energy losses. Our operational experiments on the Lurgi gasifiers (Sokolov

* E-mail address: straka@irsm.cas.cz

Coal A.S., Czech Rep.) have shown that if the content of the unburned carbon in the generator ash is for example 10.5%, then the loss of energy is 1.26% of the higher heating value introduced in the system at the input of the generator and that at a content of unburned carbon of 28.4% is already 4.39%, which are high to extreme losses. The question thus is how to lower the unburned carbon at higher shares of rubber in the gasified mixtures to an acceptable level or to the level reached at the gasification of the lignite alone. A possible solution is to lower the size of the particles of the treated rubber, e.g. from 40 mm to 15-20 mm (in the moving-bed gasifiers), and thus to increase the surface of the gasified particles, but this solution is not essential.

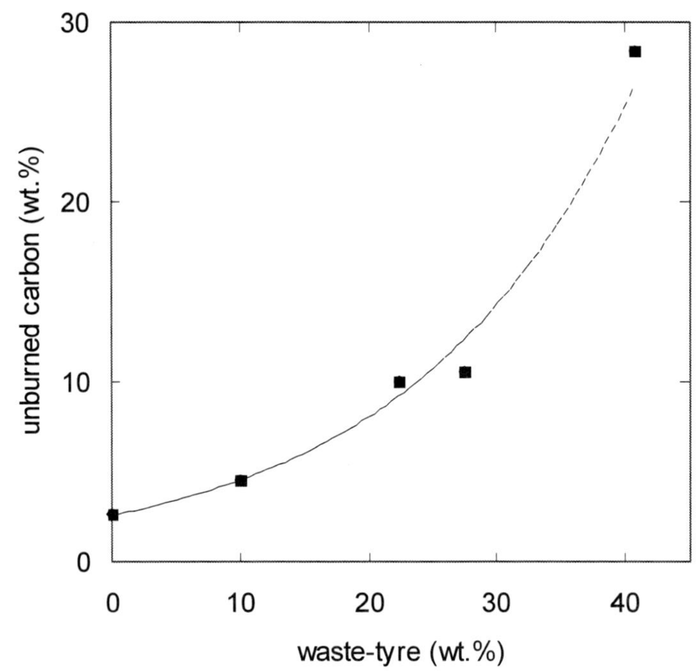

Fig. 1. The dependence of unburned carbon content in the ash on waste-tire share in the gasified mixture [3].

Another approach is to find the effective catalyst of the gasification reaction. Alkali metal salts, especially potassium salts, including those presented in a high content potassium biomass, are considered as effective catalysts for gasification by steam and CO_2 [10]. The most recently tested catalyst requires coal modified by a solvent extraction (HyperCoal), because this catalyst reacts with the mineral components of the coal [11]. It is, however, an unfeasible variant for the given case. According to our experience with the Lurgi gasification process, it is possible to consider iron-based catalysts, which are effective, available and industrially used. For our experiments as model Fe catalyst a finely ground magnetite was selected for the laboratory tests. The experiments were conceived so that the chars from lignite and rubber without magnetite and with magnetite were prepared first in a rotary kiln. The comparative chars from rubber alone, lignite/polymers, lignite/celluloses and lignite/wood mixtures, and lignite alone were prepared in a similar way. They served for the comparison of their reactivity with the reactivity of the lignite/rubber chars and further for the more general determination of the impact of magnetite on the rate of the gasification reactions. The prepared materials were then gasified by steam and CO_2 and the rates of gasification measured at isothermal conditions. With lignite/rubber chars, the effective activation energy and pre-exponential factor of the gasifying reaction with CO_2 were further evaluated using non-isothermal isoconversional kinetics, and these parameters were compared for the samples with and without magnetite.

The objective of the treatise is to describe the effect of magnetite on the rate of the gasification of the chars from the mixture of lignite and rubber, compare it with the effect determined with comparative chars and explain the impact of magnetite on the basis of non-isothermal isoconversional kinetics.

MATERIALS AND METHODS

Materials

Chars from the mixtures of lignite with waste rubber, polymers, celluloses and waste wood were used for the experiments. The lignite came from the Jiri Mine, the Sokolov Basin, Czech Rep., the waste rubber was supplied by Pneucentrum Tasy S.R.O., Czech Rep. Analyses of these key materials are shown in Tab. 1. Further, the acrylonitrile butadiene styrene (ABS) polymer, polypropylene and polyethylene, Nylon 6, methyl cellulose and cellulose (Sigma Aldrich) were used. As catalyst, the finely ground magnetite (under 5 microns, 98%, Sigma Aldrich) was used.

Table 1: Analyses of the Jiri lignite and waste rubber used. O_d-by difference

Sample	Proximate analysis			Ultimate analysis (d.a.f.)				
	Water	Ash (d.b.)	VM (d.a.f.)	C	H	N	S_o	O_d
Jiri lignite	5.55	7.64	50.72	75.03	4.81	1.22	0.43	18.50
Rubber	1.20	5.11	68.85	87.70	7.87	0.62	1.37	2.44

The preparation of the chars was conducted in a tilted rotary kiln with a retort of stainless steel with a length of 690 mm, with an inner diameter of 60 mm. The heating of the kiln was provided by three independently regulated heating sections, each with a wattage of 1 kW. The retort rotation speed was 5 rpm with a retort inclination of 3°, fed by a screw feeder at about 100-150 $g.h^{-1}$. At a temperature of 900°C, 200 g of the sample was always pyrolyzed under slight overpressure of own atmosphere to avoid the oxidation by the external air. The chars from lignite alone and rubber alone, from a mixture of lignite with waste rubber (1:1) and a similar mixture with magnetite (45% coal, 50% rubber, 5% magnetite) and further from mixtures of lignite with polymers, celluloses and waste wood were prepared in this way. Admixtures and their contents in the mixtures were as follows: ABS polymer-30%, polypropylene-30%, Nylon-30%, a mixture of polyethylene with polypropylene (1:1)-30%; further cellulose-30%, methylcellulose-30%, and wood-50%. The yield of char was always around 50%. The analyses of the key chars, i.e. those from the lignite/rubber (1:1) mixture and the similar mixture with 5% magnetite, are provided in Tab. 2, of the others in Tab. 3 (comparative chars).

Table 2: Analyses of the chars from the mixture of lignite with rubber (1:1) and a similar mixture with magnetite (45% lignite, 50% rubber, 5% magnetite)

Char from:	Proximate analysis			Ultimate analysis (dry basis)				
	Water	Ash (db)	VM (daf)	C	H	N	S	O
Lignite/rubber	3.06	11.11	3.79	84.23	0.93	0.68	1.23	1.82
Lignite/rubber with magnetite	4.85	15.60	-	80.03	0.85	0.70	0.94	1.88

Table 3: Proximate analyses of the comparative chars from rubber alone, the Jiri lignite alone, the mixtures of the Jiri lignite with polymers (30%), celluloses (30%), and wood (50%). ABS-acrylonitrile butadiene styrene, PP-polypropylene, PE-polyethylene

Char (initial admixture)	Proximate analysis		
	Water	Ash (db)	VM (daf)
Rubber (100%)	1.49	13.26	0.91
ABS (30%)	2.91	17.42	3.89
PP (30%)	2.26	14.95	4.91
PP+PE (1:1) (30%)	2.93	15.37	3.14
Nylon 6 (30%)	3.09	13.95	7.30
Cellulose (30%)	2.17	15.58	2.82
Methylcellulose (30%)	4.73	14.75	6.66
Wood (50%)	1.69	15.70	3.27
Jiri lignite (100%)	4.30	15.18	4.19

Methods

With chars, the gasification rate by steam and CO_2 was measured in the isothermal mode at temperatures of 800 and 900°C. The rate of gasification was characterized by the percentage mass loss per minute. The measurement was conducted on the MOM thermal analyzer (MOM Comp., Budapest, Hungary) with electronics and software from TERI S.R.O., Prague, Czech Republic, adapted for a sample mass of 50-100 mg. The samples with a grain size of around 0.2 mm were heated in the inert atmosphere of nitrogen to temperature of 800 or 900°C and after reaching it were dosed with 250 ml.min^{-1} of CO_2 or steam. From the moment of this dosage, the relative decrease of the mass of the sample was measured for the period of the linear decrease of the thermogravimetic curve. The mass of the sample at the moment of the beginning of the dosage of the gasification medium was taken as 100%. For measurements, it was taken into account that the gasification rate, which is the measure of reactivity, must be measured and evaluated at the initial phase of the gasification reaction, because the reactivity of the char decreases with the continuing course of the reaction. If we want to compare the reactivity of various chars, it is necessary to compare them at the initial stages of the reaction.

Since the reduction of the effective activation energy of the gasification reaction through the activity of the catalyst is important for the assessment of the catalyst's efficiency, samples of lignite/rubber without magnetite and with magnetite were further examined. Experimental data were obtained by thermogravimetric method. The samples of 10-12 mg with the grain size around 0.2 mm were tested; CO_2 with a flow rate of 12 $ml.min^{-1}$ was used as the gasification medium. The thermogravimetric curves of the samples with the heating rates of 7, 10 and 12 $K.min^{-1}$ in the temperature range of 500–1100°C were obtained, and the courses of the effective activation energy and of the pre-exponential factor for the conversion of 10–90 % were evaluated. For the calculations of the effective activation energy and the pre-exponential factor, the method of non-isothermal isoconversional analysis was selected. The thermogravimetric runs were procured on the SETARAM Setsys Evolution 18 analyzer; the AKTS Thermokinetics software, version 2.56, was used for the evaluation of the mentioned kinetic parameters.

The Raman spectra of the chars were measured by ***Jobin Yvon LabRam HR Raman d***ispersion spectrometer equipped with an Olympus confocal microscope. A laser with a wavelength of 532.2 nm and an input wattage of 50 mW served as the excitation source. The samples were measured because of the possible thermal destruction of the sample at a wattage of 5 mW, with the measurement time being 10 sec and with 30 spectra accumulations. A grating with 600 lines/mm, a split of 100, and a hole of 1000 was used. A multichannel air-cooled CCD camera served as the detector. A sample was applied without preliminary modifications on a shiny steel plate and measured with a microscope at a magnification of 50x. The measuring trace had a diameter of ca 2 micrometers. The separation of the bands in the spectrum was conducted with the Omnic 7.3 program. The spectra obtained in the region of 800-1800 cm^{-1} were separated into four bands. The G and D bands [12] were fitted using the Lorentzian function while for the other bands a Gaussian function was used.

Results and Discussion

Firstly, with the comparative chars, the rate of gasification (w) by steam and CO_2 was measured in the isothermal mode at temperatures of 800 and 900°C. These measurements served to assess an influence of polymer admixtures of lignite in the gasification rate of resulting chars. The results of the measurements are summarized in Tab. 4. From this table, the difference in the gasification rates of the char from lignite alone and from rubber alone is evident. It further arises from the data in Tab. 4 that the admixtures reduce the reactivity of the chars at 800°C in comparison with the char from lignite alone. An exception was the mixture of polyethylene with polypropylene (1:1), which on the other hand increased the reactivity of the resulting char, although only slightly in the case of gasification with CO_2.

At 900°C, the rates of gasification both by steam and CO_2 were quite similar, with the exception of the char from rubber alone. An explanation might be provided by the Raman spectra. The chars contain mainly relatively high percentages of disordered carbon structures, which is reflected in the Raman spectra as a wide band at a wavelength of 1350 cm^{-1} (disorder band, D-band), whereas the band at 1590 cm^{-1} (graphitic band, G-band) is narrower (Figs. 2

and 3). By analyzing the Raman spectra taken it was found that a) within the comparative chars the Raman spectrum of rubber differed significantly from the other chars, b) after deconvolution, in all the cases, spectra provided four bands belonging to different forms of carbon. It is obvious that the content of the various types of carbon structures affect the reactivity of chars [12]. The question is what makes the char from rubber so different from the others. An at least approximate explanation can be provided by the quantitative evaluation of Raman spectra of chars after deconvolution (Tab. 5). The chars mostly contain relatively

Table 4: The gasification rate (*w*) of the comparative chars by steam and CO_2 at temperatures of 800 and 900°C. See Tab. 3 for symbols

Char (admixture)	*w* (wt.%.min^{-1}) at 800°C		*w* (wt.%.min^{-1}) at 900°C	
	H_2O(g)	CO_2	H_2O(g)	CO_2
Rubber (100%)	2.48	0.33	4.06	1.36
ABS (30%)	4.49	0.34	16.10	6.02
PP (30%)	4.85	1.05	19.01	5.11
Nylon 6 (30%)	5.54	1.17	18.07	5.46
Methylcelulosis (30%)	5.64	1.22	14.95	4.71
Wood (50%)	5.98	1.30	19.86	-
Celulosis (30%)	7.68	1.34	19.28	6.60
Jiri lignite (100%)	8.01	1.44	18.00	5.67
PE+PP 1:1 (30%)	9.45	1.48	18.36	5.30

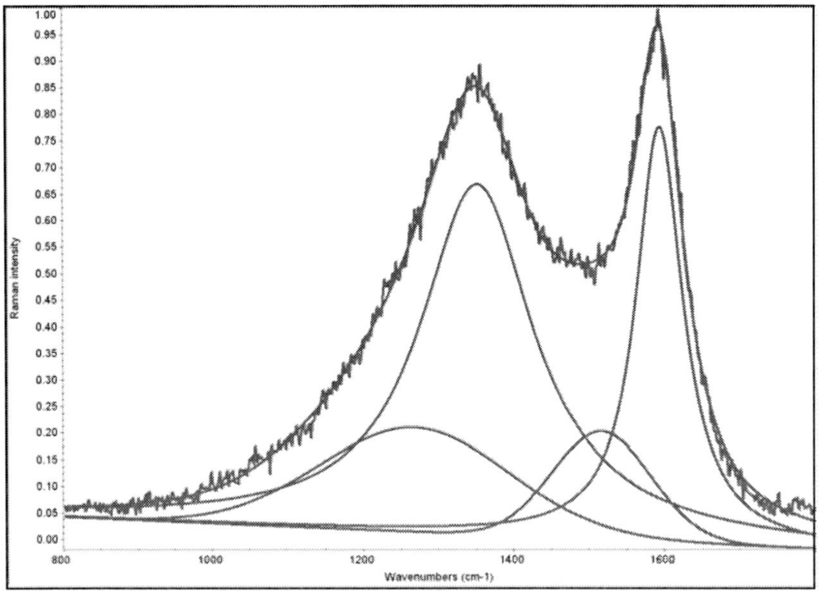

Fig. 2. The Raman spectrum of char from the rubber alone and its deconvolution.

high shares of disordered carbon structures (D-values, Tab. 5), whereas shares of the ordered structures (G-values, Tab. 5) are lower. It can be expected with the carbonaceous structures that the disordered structures will be more reactive than the ordered. The ratio of the areas of these bands (D/G) was in the chars tested of 3-3.4, with the exception of the char from rubber alone, in whose case the ratio was only 2. It means that the content of the disordered, more reactive structures was in this case significantly lower and the reactivity at steam as well as CO_2 was thus also lower in comparison with the other chars. The problem is that the more ordered, less reactive structures are still present but at an amount comparable to or even lower than the amount found with the other chars prepared from mixtures with lignite or from lignite alone, which arises from the values of the G-bands areas in Tab. 5. It is clear that the low reactivity found must be caused by yet another factor. After deconvolution, the Raman spectrum in the case of rubber significantly differs from the other spectra by the area of the band at 1100-1270 cm^{-1} (Figs. 2 and 3, Tab. 5). This band belongs to sp3 bonds, which were observed in high amounts only with the char from rubber alone. It seems that what occurs during the heating of rubber and the creation of char is not only the cracking of bonds but also the formation of new, cross-link bonds, originating from the cracking of the double bonds and from the bonds which replace the cross-link bonds of sulfur (sulfur bridges) from vulcanization. The number of sp3 bonds thus increases during heating and the rigid structures are created with cross-link bound molecular chains, whose sp3 bonds are manifested in the Raman spectra with a relatively large area of the band at 1100-1270 cm^{-1}. Thanks to these interconnected and ordered structures, the reactivity of the char from rubber is substantially lower than that of the char from lignite alone and the other chars (Tab. 4).

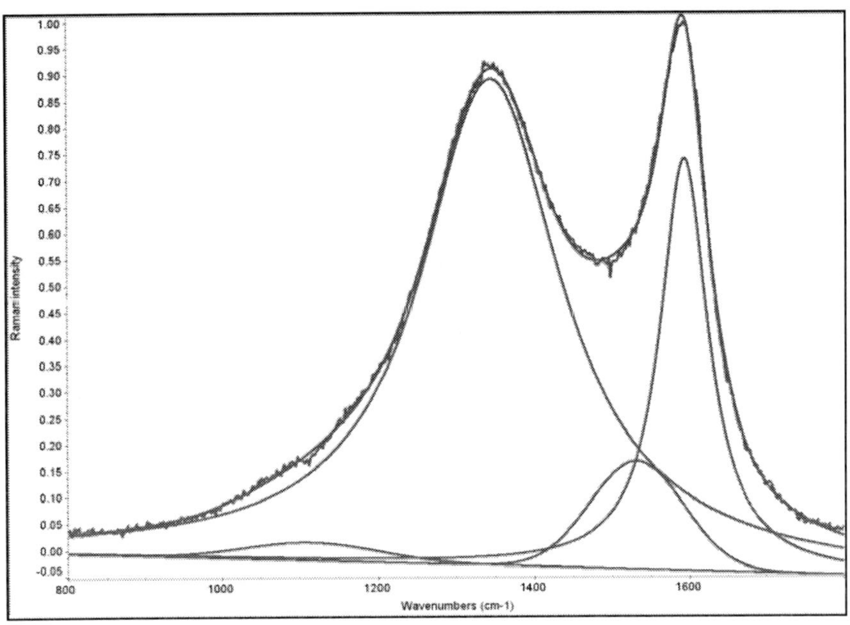

Fig. 3. The Raman spectrum of char from the lignite alone and its deconvolution.

Table 5: The areas of the bands from the Raman spectra of the chars. D-band (disorder-band), G-band (graphite band), sp3 and amorphous carbon bands

Char (admixture)	D ($1350\ cm^{-1}$)	G ($1590\ cm^{-1}$)	D/G	Band sp3 ($1100\text{-}1270\ cm^{-1}$)	Amorphous C ($1520\text{-}1560\ cm^{-1}$)
Rubber (100%)	188.61	91.90	2.05	61.46	33.42
Rubber (50%))	319.05	96.69	3.30	16.64	27.92
PP(30%)	311.60	102.10	3.05	8.10	28.88
Nylon 6 (30%)	311.16	103.71	3.00	5.13	21.05
Wood (50%)	247.27	88.08	2.81	18.96	22.53
Celulosis (30%)	312.46	93.45	3.34	10.19	31.86
Jiri lignite (100%)	313.30	92.73	3.38	7.36	31.04

Another band in the Raman spectra in the region of 800-1800 cm^{-1} was observed after deconvolution at 1520-1560 cm^{-1} and was attributed to amorphous carbon. This form of carbon is created generally in the pyrolysis of coal and carbon-containing materials and in the given case its content in the observed chars was quite similar as is evidenced by the areas of the bands shown in Tab. 5. The amorphous carbon has thus influenced the reactivity of the observed chars roughly in the same way. Of course, the content of this form of carbon was rather higher with the rubber char as compared with the others as amorphous carbon is a part of produced rubber.

Secondly, the question of activation energy was investigated. For the assessment of the efficiency of the catalyst, the reduction of the activation energy of the gasification reaction through the catalyst's activity is significant. In principle, chars undergo thermo-oxidative degradation. Thus, the effective activation energy can be calculated from the data of thermogravimetric runs by isoconversional kinetic analysis which enables evaluating a dependence of the effective activation energy (E_α) on reaction progress (extent of conversion, α) particularly for the thermally stimulated processes, whose kinetics are multi-step and/or non-Arrhenius [13,14]. Xi et al [15] used isoconversional analysis to express a catalytic effect of metal sulfates (e.g. $Fe_2(SO_4)$) on termo-oxidative degradation of poly(methyl methacrylate) on the basis of dependence of E_α on α. For virgin polymer the E_α values changed around 250 $kJ.mol^{-1}$, but in the presence of metal sulfates the E_α values decreased with α to 140-150 $kJ.mol^{-1}$. In our case, analogous procedure was used. The results achieved with CO_2 are clear from Figs. 4-7. It arises from Figs. 4 and 6 that without magnetite the values of the effective activation energy varied in the range of approximately 370-470 $kJ.mol^{-1}$ for α 10-90% (Fig. 4, simplified curve by polynomial regression), but with magnetite they dropped to values of approximately 260-270 $kJ.mol^{-1}$ in the same range of α (Fig. 6, simplified curve). The values of the pre-exponential factor (lnA) were 30-40 sec^{-1} for α 10-90% in the case without magnetite (Fig. 5, simplified curve), and 20-22 sec^{-1} when it was used (Fig. 7, simplified curve). These dependences show that a) E_α and pre-exponential factor varied with α, therefore, gasification of chars is a complex (multi-step) process as independence of E_α and pre-exponential factor on α is a sign of a single-step process, b) the decrease of E_α from 370-470 $kJ.mol^{-1}$ up to 260-270 $kJ.mol^{-1}$ indicates a significant catalytic effect of magnetite.

The above-mentioned characteristics of the order of the carbon structures derived from the Raman spectra and the isothermal measurement of the gasification rate of various chars show that the increased unburned carbon in the co-gasification of waste rubber is caused by the organized and interconnected structures of carbon, which resist the reaction with steam and CO_2. On the other hand, the catalyst works by reducing the activation energy. Therefore, finally, the mixture of lignite with rubber in a high content of 60% and a 5%-addition of magnetite catalyst was tested and the influence of the catalyst on the increase of the gasification rate was monitored.

The measurement of the gasification rate of the chars from the mentioned lignite/rubber mixture under isothermal conditions has shown that in the presence of magnetite the gasification reactions significantly accelerate both in the case of gasification by steam and when using CO_2. It is clearly demonstrated in Tab. 6, from which it arises that the rate of the gasification of the chars with magnetite is two to three times greater in comparison with the rate achieved in the gasification of chars without it. Similarly in gasification of the chosen comparative chars by CO_2 at 900°C, the rate when using a 5%-addition of a magnetite increased by almost two times. Particularly char from the mixture of lignite with polypropylene and polyethylene (1:1) was tested, where the rate of gasification in the presence of magnetite increased by more than two times (Tab. 7).

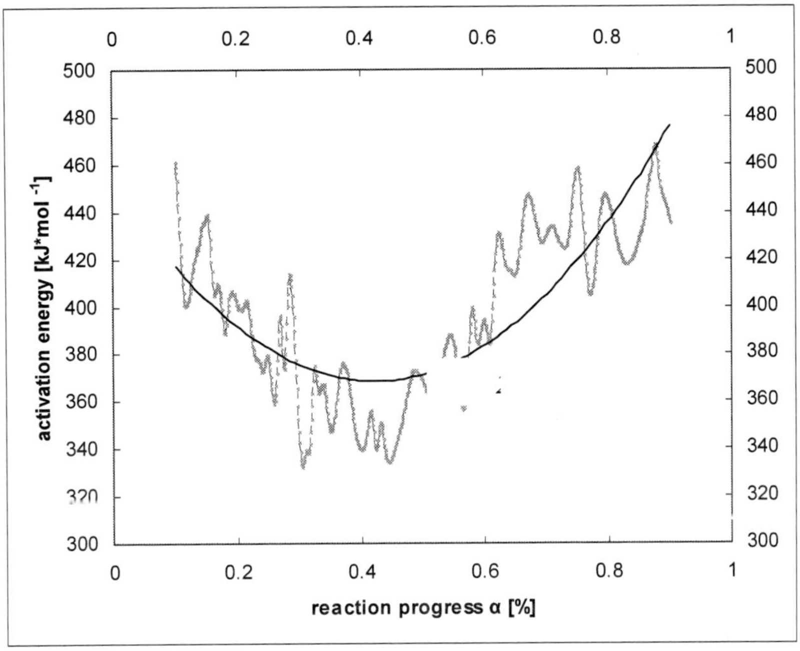

Fig. 4. The course of the effective activation energy of the gasification of char from the lignite/rubber (1:1) mixture by CO_2 for conversion of 10-90 %. Line *1*-calculated data, line *2*-simplified curve.

All of the findings listed so far show a favorable influence of the discussed catalyst on the rate of the gasification reactions and thus also a reduction of unburned carbon. Using this catalyst the gasification rates of chars in question are fully comparable with those obtained with char from lignite alone (Tab. 4, Tabs. 6 a 7).

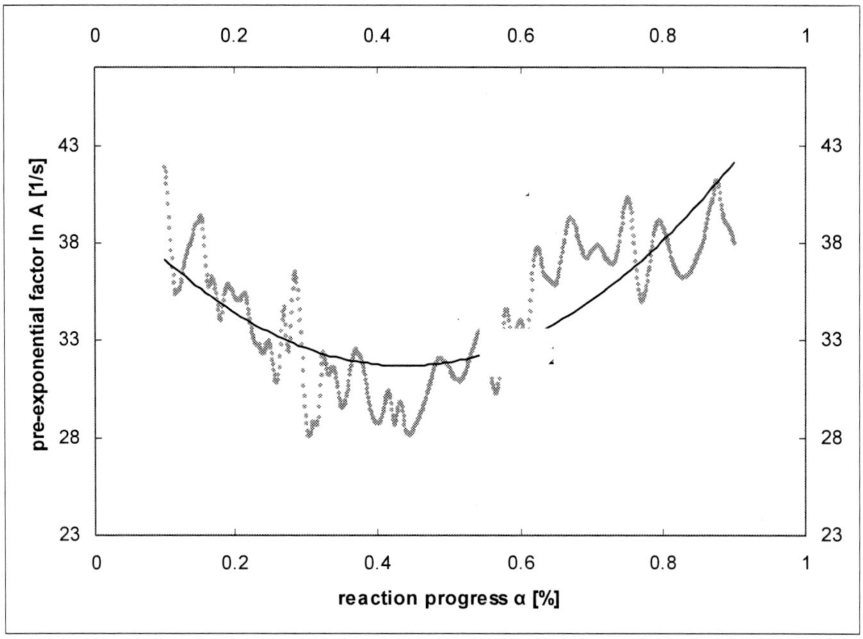

Fig. 5. The course of the pre-exponential factor of the gasification of char from the lignite/rubber (1:1) mixture by CO_2 for conversion of 10-90%. Line *1*-calculated data, line *2*-simplified curve.

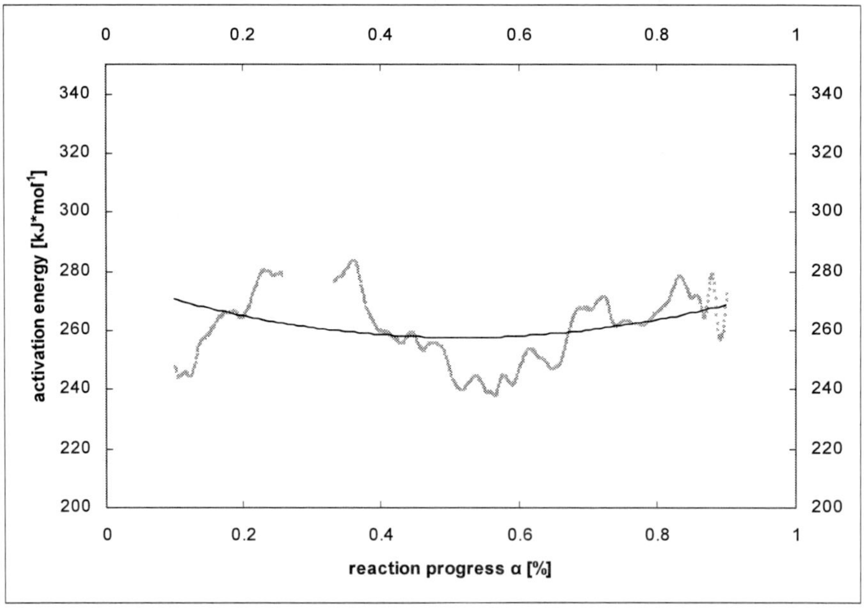

Fig. 6. The course of the effective activation energy of the gasification of char from the lignite/rubber (45:50) mixture using a 5%-addition of magnetite by CO_2 for conversion of 10-90 %. Line *1*-calculated data, line *2*-simplified curve.

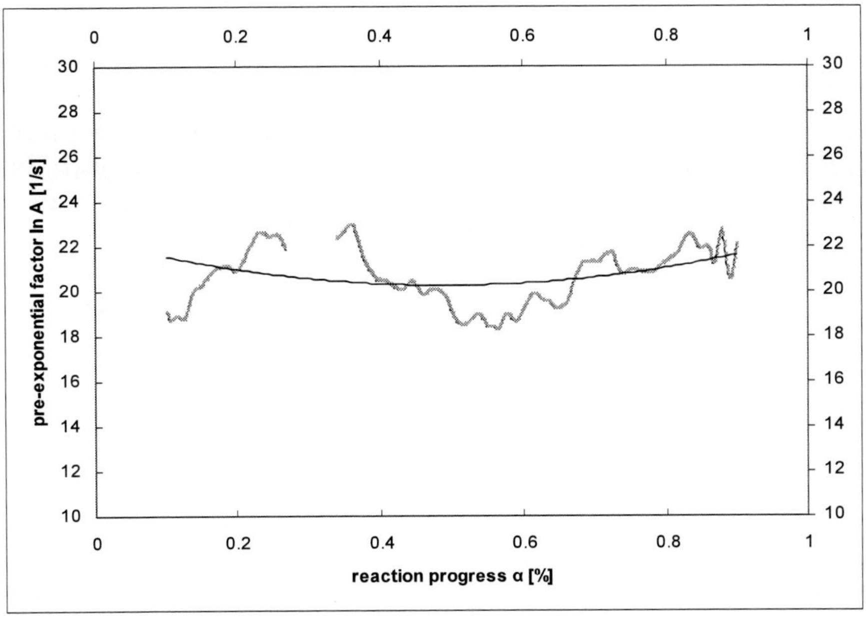

Fig. 7. The course of the pre-exponential factor of the gasification of char from the lignite/rubber mixture (45:50) using a 5%-addition of magnetite by CO_2 for conversion of 10-90%. Line *1*-calculated data, line *2*-simplified curve.

Table 6: The rate of the gasification of the chars from the lignite/rubber mixture (35:60) with and without magnetite (5%-addition) by steam and CO_2

Char from:	*w* (wt.%.min^{-1}) at 800°C		*w* (wt.%.min^{-1}) at 900°C	
	H_2O(g)	CO_2	H_2O(g)	CO_2
Lignite/rubber	2.94	0.33	4.98	2.65
Lignite/rubber with magnetite	5.31	1.05	11.71	5.89

Table 7: The rate of the gasification of the chosen comparative chars without and with magnetite by CO_2 at 900°C (5%-addition of magnetite)

Char (admixture)	*w* (wt.%.min^{-1}) at 900°C	
	Without	With
ABS (30%)	6.02	11.73
PP (30%)	5.11	9.70
Methylcellulose (30%)	4.71	8.97
Cellulose (30%)	6.60	11.88
PE+PP 1:1 (30%)	5.30	11.94

Conclusion

Through the measurement of the gasification rate of the chars from lignite with admixtures by steam or CO_2, it was discovered that the admixtures of polymers, celluloses, wood and rubber reduce their reactivity. This phenomenon was more closely examined with rubber chars, with which the decrease of reactivity very significantly manifests itself both with chars from rubber alone and from the lignite/rubber mixtures at 800°C and 900°C. This is caused by the ordered and interconnected structures of carbon which are imported into the chars by rubber and which resist the gasification reactions. In practice, this reduced reactivity is manifested in an increased amount of unburned carbon during co-gasification of waste rubber with lignite. This negative phenomenon can be compensated by an addition of magnetite, which acts as a catalyst of the reactions of chars with steam as well as CO_2 and accelerates the reactions. Through the isoconversional analysis, it was proven that the presence of magnetite significantly lowers the effective activation energy of these reactions, thus accelerating their course. The measurements of the reaction rate showed that it leads to a two- or three-fold acceleration of gasification. The catalyst activity of magnetite showed itself also during the gasification of chars from the lignite/polymers and lignite/celluloses mixtures, particularly from the lignite/polyethylene-polypropylene mixture, when the rate of the gasification with CO_2 at 900°C increased by more than two times.

Acknowledgment

This work was supported by the Czech Science Foundation (GA ČR) as the project No. 105/07/1407. In particular, I thank Dipl.-Eng. V. Káš, CSc. for valuable consultation and assistance in experiments.

References

[1] E. Murillo, E. Aylón, M.V. Navarro, M.S. Gallén, A. Aranda, A.M. Mastral: The application of thermal processes to valorise waste tyre, Fuel Processing Technology 87, Issue 2 (2006) 143-147.

[2] S. Ucar, S. Karagoz, J. Yanik, M. Saglam, M. Yuksel: Copyrolysis of scrap tires with waste lubricant oil, Fuel Processing Technology 87, Issue 1 (2005) 53-58.

[3] P. Straka, Z. Bučko: Co-gasification of lignite/waste-tyre mixture in a moving bed, Fuel Processing Technology 90, Issue 10 (2009) 1202-1206.

[4] V. Kříž, Z. Brožová, O. Přibyl, I. Sýkorová: Possibility of obtaining hydrogen from coal/waste-tyre mixture, Fuel Processing Technology 89, Issue 11 (2008) 1069-1075.

[5] C. Troca-Torrado, M. Alexandre-Franco, C. Fernández-González, M. Alfaro-Domínguez, V. Gómez-Serrano: Development of adsorbents from used tire rubber: Their use in the adsorption of organic and inorganic solutes in aqueous solution, Fuel Processing Technology, in press (Abstract).

[6] I. de Marco Rodrigues, M.F. Laresgoiti, M.A. Cabrero, A. Tores, M.J. Chomón, B. Caballero: Pyrolysis of scrap tyres, Fuel Processing Technology 72, Issue 1 (2001) 9-22.

[7] K. Unapumnuk, M. Lu, T.C. Keener, Carbon Distribution from the Pyrolysis of Tire-Derived Fuels, Ind. Eng. Chem. Res. 45 (2006) 8757-8764.

[8] D.Y.C. Leung, C.L. Wang: Fluidized-bed gasification of waste tire powders, Fuel Processing Technology 84, Issues 1-3 (2003) 175-196.

[9] R. Murillo, M.V. Navarro, J.M. López, E. Aylón, M.S. Callén, T. García, A.M. Mastral: Kinetic Model Comparison for Waste Tire Char Reaction with CO_2, Ind. Eng. Chem. Res. 43 (2004) 7768-7773.

[10] W. Zhu, W. Song, W. Lin: Catalytic gasification of char from co-pyrolysis of coal and biomass, Fuel Processing Technology 89 (2008) 890-896.

[11] A. Sharma, T. Takanohashi: Controlling the Synthesis Gas Composition from Catalytic Gasification of HyperCoal and Coal by Changing the Gasification Parameters. Proceedings – Twenty-Seventh Annual International Pittsburgh Coal Conference, October 11-14, Istanbul 2010, Session 2, 2-4.

[12] Y. Sekine, K. Ishikawa, E. Kikuchi, M. Matsukata, A. Akimoto: Reactivity and structural change of coal char during steam gasification, Fuel 85 (2006) 122-126.

[13] S. Vyazovkin, A.I. Lesnikovich: On the method of solving the inverse problem of solid-phase reaction kinetics, J. Thermal Anal. 36 (1990) 599-615.

[14] S. Vyazovkin, N. Sbirrazzuoli: Isoconversional Kinetic Analysis of Thermally Stimulated Processes in Polymers, Macromol. Rapid Commun. 27 (2006) 1515-1532.

[15] G.X. Xi, S.L. Song, Q. Liu: Catalytic effect of sulfates on thermal degradation of waste poly(methyl methacrylate), Thermochim. Acta 435 (2005) 64-67.

In: Gasification: Chemistry, Processes and Applications
Editor: Michael D. Baker
ISBN: 978-1-61209-681-0

Chapter 12

THE CHARACTERIZATION OF NON-VOLATILE IMPURITIES IN MIXED CARBON FEEDSTOCK AND THEIR INTERACTIONS WITH GASIFIER LINER MATERIALS

Jinichiro Nakano[1,2], Kyei-Sing Kwong[1], James Bennett[1], Tetsuya Kaneko[1,3] and Seetharaman Sridhar[1,3]

[1] US Department of Energy National Energy Technology Laboratory, Albany, OR, U.S.A.

[2] URS Corp, Albany, OR, U.S.A.

[3] Department of Materials Science and Engineering, Carnegie Mellon University, Pittsburgh, PA, U.S.A.

ABSTRACT

Carbon, slag chemistry, temperature, and partial pressure of oxygen affect material reactions in a gasification chamber. Corresponding thermodynamic parameters are critical as they determine the types of crystals that form in slags, weight fractions of the phases, and corrosive interactions between slags and refractory materials. The crystal formation then effectively changes physical characteristics of slags, impacting slag viscosity. In order to study interactions between slags and refractory materials, it is important to assess thermodynamic data for the slags and the timing of crystal formation as opposed to slag residence time in the gasification chamber. The current study focuses on the thermodynamic analysis of impurities contained in carbon feedstock used in gasification, its kinetic behavior, and the effects of carbon feedstock chemistry on the degradation of gasifier refractory materials. Utilizing qualitative and quantitative techniques (XRD, SEM-EDX/WDX, TEM-EDX and ICP-OES); a thermodynamic phase diagram encompassing industrial slags based upon coal/petcoke ash mixture compositions was constructed using equilibration conditions obtained in samples after 72 hours of exposure at 1500 °C in an atmosphere of 64%CO/36%CO_2. This CO/CO_2 ratio corresponded to a simulated gasification environment of $Po_2 = 10^{-8}$ atm. Mullite crystal was found in Al_2O_3 rich slags, while V_2O_3 crystal was present in slags with high V_2O_3

(content corresponding to slags with high petcoke content). No crystalline phases were observed in SiO_2 rich slags at low V_2O_3 content (slag with high content of coal ash).

The formation kinetics of crystals in coal/petcoke slags was also studied from 1200 – 1700 °C in an atmosphere of 64%CO/36%CO_2, and a Time-Temperature-Transformation diagram constructed. This diagram exhibited two distinct crystallization events; with higher than 10 wt.% additions of petcoke slag to coal slag promoting V_2O_3 crystallization in the slag above 1350 °C, and with additional crystalline phases such as V-rich spinel forming at temperatures lower than 1350 °C. Experimental results indicated that the V_2O_3 crystal would form in the slag as fast as 35 seconds.

Finally, an evaluation of the interfacial reactions between synthetic slags (average coal and petcoke slag chemistry) and refractory materials (90 wt.% Cr_2O_3 – 10 wt.% Al_2O_3 and 100 wt.% Al_2O_3) was conducted utilizing a Confocal Scanning Laser Microscope. Ground slag samples at specific locations on refractory coupons were heated at 1500 °C in a 64%CO/36%CO_2 gas mixture using a gold-image heating chamber. Characterization of the slag/refractory interface by the SEM-EDX indicated that slag penetration occurred into porous areas of the refractory microstructure, and that refractory dissolution into slag occurred through the refractory surface and in the porous areas penetrated by slag, which promoted de-bonding of refractory material. It was noted that VO_x-containing crystalline phases precipitated at the slag/refractory interface in coal/petcoke slag mixtures. Chemical spalling of the Cr-containing refractory grains along with chemical dissolution of the grain in the slag facilitated surface degradation of the refractory.

1. INTRODUCTION

Advanced gasification systems such as the Integrated Gasification Combined Cycle (IGCC) process, use coal, petroleum coke (petcoke), or combinations of them as the carbon feedstock, with biomass being considered as a potential future feedstock material. During gasification in slagging gasifiers, non-volatile impurities from the feedstock coalesce and form a viscous liquid of mixed oxides (slag), which flow down the sidewalls of the gasification chambers and interact with the liner materials, causing their failure. High Cr_2O_3 refractory materials are the liner material of choice to protect the gasifier vessel against the severe service environments at high temperature. The gasification temperature is controlled so the slag flows down the gasifier sidewalls, while keeping it low so as to minimize degradation of the refractory materials.

Petcoke is a waste byproduct from petroleum refining and is used as a feedstock material. It contains vanadium oxide as a major impurity and causes a different mechanism of refractory degradation. Coal used in gasification contains approximately 10 wt.% non-volatile impurities, whereas petcoke averages approximately 1 wt.%. When coal and petcoke are mixed, the overall slag amount decreases as more petcoke is used (Figure 1). Coal/petcoke mixed feedstock has been used as a carbon source, e.g., at Tampa Power and Electric (TECO), where up to 55 wt.% petcoke has been mixed with coal [1, 2]. If a 45 wt.% coal – 55 wt.% petcoke mixture is used, it would produce approximately 5 wt.% slag content (assuming 10 wt.% ash in coal and 1 wt.% ash in petcoke). As the petcoke content increases, the vanadium oxide content increases in the slag, which can drastically affect the chemical and physical properties of the resulting slag. The variation of slag constituents with petcoke addition to coal feedstock (averaged) is shown in Figure 2.

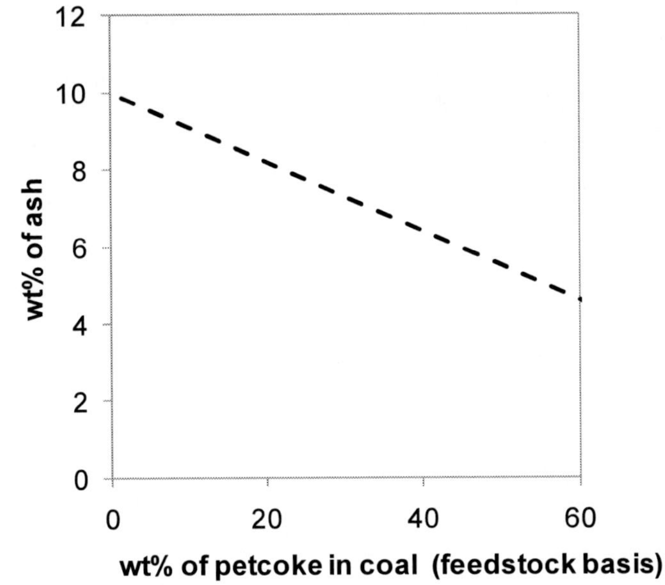

Figure 1. Change in slag quantity with petcoke addition (Note: coal has 10 wt.% ash and petcoke has 1 wt.% ash.).

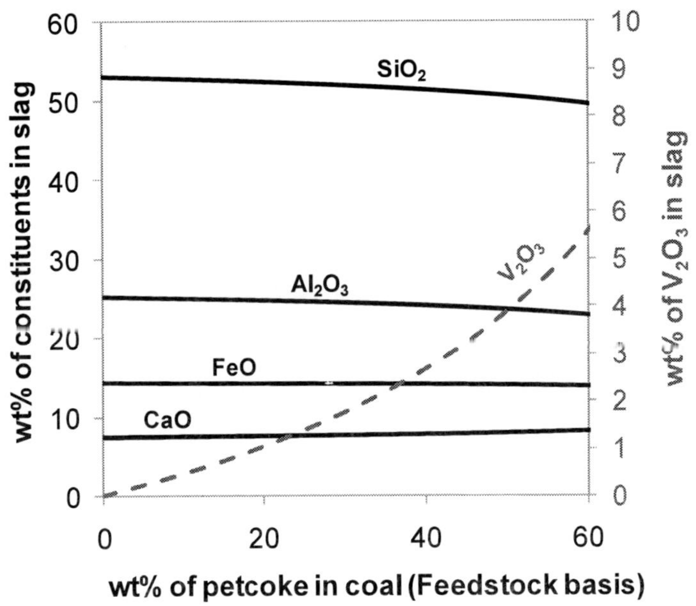

Figure 2. Chemical changes in slag constituents with petcoke addition.

Vanadium is a transition metal, with its valence varying with temperature, slag chemistry, and partial pressure of oxygen [3, 4, 5, 6, 7]. A stability phase diagram in Figure 3 (calculated using FactSage[TM] 6.1 [8]), indicates that the equilibrium valence of vanadium in pure vanadium oxides (i.e., no other oxides considered) is 3+ under the practical gasification conditions (T = 1325 °C ~ 1575 °C, $Po_2 = 10^{-9} \sim 10^{-7}$ atm, and P_{total} = 20 ~ 68 atm). The low partial pressure of oxygen inside a gasifier causes vanadium to be present in the form of karelianite crystal (V_2O_3). If this phase is present in the slag, it will impact slag visicosity.

Vanadium oxides also have different chemical interactions with refractory materials, depending on the oxidation state. According to FactSage calculations, the V_2O_5 phase in air ($Po_2 = 10^{-0.68}$ atm) melts at 669 °C in air, while the V_2O_3 crystal would remain as solid below 1938 °C at an oxygen partial pressure of 10^{-8} atm. Figure 3 assumes unit activities of vanadium oxides across the diagram. The activities of vanadium oxides in a slag solution are, however, not available in publicly available databases, and thus corresponding phase diagrams cannot be calculated using commercial thermodynamic programs or databases. The development of such equilibrium data on slag systems containing vanadium is necessary to understand the related thermodynamic systems, and is the goal of part of this research.

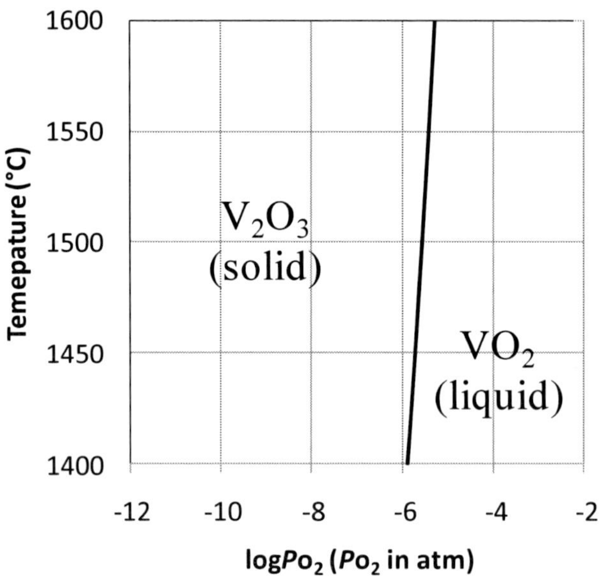

Figure 3. A stability phase diagram of pure vanadium oxide.

In a constant partial pressure of oxygen, the crystallization of vanadium oxide is thermally and chemically influenced by slag constituents, and the resulting system directly affects slag viscosity in a gasification system with varying temperature and oxygen partial pressure. If petcoke is used as a feedstock in a gasifier, readjustment of the gasification temperature may be required for good slag flow. If the slag is too viscous, it may clog the exit at the bottom of the gasifier chamber, causing premature shutdown. Slag chemistry and the resulting properties also influence the rate of refractory wear. If crystalline phases form on a surface or in the pores of a refractory liner, they may also alter the liner's properties such as thermal conductivity and thermodynamic interactions with refractory. Slag penetration also causes structural spalling of the top surface layers, which is shown in Figure 4. In this figure, a commercial brick sampled after a long service life indicates structural spalling occurring at the hot face of the refractory caused by slag infiltration of the refractory surface [9].

Figure 4. Slag penetration and structural spalling in a brick collected after commercial service.

In industrial gasification practice, complex metastable conditions are created inside the gasifier as 'fresh' slag continuously lands on the gasifier sidewalls and flows down. Thermodynamic predictions of the solid phases (to be formed) in slags originating from coal, petcoke or mixtures are essential to understanding refractory degradation and to gasifier operation. A few studies exist in the literature on phase equilibria in V_2O_3-containing systems [10, 11, 12, 13] Körber and Oelson [11] investigated the V_2O_3-FeO system, while Lopatko et al. [14] evaluated the V_2O_3-CaO system at temperatures between 1300 ~ 2080 °C in an inert atmosphere by utilizing Differential Thermal Analysis (DTA), Electron Probe Micro-Analysis (EPMA) and X-ray Diffractometry (XRD). Muan and Najjar [12, 13] reported phase equilibria of the V_2O_3-CaO system between 700 ~ 1600 °C, and the V_2O_3-CaO-Al_2O_3 system at 1300 °C in a CO_2/H_2 gas mixture (10/1 ~ 1/1 volume ratios). None of these studies, however, describe the equilibrium stability of phases in the system related to coal/petcoke slag chemistry under gasification conditions.

The present contribution discusses the following three main sections based on key components in the gasifier material studies as described so far – coal and petcoke slag thermodynamics, crystallization behavior, and the interactions of these slags with refractory materials.:

- Section 0: Phase Equilibria Studies in Coal/Petcoke Slag Systems
- Section 0: The Formation Kinetics of Crystals in Coal/Petcoke Slag Mixtures
- Section 0: Interactions Between Slags and Refractory Materials

Firstly, thermodynamic equilibria in synthetic coal/petcoke slags were evaluated in a simulated gasification environment. The phases identified in the coal/petcoke synthetic slags were used to develop an equilibrium phase diagram for the composition range covering industrial gasification operation. Secondly, the formation kinetics of crystalline phase precipitation in the coal/petcoke synthetic slags was also investigated. A Time-Temperature-

Transformation diagram was then constructed. Finally, interactions between synthetic slags (coal and petcoke) and refractory materials (90 wt.% Cr_2O_3 – 10 wt.% Al_2O_3 and 100 wt.% Al_2O_3) were studied in situ under the simulated gasification conditions using a confocal microscope with a hot stage in which the spreading of molten ash mixtures on refractory surfaces could be imaged in real time.

2. Phase Equilibria Studies in Coal/Petcoke Slag Systems

A gasifier environment is affected mainly by slag chemistry, operation temperature, and partial pressure of oxygen, which modifies types of crystals forming in slags, the amount of crystals, and so on. The presence of crystal drastically increases slag viscosity, creating more potential for clogging issues. The crystalline formation also affects the slag/refractory interaction. It is therefore inevitable to study the stability of crystal phases in slags. In this section, thermodynamic phase equilibria in synthetic coal/petcoke slag mixtures covering up to 72 wt.% of petcoke addition were evaluated under simulated gasification conditions (1500 °C and $Po_2 = 10^{-8}$ atm) and an equilibrium phase diagram encompassing the industrial composition range was constructed.

2.1. Experimental

2.1.1. Materials

The coal and petcoke ash mixtures selected in this study represent averages of feedstock mixtures from 0 ~ 72 wt.% petcoke additions, with ash quantities representing 10 wt.% in coal and 1 wt.% in petcoke. The coal slag chemistry was determined from a statistically averaged composition of over 300 U.S. coal samples reported by Selvig and Gibson [14]. The petcoke slag composition was based on averaged and normalized data from Bryers [15] and Swanson [16], which represented various types of petcoke. SiO_2, Al_2O_3, FeO, CaO and V_2O_3 were the major constituents in the coal/petcoke slags impacting viscosity and corrosion, and are the only constituents evaluated in this study. Variations of slag chemistry with respect to petcoke additions evaluated in the research are shown in Figure 2. The amount of V_2O_3 increases with petcoke additions, while the amounts of FeO and CaO remain rather constant. Because of that, FeO and CaO were fixed as constants across the present sample matrix at 13.5 wt.% and 7.0 wt.%, respectively. Through Selvig and Gibson's report on coal ash, the ratio of Al_2O_3/SiO_2 in coal slags was found to range from 0.25 to 1. Therefore, in the present five component oxide mixtures, SiO_2, Al_2O_3 and V_2O_3 were thus considered as variables.

Coal/petcoke mixtures (20 g each) were made using reagent grade powders of Al_2O_3, CaO, Fe_2O_3, SiO_2 and V_2O_5, then mixed in Nalgene® containers by dry-turning for 48 hours. During mixing, the bottles were manually stirred periodically shaking to avoid possible agglomerates. Chemical compositions of the mixtures were analyzed using Inductively Coupled Plasma Optical Emission Spectrometry (ICP-OES) after each equilibration treatment (shown in Table 3). FeO and CaO contents were confirmed to be constant (13.5 wt.% ± 2.4 and 7.0 wt.% ± 0.7, respectively). Uncertainty of the ICP values was within ± 0.2 %.

Table 3. Experimental sample compositions by ICP-OES (after equilibration)

	Bulk slag composition (wt.%)					XRD, WDX and TEM
Sample	Al_2O_3	CaO	FeO	SiO_2	V_2O_3	Equilibrium phases
1	14.7	6.9	12.8	65.6	-	Slag
2	18.8	6.9	12.8	61.4	-	Slag
3	23.3	7.2	13.4	56.0	-	Slag
4	14.7	6.7	12.8	64.2	1.7	Slag
5	25.1	6.6	12.7	55.7	-	Slag + Mullite
6	36.5	6.6	15.8	41.0	-	Slag + Mullite
7	38.9	7.0	13.5	39.0	1.6	Slag + Mullite
8	35.4	7.0	14.3	39.8	3.5	Slag + Mullite
9	33.6	7.5	14.0	38.9	6.0	Slag + Mullite
10	32.8	6.9	13.7	44.4	2.1	Slag + Mullite
11	26.9	6.7	14.3	49.0	3.1	Slag + Mullite
12	14.2	6.8	14.0	62.0	3.0	Slag + V_2O_3
13	15.0	7.0	12.1	60.9	4.9	Slag + V_2O_3
14	11.7	6.4	12.3	60.9	8.7	Slag + V_2O_3
15	20.8	6.6	15.9	49.6	7.0	Slag + V_2O_3
16	18.7	6.9	13.2	55.8	5.5	Slag + V_2O_3
17	25.1	6.9	14.2	46.4	7.4	Slag + V_2O_3+ Mullite
18	28.7	7.6	13.1	42.8	7.7	Slag + V_2O_3+ Mullite
19	33.9	7.7	13.8	36.6	8.0	Slag + V_2O_3+ Mullite
20	24.0	7.0	12.4	51.4	5.2	Slag + V_2O_3+ Mullite

2.1.2. Sample Treatment Procedure

Each mixed powder of approximately 0.5 g was placed in a cylindrical platinum crucible (99.99%, 10D×7.5H×0.05T in mm). The platinum crucibles with samples were then set in a vertical resistance tube furnace. Sample temperature was monitored by a B-type thermocouple positioned close to the sample inside the furnace. The furnace was initially heated to 1200 °C at a rate of 191 °C/hour, followed by heating to 1500 °C at a rate of 120 °C/hour.

The time of hold was determined through an evaluation of the work in the literature. Dong et al. [17] determined from chemical uniformity changes across low V_2O_3 samples, that 20 hour holds at 1500 °C were long enough to attain equilibrium in $Po_2 - 10^{-10} \sim 10^{-8}$ atm for high CaO oxide in the SiO_2-Al_2O_3-MgO-CaO-V_2O_3 system. Johnston [18] found 70 hours was required at 1100 °C for $Po_2 = 10^{-17} \sim 10^{-1}$ atm to achieve phase equilibrium in the SiO_2 rich SiO_2-Na_2O-Fe_2O_3-V_2O_3 compositions. Considering the time variations used by past researchers to approach equilibrium, 72 hours was assumed to be enough to reach the equilibrium at 1500 °C and $Po_2 = 10^{-8}$ atm in this study.

To simulate a gasifier atmosphere of $Po_2 = 10^{-8}$ atm at 1500 °C, a 64%CO/36%CO_2 gas mixture was introduced in the furnace at a rate of 30 ml/min. In order to minimize thermal segregation of the gas before contacting samples, the gas was delivered directly to the sample surface through an alumina tube. After a 72 hour hold at 1500 °C, slag samples were withdrawn from the furnace and quenched in water. Quenched samples were then characterized for phases formed by X-ray diffractomery (XRD): Rigaku Ultima III XRD and Scanning Electron Microscope (SEM) with Wavelength Dispersive X-ray (WDX): JEOL-

7000F equipped with an INCA X-ray detector. Samples representing each proposed phase region were selected for Transmission Electron Microscopy (TEM). To confirm phase structures, TEM samples were prepared using a Helios Nano Lab Dual Beam Focus Ion Beam (DB-FIB). TEM examinations were employed using a Phillips CM200 at 200 kV with Energy Dispersive X-ray (EDX). Electron diffraction patterns were analyzed from simulated patterns using the JEMS software [Version 3.4008U2009, CIME-EPFL] to confirm phases observed.

2.2. Results and Discussion

2.2.1. The Formation of Crystalline Phases in Slags

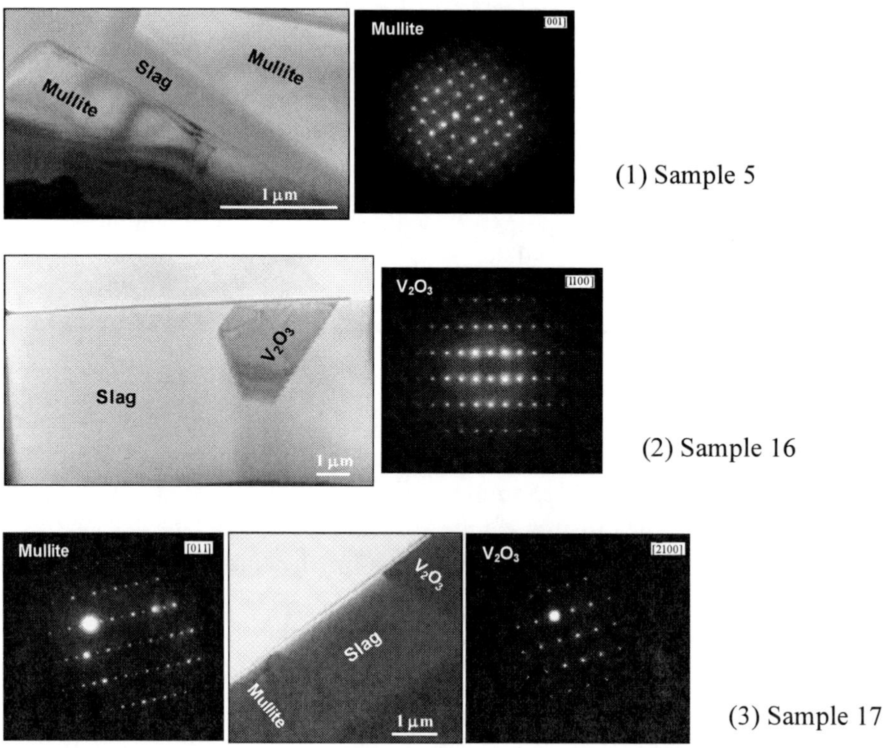

Figure 5. TEM bright field image diffraction patterns of selected samples representing different equilibria: (1) Sample 5 containing Slag and mullite; (2) Sample 16 containing Slag and V_2O_3 phases; and (3) Sample 17 containing Slag, mullite and V_2O_3 phases.

Based on results from XRD and SEM-WDX performed on samples after 72 hours of temperature hold at 1500 °C, phases identified in different samples are presented listed in Table 3. No crystalline phase was identified when Al_2O_3 and V_2O_3 contents were low. Two crystalline phases were found to be in equilibrium with slag; mullite ($3Al_3O_3 \cdot 2SiO_2$) in the Al_2O_3 rich samples; and karelianite (V_2O_3) if sufficient amounts of V_2O_3 crystals existed in the slag. The mullite formation was observed at 25.1 wt.% or higher Al_2O_3, and did not appear to be influenced by V_2O_3 content in slags in the compositions evaluated. V_2O_3 precipitation was favored in SiO_2 rich slags, but higher V_2O_3 content was required to precipitate V_2O_3 crystals in Al_2O_3 rich slag samples. Bright field TEM images of samples in Figure 5 show coexisting three phases. The mullite phase was found to have a rhombohedral structure, while the V_2O_3 crystal had an orthorhombic structure. TEM-EDX analysis indicated certain levels of Al and Fe dissolved in the V_2O_3 phase, with some solubility of V in the mullite phase. Vanadium atoms seemed to have limited substitution for aluminum in these phases. The constant SiO_2 concentration in the mullite phase of Samples 5 and 17 indicates vanadium ions substituted directly for alumina ions in the rhombohedral structure.

2.2.2. Construction of Phase Diagram

A series of XRD patterns for the different samples and results of WDX were analyzed to determine phase equilibria across the composition region studied. Using information on the phases determined, an isothermal section of the Al_2O_3-CaO-FeO- SiO_2-V_2O_3 system at 1500 °C in an oxygen partial pressure of 10^{-8} atm was constructed (Figure 6), where FeO and CaO contents were fixed at 13.5 wt.% and 7.0 wt.%, respectively. In the proposed phase diagram, directions of homogeneity boundaries were found to strictly follow Schreinemakers' rule [19], where the extrapolations of two phase boundaries that intersect, fall into or out of the highest-order phase field. In Figure 6, the extrapolation of a phase boundary between the 'slag' region and the 'slag + mullite' region moves into the 'slag + V_2O_3' region, while that between the 'slag' region and the 'slag + V_2O_3' region falls into the 'slag + mullite' region.

A general morphology variation of the V_2O_3 phase with Al_2O_3 (or SiO_2) content variations was noted in Figure 6. Needle-like V_2O_3 crystals formed in SiO_2 richer slags, whereas round particles predominant in Al_2O_3 richer slags. Lower viscosity in the Al_2O_3 richer slags may have kinetically facilitated the formation of mullite, which then may have influenced the V_2O_3 morphology.

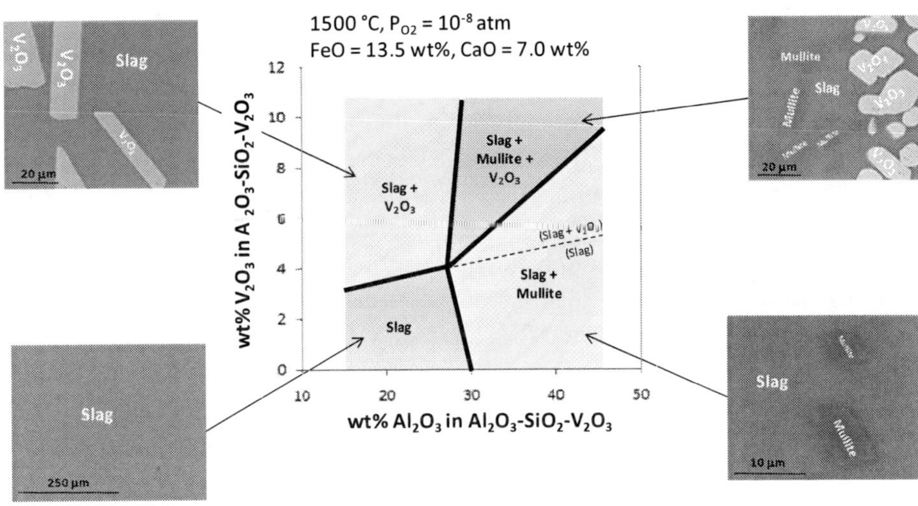

Figure 6. An experimental isothermal section of the Al_2O_3-CaO-FeO- SiO_2-V_2O_3 system at 1500 °C in an oxygen partial pressure of 10^{-8} atm. The dashed line represents a proposed metastable phase boundary in the absence of mullite.

A single phase region (slag) was found to exist in the SiO_2 rich area of Figure 6. The homogeneity region of mullite slightly expanded slightly with increasing V_2O_3 content up to approximately 4 wt.% V_2O_3; where slag, karelianite (V_2O_3) and mullite coexisted at the same time. The stability of karelianite (V_2O_3) is weakened by the formation of mullite in Al_2O_3 rich slags. The V_2O_3 phase has the rhombohedral crystal structure, and can form a limited solid solution with Al_2O_3 [14]. The limited solid solution may increase the availability of Al_2O_3 in the slag; if a slag exceeds the solubility limit of Al_2O_3, the Gibbs free energy of mullite decreases with respect to that of the V_2O_3 phase in Al_2O_3 richer regions. As a result, more V_2O_3 content is required to form the V_2O_3 crystal in Al_2O_3 rich regions.

2.2.3. Coal-Petcoke Slag Equilibria Using the Proposed Diagram

In Figure 7, the proposed phase diagram is superimposed with normalzied compositions of industrial coal slags from Selvig and Gibson [15] and mixtures of coal/petcoke slags from Bryers [16] and Swanson [17] (at a 45 wt.% coal / 55 wt.% petcoke feedstock ratio). The 100 wt.% coal slag average composition is located in the equilibrium phase region where slag and mullite coexist. The average composition change in the mixtures is shown in Figure 2. For the 100 wt.% coal slag, the proposed diagram indicates that mullite is not thermodynamically favored to precipitate if the composition is SiO_2 richer (i.e., < 30 wt.% Al_2O_3 in Figure 7) than the average coal slag composition. Mullite may form in coal slags with certain compositions (e.g., Al_2O_3 rich in the diagram) during gasification operations if it is kinetically allowed. It should be reminded that the phase diagram in Figure 7 is an isothermal section at 1500 °C. If more petcoke is added to a coal feedstock at 1500 °C, for example, a slag would become supersaturated with vanadium, causing the relative thermodynamic stability of the V_2O_3 crystal to increase (with respect to 'slag' or 'mullite and slag'). When V_2O_3 content in the slag exceeds certain amounts in the Al_2O_3 rich region (approximately 5 wt.% for 30 wt.% Al_2O_3 or 8 wt.% for 40 wt.% Al_2O_3 in Figure 7), the Gibbs free energy of the V_2O_3 phase becomes low enough to extend beyond the equilibrium tangential surface drawn through slag and mullite. When a coal feedstock is mixed with 55 wt.% petcoke (on a feedstock basis), the resulting slag compositions vary drastically. Note that the formation of the V_2O_3 crystals is thermodynamically favored in V_2O_3 rich slags. Depending on temperature, chemistry, and viscosity; V_2O_3 crystal formation may be kinetically favored over mullite [20]. If the formation of mullite is somehow suppressed, the thermodynamic stability of the V_2O_3 phase becomes more favorable, and it may be able to form in the slag at much lower V_2O_3 content than the global equilibrium phase diagram suggests. This is illustrated by a dashed line in Figure 6. It is important to note that the formation of V_2O_3 crystals could result in an increase in slag viscosity. If precipitation occurs in a short time, it would most likely take place on refractory surfaces.

In practice, changes in bulk slag chemistry, which may be altered by interacting with refractory materials and/or incoming fresh ash, also need to be considered. Gasifier process parameters such as temperatures and oxygen partial pressures are variable during gasification, causing equilibrium conditions to change with respect to location and time. If global equilibrium (i.e., long-term equilibrium) within the gasifier is assumed (

Figure 7), it may be necessary to maintain V_2O_3 content below 3 wt.%, and the Al_2O_3 content less than 21 wt.% in the coal/petcoke slag mixtures gasified at 1500 °C so that no crystal formation would occur during gasifier operation in order to ensure sufficient fluidity of the slag. Based on Figure 8, flux would be required for the 100% coal slag and the 45 wt.% coal slag – 55 wt.% flexicoke slag to remain in phase regions where there would be no crystalline phase present. This was estimated assuming a long-term equilibrium case in the mixed slags. Fluxing slags with SiO_2 (and Al_2O_3) in Figure 8 simulates how the slags can be modified if the system needs to be a homogeneous liquid solution without any crystallization at equilibrium. Each slag composition needs to be evaluated individually to determine phase conditions existing in the gasification environment. It is possible a metastable equilibrium (i.e., short-term equilibrium) may exist where the mullite formation is kinetically delayed as indicated by the dashed line in Figure 6, pointing out the importance of reaction kinetics (i.e., 100 % coal slags and coal/petcoke slag mixtures may remain as homogeneous solutions without any crystallization).

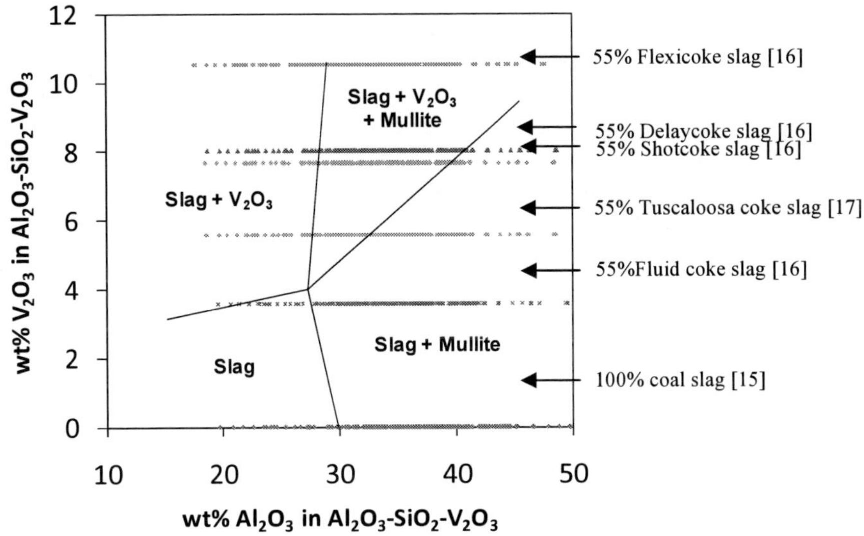

Figure 7. A proposed phase diagram with industrial slag compositions at 100 wt.% coal (~ 300 different compositions) and at 45 wt.% coal – 55 wt.% petcoke (feedstock basis).

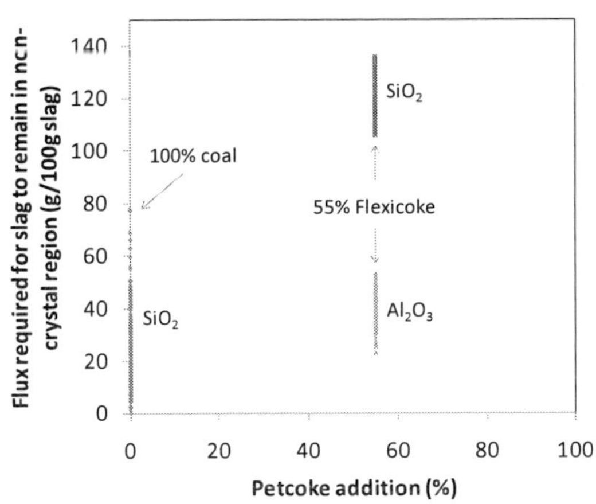

Figure 8. Flux required for slags to remain in a region where no crystal forms determined from Figure 8.

To this point, the thermodynamic aspects of slags with various compositions resulting from mixtures of coal and petcoke feedstocks used in gasifiers have been discussed. Kinetic behaviors of the crystalline phases in the mixed slags will be discussed next.

3. The Formation Kinetics of Crystals in Coal/Petcoke Slag Mixtures

Crystal phase formation impacts both slag viscosity and slag chemistry. Changes in slag chemistry and the solid phases crystallized affect the wear of interfaces between slag and refractory materials. In this section, the kinetics of crystal phases forming in coal/petcoke slags with various compositions was studied. A hot stage attached to a confocal microscope allowed a rapid heating rate and real time observations of crystal formation. A Time-Temperature-Transformation diagram was constructed based on the isothermal observations with various compositions and temperatures.

3.1. Experimental

3.1.1. Materials

Chemical compositions of 100% coal slag and 100% petcoke slag were adopted from Selvig and Gibson [15], and Bryers [16] and Swanson [17], respectively. Note that K_2O was included in this section of the study, which is only different from sample compositions used in Section 0. The coal slag and petcoke slag samples were separately produced by blending appropriate amounts of reagent grade oxide powders in a 500 ml nalgene® plastic bottle filled with ethanol, which were then mixed by rotation at 55 rpm with alumina balls (10 mm diameter) for 48 hours. Six slag mixtures of 200 g each were prepared from the above slag samples by varying petcoke slag amounts from 0 wt.% to 100 wt.%. to study the effect of different slag chemistry. Final chemical compositions of the mixtures analyzed by ICP-OES are shown in Table 4. A drastic increase in vanadium content by the petcoke addition should be noted.

Table 4. Chemical compositions of coal/petcoke mixtures (wt.%) used in crystallization kinetics studies

Coal/petcoke	Coal/petcoke	Constituents in slag mixtures (ICP)						Reference
(Slag basis)	(Feedstock basis)	Al_2O_3	CaO	Fe_2O_3	K_2O	SiO_2	V_2O_5	
100/0	100/0	24.0	6.4	14.5	2.9	52.0	-	[15]
90/10	47/53	22.4	6.8	14.1	2.7	49.3	4.7	
70/30	19/81	19.2	7.5	13.2	2.4	43.9	13.7	
50/50	9/91	15.9	8.3	12.3	2.0	38.4	23.1	
30/70	4/96	12.6	9.0	11.5	1.7	32.9	32.5	
0/100	0/100	7.8	10.1	10.2	1.2	25.0	46.0	[16, 17]

3.1.4. Procedure

A Confocal Scanning Laser Microscope (CSLM) attached to a gold-image heating chamber (Lasertec 1LM21H) was utilized for the in situ observations of crystallization as it occurred. Details of the CSLM are widely available in the literature [21]. A schematic of the CSLM heating chamber is shown in Figure 9. An He-Ne laser applied energy to the sample surface where the focus point returns the brightest contrast. Real-time images with up to 0.25 μm resolution were obtained using the confocal optics. A sample was placed at one of the two focal points in the gold-coated elliptical furnace, with a halogen heating lamp located at the other. In the experimental setup, cylindrical platinum crucibles (99.99%, 5D×5H×0.4T in mm) containing approximately 30 mg of each slag mixture was set on a platinum sampler pan (10 mm diameter), as indicated in Figure 9. The experiment was carried out in an atmosphere simulating a commercial gasifier environment (Po_2 = 10^{-8} atm at 1500 °C), using a 64%CO/36%CO_2 gas. The average heating rate from room temperature to 1500 °C was determined to be 77 °C/sec. The details of the experimental technique used are described elsewhere [21].

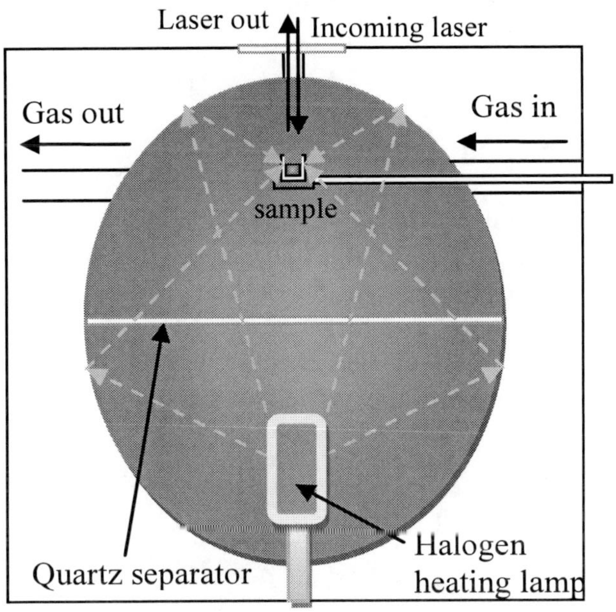

Figure 9. A schematic of CSLM heating chamber.

In order to ensure slag's homogeneity prior to each run, mixtures were premelted in air at 1500 °C (or higher) in the CSLM heating chamber. The chamber was then lowered to the designated test temperature, where the sample was held for two minutes, after which the chamber atmosphere was isothermally switched to the 64%CO/36%CO_2 mixture. The gas flow rate was kept constant at 616 ml/min during experiment.

When the atmosphere was changed from air to the CO/CO_2 gas, the oxygen partial pressure change causes liquid vanadium oxide (V_2O_5 or VO_2), to transform into a crystalline V_2O_3. In this study, the precipitation event was optically measured in situ using the CSLM, and the beginning time of crystal formation on the surface of the slag liquid was recorded for each slag mixture. Temperature of the sample was controlled by a computer-based program, and was continuously adjusted to compensate for exothermic and endothermic reactions over the atmospheric switch.

The crystalline phases formed in the CSLM studies were identified by powder XRD (Rigaku GeigerFlex). For XRD evaluation, separate samples were produced by individually heating material at 1350 °C and 1450 °C for a longer time of period (one hour), followed by He gas quenching (99 °C/sec). It was visually confirmed under the CSLM that phases which first crystallized in the isothermal reactions grew and remained stable during studied time periods. XRD samples at 1200 °C were prepared separately as the complex multi-phase system at 1200 °C required a larger sample volume for distinguishable XRD peak patterns. In order to produce a sufficient quantity of sample, a high temperature resistance tube furnace was utilized for thermal treatment (3 days) under the same gas condition as the CSLM study, followed by water-quenching. The samples were also analyzed using SEM-EDX (Philips XL-30 FEG-SEM with an INCA EDX detector) to verify phases formed. The results from 1200 °C must be carefully interpreted because it is uncertain how the resulting phases observed after 3 days are correlated with the phases that formed at an early stage corresponding to the TTT diagram (0), but should return an indication of expected phase formation.

3.2. Results and Discussion

3.2.1. Crystalline Formation in the Slag Mixtures

A series of consecutive CSLM images in Figure 10 show a typical crystallization sequence for the V_2O_3 phase at the gas/liquid interface. This sequence occurred shortly after the atmospheric gas was switched to the 64%CO/36%CO_2 mixture. A time delay before the CO/CO_2 gas mixture reached the sample was estimated to be approximately 20 seconds in the experimental setup used. Because of the time lag for gas changes, time offsets were therefore taken into account throughout this study. Particles formed at the gas/liquid interface grew and agglomerated, which eventually slowed down as the system was approaching equilibrium (Figure 10-(4)).

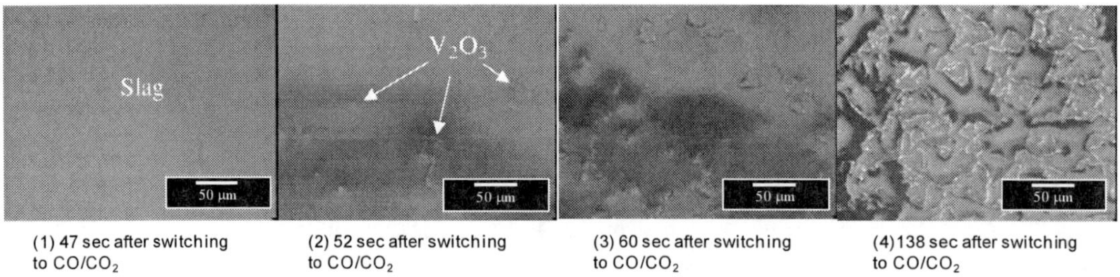

Figure 10. Consecutive CSLM images of the karelianite (V_2O_3) formation in 100% petcoke slag at 1500 °C: (1) before any crystal formed; (2) first crystal appearance; (3) crystal growth; and (4) further growth and agglomeration.

3.2.2. Time-Temperature-Transformation Diagram

The time for the initial crystal formation observed in situ using the CSLM was recorded for various temperatures and coal/petcoke slag mixtures. The corresponding Time-Temperature-Transformation (TTT) diagram was then constructed, as shown in Figure 11. Two distinctive noses were observed in the 30, 50 and 70% petcoke slag samples, implying different phase development above and below 1350 °C. The present TTT diagram indicates that crystallization would occur as fast as 35 seconds, even homogeneously (i.e., without aid of significant surface energy increase) in the liquid slag. No precipitation was observed in the 10% petcoke / 90% coal slag above 1400 °C during the time evaluated (up to two hours). This observation is supported by the absence of crystalline vanadium compounds in slag or on the refractory liner by commercial gasifier operators with low vanadium slags. Also crystallization of any solid phase was not found in the 100% coal slag in the present study at temperatures between 1250 °C and 1550 °C.

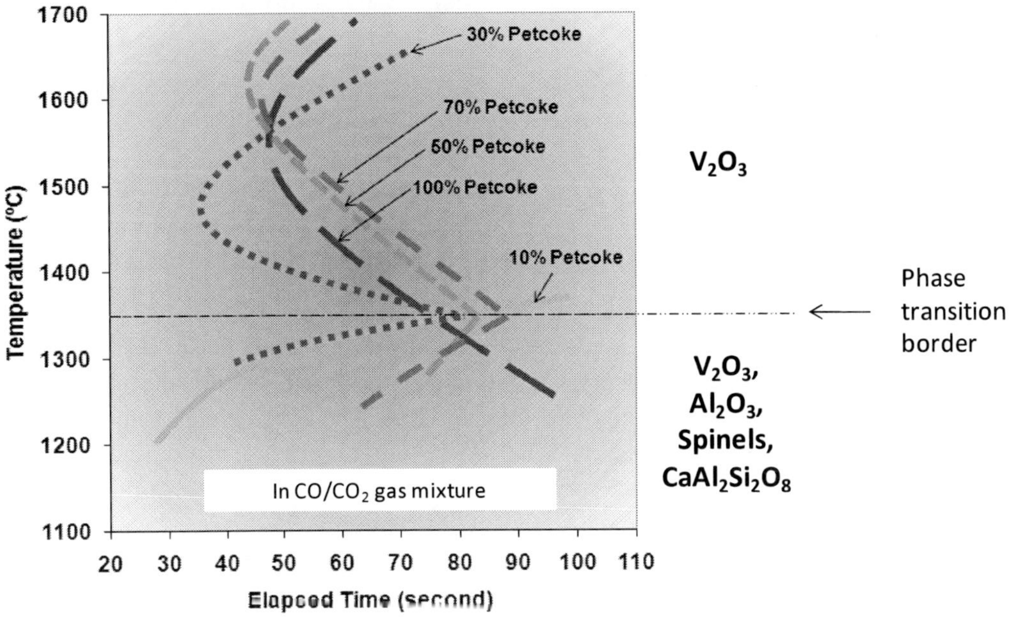

Figure 11. TTT diagram for the crystal formation in coal/petcoke slag mixtures in a CO/CO_2 gas.

Distinct V_2O_3 crystal formation morphologies existed between the upper and lower portions of the TTT curve, with fine crystals having angular characteristics above the nose of the curve and with columnar morphology below the nose. Examples of these two types of crystal structures are shown in Figure 12. These crystallization tendencies were most apparent in the petcoke rich coal/petcoke slag mixtures, especially with the 70% and 100% petcoke slags. Once nucleated, the structures quickly grew across the slag surface. Agglomerate layers that formed at the gas/liquid interface moved into the slag. No crystalline phases were observed when petcoke additions were reduced to 10 or 0 %.

As more petcoke slag was added to the coal slag, the basicity (CaO/SiO_2 ratio) of the resulting slag increased, and its viscosity decreased drastically as the degree of SiO_2 polymerization was altered. It was also noted that oxygen diffusion through the slag solution in the reducing atmosphere affected the kinetics of the V_2O_3 crystal formation above the nose (30% petcoke slag vs. 50, 70 and 100% petcoke slags). The smaller undercooling lowers the driving force for precipitation. The lower nose location for the 30% petcoke slag may be induced by a lower liquidus temperature of this slag system. It is noteworthy that the precipitation was recorded through the 'direct' observation under the CSLM rather than an 'indirect' quench experiment, minimizing potential cooling artifacts.

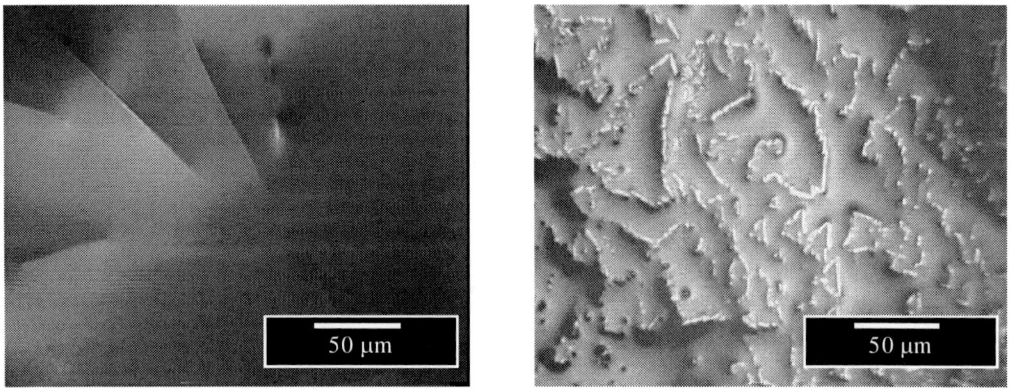

(1) 100% 1450 °C (2) 100% 1600 °C

Figure 12. Two distinct morphologies of the V_2O_3 crystal formation in 100% petcoke slag (1) below the nose and (2) above the nose.

Depending on thermodynamic conditions, crystallization would occur in coal/petcoke slags while on gasifier sidewalls. If crystallization occurs before exiting the gasifier, slag viscosity may increase and clogging may become an issue. The cooling rate of the slag while moving down the refractory walls should be small, but may cause crystallization if the cooling rate is slower than the critical cooling rate for the glass transition. Liquid V_2O_5 can solidify in air below 669 °C without modifying its composition.

3.2.3. Crystalline Phases Present in the Slag

XRD analysis of samples heated at 1350 °C and 1450 °C indicated that a corresponding crystalline phase for the upper curves in the TTT diagram was the V_2O_3 phase. XRD patterns at 1200 °C indicated the presence of anorthite in all compositions except for 100% petcoke slag, where goldmanite ($Ca_3(Al,Fe^{3+},V^{3+})_2(SiO_4)_3$) was found. Spinel phases of various compositions were also found in all the coal/petcoke mixtures heated at 1200 °C. For example, $FeAl_2O_4$ was identified in the 100% coal slag, while VFe_2O_4 was found in the 10% through 100% petcoke/coal slag mixtures. XRD peaks of the V_2O_3 phase grew larger for the higher petcoke content at 1200 °C, implying greater quantities of the V_2O_3 crystal in the petcoke-rich mixtures.

Crystalline phases identified by the XRD are summarized in Figure 13 for different coal/petcoke slag compositions. V_2O_3 was the only phase identified in equilibrium with slag at1350 °C and higher. At 1200 °C, two types of spinel phases ($FeAl_2O_4$ and VFe_2O_4) were found. The types of spinel varied with each coal/petcoke mixture when evaluated using the same CO/CO_2 gas mixture. For the 100% coal slag and 10% petcoke slag, Al_2O_3 was identified by XRD, with $FeAl_2O_4$ or VFe_2O_4 identified at higher petcoke additions. Corundum (Al_2O_3) was found to form in the absence of the thermodynamically stable mullite phase, a phase which was not found at any compositions.

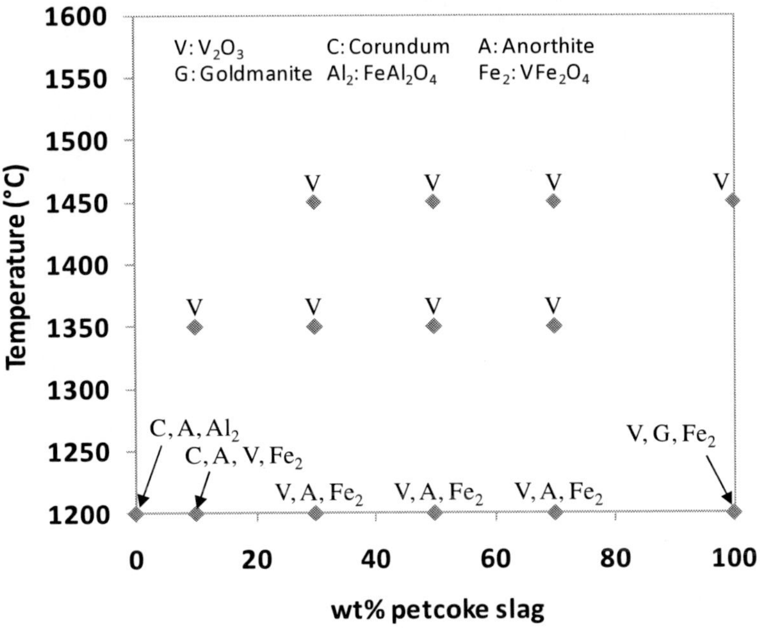

Figure 13. Crystal phases identified by XRD at various mixtures and temperatures.

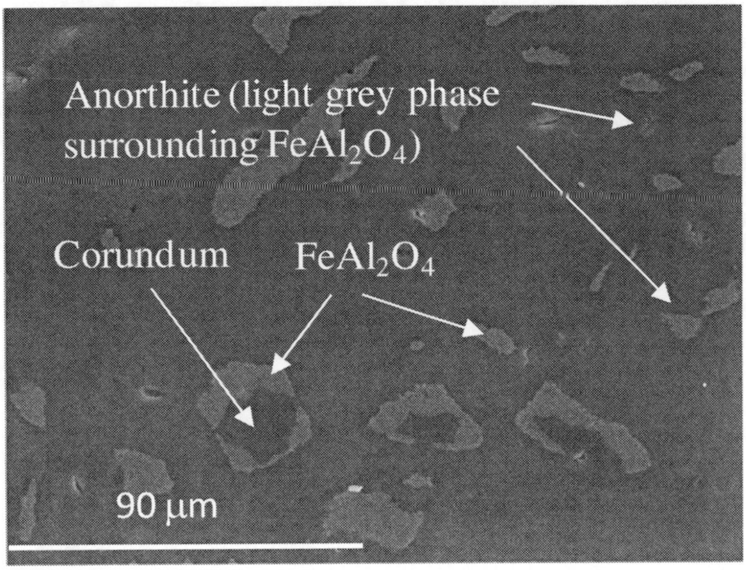

Figure 14. SEM image of 100% coal slag heated at 1200 °C.

At 1200 °C, crystal phases of corundum, anorthite, and $FeAl_2O_4$ formed in the 100% coal slag (Figure 13); while no crystallization was noted to occur above that temperature (up to 2 hours). It is thought that a sufficient increase in viscosity occurred that slowed crystallization in this sample. In a relation between ion diffusion and viscosity in polymeric melts silicate species would be coupled to viscosity. However, the present result refers more to homogeneous nucleation due to the nature of the experiment. In practice, faster kinetics may be observed with a favorable heterogeneous nucleation site like a refractory surface. Similarly, slower forming phases could form on particles that nucleated earlier. An example of this is shown in Figure 14, where a 100% coal sample heated at 1200 °C had $FeAl_2O_4$ around the corundum (Al_2O_3) particles, with anorthite encasing that phase. In this case, it seems that corundum formed at an early stage acted as a nucleation site for $FeAl_2O_4$, which also aided the formation of anorthite. It is not certain if solid state diffusion was involved in these phase formations.

When the 10% petcoke was added to coal slag, VFe_2O_4 was observed instead of $FeAl_2O_4$ (found in 100% coal slag), as shown in Figure 15. VFe_2O_4 was found to form on the V_2O_3 particles for 30, 50, and 70% petcoke additions to coal slag; but the trend weakened for petcoke richer samples. Anorthite formation was independent of the spinel particles for the 30, 50 and 70% petcoke slag; implying easier formation of anorthite in the higher petcoke mixtures. As a volatile substance with high activity in the slag, vanadium appears to play a catalytic role, promoting slowly forming phases to crystallize sooner by reacting with other species, and to produce heterogeneous nucleation sites.

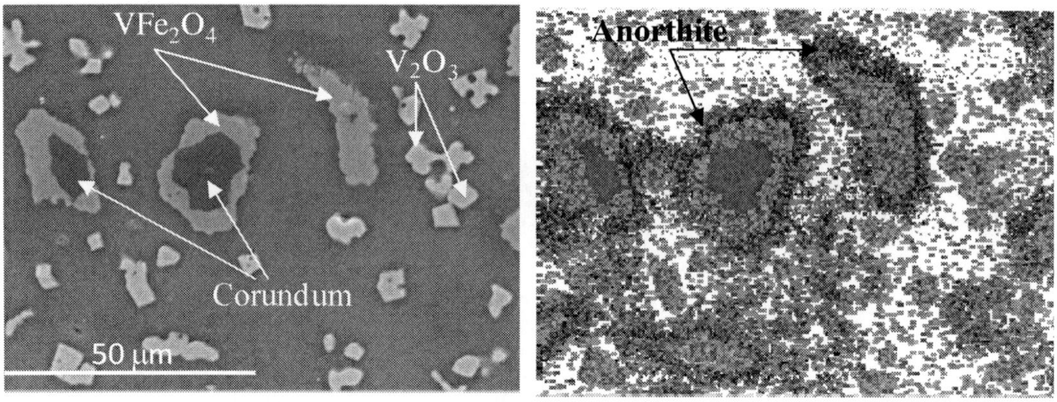

Figure 15. SEM image (left) and corresponding EDX elemental map for Al and Fe (right) of 10% petcoke/coal mixture heated at 1200 °C.

In section 0, interactions between slags and refractories will be discussed. During gasification operation, impurities in a carbon feedstock create viscous slag solution on the surface of the refractory sidewalls in the gasifier, leading to slag/refractory interactions. An understanding of how the slag interacts with the refractory is important in order to determine mechanisms of materials degradation under gasification conditions. This information can then be used to develop new, improved refractory materials.

4. Interactions Between Slags and Refractory Materials

Interactions between synthetic slags (coal and petcoke) and refractory materials (90 wt.% Cr_2O_3 – 10 wt.% Al_2O_3 and 100 wt.% Al_2O_3) were studied in situ under the simulated gasification conditions using a CSLM previously described in Section 0. Real time interactions between slags and refractories were investigated at a microscopic level.

4.1. Experimental

4.1.1. Materials

Two slag samples simulating coal and petcoke slags were prepared for this investigation. Compositions of the 100% coal slag and 100% petcoke slag were described in 0. Mixtures of oxide powders in pure graphite crucibles were melted in a high purity Ar gas (99.998 wt.%) using an induction furnace. The samples were carefully separated from the graphite crucibles and decarburized at 580 °C in air for 18 hours. At the beginning of each experiment, a sessile drop (0.02 ~ 0.04 mg) of the premelted slag was melted a second time in 64%CO/36%CO_2 (producing an oxygen partial pressure of 10^{-8} atm) at 1500 °C, to minimize potential reducing reactions during the experiment and to simulate gasification conditions. The resulting slag chemical compositions were analyzed by ICP-OES (Table 5).

Table 5. Experimental compositions of coal and petcoke slags used in the present study (wt.%)

Slag constituents	Coal slag	Petcoke slag
Al_2O_3	24.0	7.8
CaO	6.4	10.1
Fe_2O_3	14.5	10.2
K_2O	2.9	1.2
SiO_2	52.0	25.0
V_2O_5	-	46.0
Total	99.8	100.3

Refractory samples used were commercial refractory bricks whose compositions and porosity were taken from product data sheets and are listed in

Table 6. There are two structurally distinct regions observed in the refractory: dense grain agglomerate regions (individual grain size up to 3 mm), and fine particle size matrix regions. This microstructure is engineered by refractory manufacturer and is important for thermal shock resistance and for bonding the refractory grains. The porous matrix regions increase thermal shock resistance, but are susceptible to chemical dissolution, mechanical erosion, and slag penetration. SEM images of the two refractory materials shown in Figure 16 indicate typical refractory grain agglomerate and matrix regions.

Table 6. Select chemical and physical properties of refractory samples used in this study

Refractory type		$90Cr_2O_3$-$10Al_2O_3$	Al_2O_3
Chemical composition (wt.%)	Al_2O_3	10.2	99.6
	Cr_2O_3	89.0	-
	CaO	0.3	Trace
	SiO_2	0.2	0.2
	TiO_2	0.2	Trace
	Fe_2O_3	0.1	0.1
	MgO	0.1	Trace
	NaO + K_2O	-	0.1
Porosity (%)		16.7	19.4

(1) $90Cr_2O_3$-$10Al_2O_3$ refractory (2) Al_2O_3 refractory

Figure 16. SEM images of refractory surfaces showing fine matrix regions and dense grain agglomerate areas.

4.1.2. Procedure

Slag/refractory interactions were studied as sessile drop interactions using the CSLM. A premelted slag particle was placed on the selected microstructure areas of a refractory coupon and heated at 1500 °C for 0 minutes and 10 minutes in the CO/CO_2 gas. The surface of the sample where the slag particle was placed had previously been polished through 1 µm. The heating rate of the CSLM used in this study was approximately 77 °C/sec. Wetting behaviors and interactions between the slag and refractory surface were video recorded while holding at 1500 °C for the designed amount of time. Test samples were then rapidly quenched by blowing room temperature He gas onto sample surfaces, causing a cooling rate of 99 °C/sec from 1500 °C to 900 °C. The slag/refractory interface was then analyzed by SEM-EDX. A total of 16 slag/refractory interaction experiments were conducted using two types of slags and two refractory microstructures. Samples were held for 0 and 10 minutes at 1500 °C before the microstructures were frozen.

Table 7. Experimental configurations in the present study

Experiment	slag	Refractory (wt.%)	Microstructure location	Holding time (min)
1	Coal	$90Cr_2O_3$-$10Al_2O_3$	Grain	0
2	Coal	$90Cr_2O_3$-$10Al_2O_3$	Grain	10
3	Coal	$90Cr_2O_3$-$10Al_2O_3$	Matrix	0
4	Coal	$90Cr_2O_3$-$10Al_2O_3$	Matrix	10
5	Coal	Al_2O_3	Grain	0
6	Coal	Al_2O_3	Grain	10
7	Coal	Al_2O_3	Matrix	0
8	Coal	Al_2O_3	Matrix	10
9	Petcoke	$90Cr_2O_3$-$10Al_2O_3$	Grain	0
10	Petcoke	$90Cr_2O_3$-$10Al_2O_3$	Grain	10
11	Petcoke	$90Cr_2O_3$-$10Al_2O_3$	Matrix	0
12	Petcoke	$90Cr_2O_3$-$10Al_2O_3$	Matrix	10
13	Petcoke	Al_2O_3	Grain	0
14	Petcoke	Al_2O_3	Grain	10
15	Petcoke	Al_2O_3	Matrix	0
16	Petcoke	Al_2O_3	Matrix	10

4.3. Results and Discussion

4.3.1. Wetting Behaviors of Slags on the Refractory Surface

The high temperature spreading and penetration behaviors of slag on a refractory surface are affected by: surface topology (including production defects and surface smoothness) and geometry (pores and microcracks), types of chemical interactions between the slag and refractory surface (formation of reaction layers and refractory dissolution), interfacial energy (grain agglomerates or matrix), pore connectivity, etc. The types of chemical reactions depend on slag and refractory chemistry, temperature and atmosphere, and how the test materials were prepared. Capillarity pressure within pores and slag viscosity should also influence penetration kinetics. An example of coal slag wetting a matrix region of the $90Cr_2O_3$-$10Al_2O_3$ refractory is shown in Figure 17. On heating, the coal slag particle melted and formed a semispherical cap, followed by spreading and penetration of the refractory surface. The majority of coal slag penetrated the pores and microcracks of the refractory substrate at 1500 °C, having melted at a lower temperature. Petcoke slag, on the other hand, started melting at lower temperatures than the coal slag, and started rapidly spreading without forming a semispherical shape (Figure 18). The formation of VO_x-based crystal phase was observed in the petcoke slag and a spreading radius (relative to its original size) of the petcoke slag was smaller than that of the coal slag. If slags were placed on grain agglomerates, a coal slag cap remained on a surface of both refractories after 10 minutes (Figure 19), whereas petcoke slag irregularly spread over the surface leaving VO_x-based particles (Figure 20).

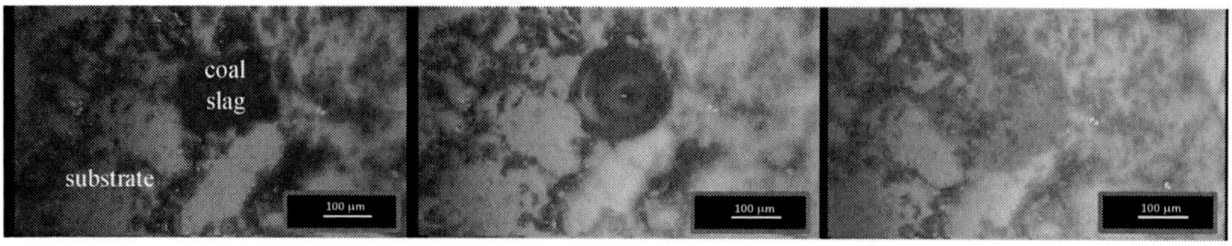

(1) Coal slag particle (10 sec, 780 °C) (2) The sphere is expanding (15 sec, 1300 °C) (3) The majority penetrated (40 sec, 1500 °C)

Figure 17. Spreading of coal slag on a matrix region of $90Cr_2O_3$-$10Al_2O_3$ refractory in an oxygen partial pressure of 10^{-8} atm.

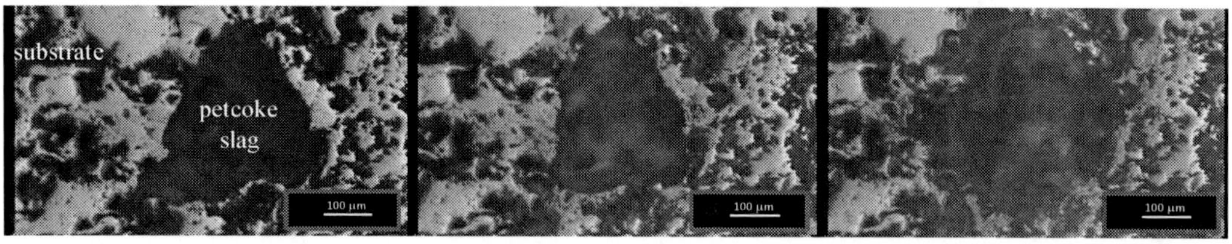

(1) Petcoke slag particle (10 sec, 900 °C) (2) Slow spreading (15 sec, 1280 °C) (3) Rapid bubbling over penetration (40 sec, 1500 °C)

Figure 18. Spreading of petcoke slag on a matrix region of $90Cr_2O_3$-$10Al_2O_3$ refractory in an oxygen partial pressure of 10^{-8} atm.

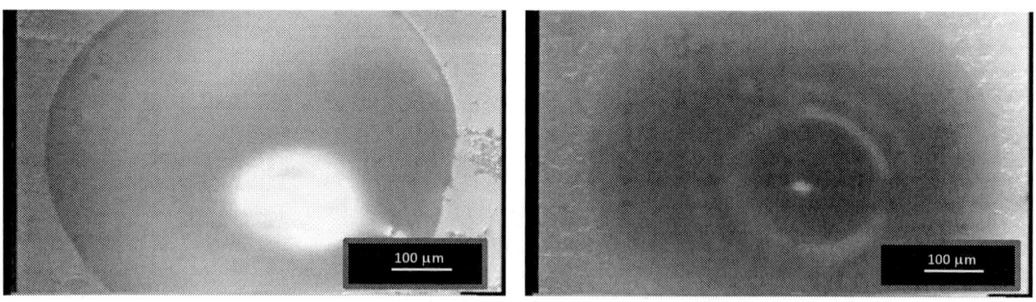

(1) Coal slag on $90Cr_2O_3$-$10Al_2O_3$ (2) Coal slag on Al_2O_3

Figure 19. CSLM images showing a wetting behavior of coal slag on a grain agglomerate of (1) $90Cr_2O_3$-$10Al_2O_3$ and (2) Al_2O_3 in an oxygen partial pressure of 10^{-8} atm.

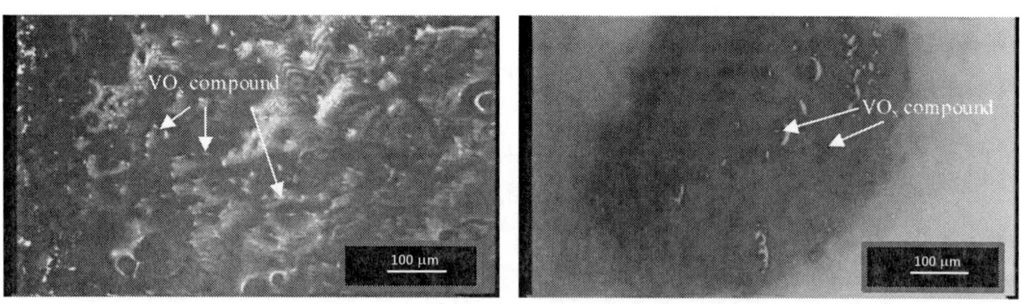

(1) Petcoke slag on $90Cr_2O_3$-$10Al_2O_3$ (2) Petcoke slag on Al_2O_3

Figure 20. CSLM images showing a wetting behavior of petcoke slag on a grain agglomerate of (1) $90Cr_2O_3$-$10Al_2O_3$ and (2) Al_2O_3 in an oxygen partial pressure of 10^{-8} atm.

4.2.2. Penetration of Slag Into Refractory

Based on SEM-EDX analysis of cross-sectioned slag/refractory samples, coal and petcoke slag penetrated the porous matrix regions. The effects of refractory microstructure on slag penetration behavior seem constant at any depth, i.e., slag liquid penetrated through fine porous regions first, even at the deepest penetrated areas. Figure 21 shows the average penetration depths of slag constituents in the refractory surface, which indicated the general penetration behavior depended on the type of slag and refractory for the same exposure conditions. For a given slag, refractory, and time, each curve represents a different element at a certain depth from a refractory surface, as measured by SEM-EDX. From the data in Figure 21, the penetration depth for a given element was found to be shallower for the $90Cr_2O_3$-$10Al_2O_3$ substrate, when compared to the Al_2O_3 refractory. Also, petcoke slag penetrated deeper than coal slag in both high chromia and high alumina refractories and in both microstructures (grain agglomerates and fine grain matrix regions). In some cases, the penetration depth became smaller for the 10-minute samples with respect to the 0-minute samples, which may be explained by slag also scattering sideways (Exp. 1 and 2, 5 and 6, 9 and 10, 11 and 12). The petcoke slag penetrated much faster into the porous matrix regions of Al_2O_3 refractory (no change in concentration profile between 0-minute and 10-minute samples) than those of $90Cr_2O_3$-$10Al_2O_3$ substrates (drastic change in profile). While the amount of slag may be a limiting factor, the difference could also be due to average pore diameters. Generally, slag penetration was lower in the $90Cr_2O_3$-$10Al_2O_3$ versus the Al_2O_3 refractory, especially when slag was placed on a grain agglomerate. The exception was that, compared to the high chromia refractory, the Al_2O_3 refractory resisted more efficiently against petcoke slag penetrating into the matrix regions after 10 minutes (Exp. 16 vs. Exp. 12). The majority of V-species from petcoke slag remained as VO_x-based crystals precipitated above the surface. Penetrated V-species in the matrix regions of the Al_2O_3 substrate stayed at shallow regions after 10 minutes.

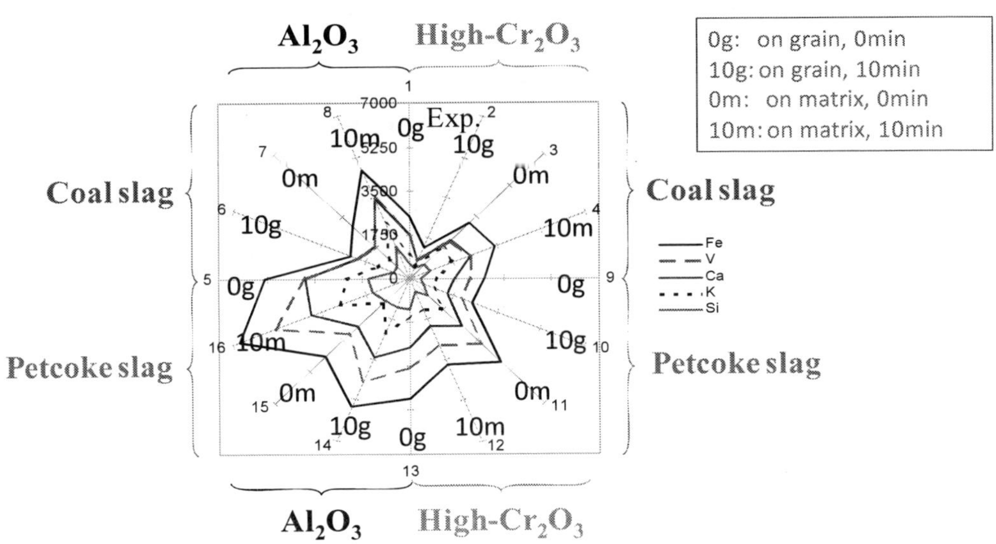

Figure 21. Selective penetration of elemental slag species into refractories at 0 min and 10 min (High-Cr_2O_3 = $90Cr_2O_3$-$10Al_2O_3$). An experiment number is indicated for each axis (1 – 18).

4.2.3. Degradation of Refractory By Slag Interactions

De-bonding of Al_2O_3 refractory grains occur when slag first attacks the porous matrix regions, dissolving refractory particles and bond material. If slag penetration is fast enough, alumina grains remain on the surface, exposing the grain boundaries (or agglomerate surfaces). If slag penetration is impeded, however, slag liquid remains on the refractory coupon surface, and the Al_2O_3 grains are released into the slag. This is shown in Figure 22 for coal slag / Al_2O_3 refractory interactions, along with slag penetration into the pores and microcracks.

When petcoke slag contacted the high chromia substrate, rapid slag/refractory interactions resulted in the formation of the continuous (Cr, Fe, V)O_x spinel crystal layer at the slag/refractory interface, impeding further slag penetration. A SEM-EDX elemental mapping of Si and V in a high chrome oxide refractory (Figure 23) shows a spinel layer formed along the slag/refractory interface immediately after slag contacted the refractory. This spinel interfacial layer grew into the molten slag, in some cases, breaking off into the slag, causing material loss. The spinel layer remained continuous during the 10 minute hold at 1500 °C. V was found at the bottom of the refractory in most of the cases, implying penetration of the V-containing slag into the refractory occurred before the interfacial layer formed. If the VO_x-based crystal layer formed over the matrix regions, the layer started flaking off into the liquid, creating "fresh" refractory surface for more interaction to occur. This agreed with the result for porous regions of the $90Cr_2O_3$-$10Al_2O_3$ refractory, which allowed petcoke slag to penetrate deeper over 10 minutes.

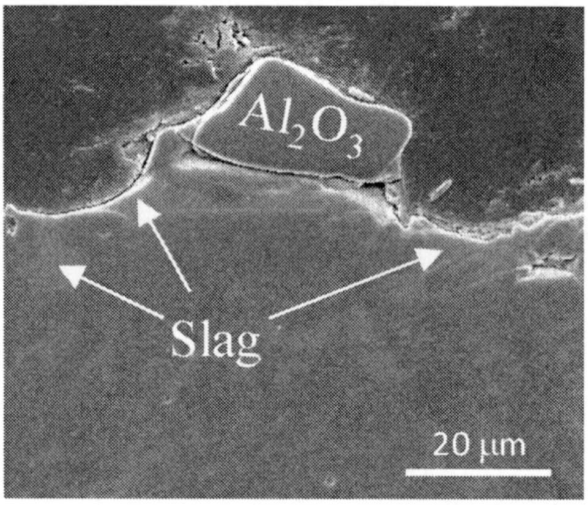

Figure 22. Cross-sectional SEM image showing coarse Al_2O_3 grains eroded and released by coal slag attack at 1500 °C in an oxygen partial pressure of 10^{-8} atm after 10 minutes of exposure.

Under the assumption that a metastable local equilibrium between the petcoke slag and $90Cr_2O_3$-$10Al_2O_3$ can exist, then the solubility of Cr_2O_3 in the slag (as at the contact point) can be greater than the global equilibrium value expected from the phase diagram, and therefore, that area soon experiences supersaturation with Cr_2O_3 (and Al_2O_3). On first contact between the petcoke slag and the high chromia refractory, therefore, a larger amount of Cr_2O_3 (and Al_2O_3) can chemically dissolve into the slag to the extent it is thermodynamically allowed. This supersaturation in the liquid must be compensated by eventually forming a globally stable phase(s).

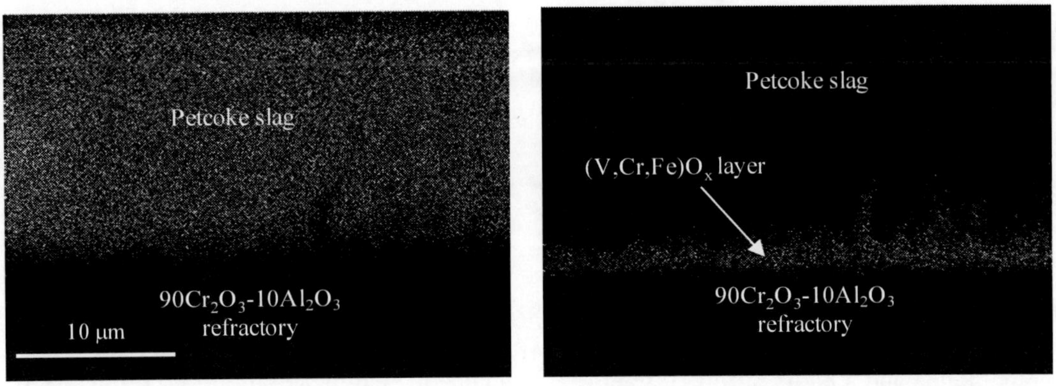

(1) Si (2) V

Figure 23. A SEM-EDX elemental map for Si and V showing the formation of an intermediate layer at the interface between petcoke slag and a $90Cr_2O_3$-$10Al_2O_3$ substrate at 1500 °C in an oxygen partial pressure of 10^{-8} atm.

CONCLUSION

5.1. Phase Equilibria Studies in Coal/Petcoke Slag Systems

Phase equilibria in the Al_2O_3-CaO-FeO-SiO_2-V_2O_3 system representing coal/petcoke slag mixtures at 1500 °C in an oxygen partial pressure of 10^{-8} atm were analyzed by a series of quench experiments. Results indicated four homogeneity regions: (1) slag; (2) slag + mullite; (3) slag + V_2O_3; and (4) slag + mullite + V_2O_3. TEM-EDX indicated that vanadium ions dissolved in the mullite phase formed to a limited extent; while karelianite (V_2O_3) formed a solid solution with Al_2O_3. A corresponding isothermal phase diagram where FeO and CaO contents were fixed at 13.5 wt.% and 7.0 wt.%, respectively, was proposed. Industrial coal slag, petcoke slag, or their mixtures (averaged) fall well within the proposed phase diagram. The diagram suggested that under gasification conditions, no crystal phase would precipitate if the slags are high (above 55.7 wt.%) in SiO_2 and low (below 3 wt.%) in V_2O_3. Slags may be in equilibrium with mullite and karelianite at the same time if enough petcoke is in the carbon feedstock.

5.2. The Formation Kinetics of Crystals in Coal/Petcoke Slag Mixtures

The formation kinetics in molten slags corresponding to coal/petcoke slag mixtures was studied using a CSLM in a gasifier-simulated atmosphere. Through in situ observations, a TTT diagram was constructed based on the time for initial crystallization to occur between 1200 °C ~ 1700 °C. Two distinct curves were found in the diagram, corresponding to 30, 50 and 70% petcoke slag mixtures. High temperature curves contained V_2O_3 formation in all cases. A single curve was found for a coal/petcoke slag containing 10% petcoke slag. A TTT diagram corresponding to 100% coal slag could not be constructed since crystallization was found only at 1200 °C in small amounts. The present TTT diagram indicates that the V_2O_3 crystallization in slag would occur as fast as 35 seconds even homogeneously (nucleated) in the slag liquid.

5.3. Interactions Between Slags and Refractory Materials

Interactions between synthetic slags and refractories were studied using sessile drop interactions at the slag/refractory interface utilizing a CSLM. Interactions and slag penetration were studied using synthetic coal and petcoke slags on commercial refractory materials (a high alumina and a high chromia refractories). In general, slag penetration into refractory was deeper and more rapid for petcoke slag than coal slag. The formation of the VO_x-bearing phase on the refractory surface limited the V penetration as time increased. Areas of initial slag interaction of the refractory were fine matrix regions, pore networks, and grain boundaries. The chemical dissolution of the alumina particles caused release of alumina grains from the high alumina refractory matrix. Refractory loss was observed when product layers of (Cr, Fe, V)O_x at the interface between the petcoke slag and high chromia refractory formed and were detached into the slag.

The information obtained through the studies summarized in 0, 0, and 0 will lead to a better understanding of the phases existing in a gasifier environment and how they interact with materials in a gasifier. This information can be used to develop improved liner materials, or to a better understanding of how to operate a gasifier.

ACKNOWLEDGMENT

This technical effort was performed in support of the National Energy Technology Laboratory's ongoing research in Control of Carbon Feed stock and the Impact on Gasifier under the RES contract DE-FE0004000. Dr. Thomas Lam (NETL, Albany, OR) is acknowledged for TEM microanalysis. The authors also thank Dr. Il Sohn (US Steel, Pittsburgh, PA, currently at Yonsei University, South Korea) for ICP analysis and Dr. Paolo Nolli (Vesuvius, Pittsburgh, PA) for machining the refractory samples. In NETL, SEM-WDX analysis of Mr. Keith Collins and Mr. David Smith for XRD crystalline phase analysis is appreciated.

REFERENCES

[1] Tampa Electric Company. *Biomass Test Burn Report – Polk Power Station Unit 1* 2002; April.

[2] Tampa Electric Polk Power Station – Integrated Gasification Combined Cycle Project, Final Report for US DOE NETL 2002, August *(http://www.tampaelectric.com/data/files/PolkDOEFinalTechnicalReport.pdf)*.

[3] Johnston WD. *J American Ceramic Society* 1965; 48(12): 608-611.

[4] Wakihara M, Katsura T. *Bulletin of Chemical Society of Japan* 1971; 44: 3043-3046.

[5] Rovnushkin VA, Nikiforov SV, Toporishchev GA. *Izvestiya Akademii Nauk SSSR. Metally* 1989; 5: 1-7.

[6] Mittelstädt R, Schwerdtfeger K. *Metallurgical Transactions* B 1990; 21B: 111-120.

[7] Farah H. Journal of Materials *Science* 2003; 38: 727-737.

[8] Bale CW, Chartrand P, Degterov SA, Eriksson G, Hack K, Ben Mahfoud R, Melançon J, Pelton AD, Petersen S. *CALPHAD* 2002; 26(2): 189-22.

[9] Kwong K-S, Petty A, Bennett J, Krabbe R, Thomas H. *Applied Ceramic Technology* 2007; 4(6): 503-513

[10] Körber vF, Oelsen W. *Stahl und Eisen* 1940; 60(43): 948-955.

[11] Muan A, Najjar MS. *US Patent no.5070065*, Dec. 3, 1991.

[12] Muan A, Najjar MS. *US Patent no.5070260*, Dec. 31, 1991.

[13] Lopatko NN, Mikhailov GG, Archugov SA. *Inorganic Materials* 1996; 32(1): 71-73.

[14] Selvig WA, Gibson FH. *Analyses of Ash from United States Coals. Bureau of Mines Bulletin 567*. Washington: United States Government Printing Office;1956.

[15] Bryers RW. *Fuel Processing Technology* 1995; 44: 121-141.

[16] Swanson ML, Hajicek DR. Proc. Int. Conf. Fluid. *Bed Combust*. 2001; 16: 496-507.

[17] Dong P, Wang X, Seetharaman S. *Steel Research Int* 2009; 80(4): 251-255.

[18] Johnston WD. *J American ceramic society* 1965; 48(12): 608-611.

[19] Schreinemakers FAH. Die Ternäre Gleichgewichte. *Braunschweig: Vol. III, part II;*1912.

[20] Nakano J, Sridhar S, Moss T, Bennett J, Kwong K-S. *Energy Fuels* 2009; 23: 4723-4733.

[21] Yin H, Emi T, Shibata H. *Acta Mater*. 1999; 47: 1523-1535.

INDEX

C

D

E

F

G

H

I

J

K

L

M

N

O

P

Q

R

S

T

U

V

W

X

Y

Z